Phase Equilibria
*Basic Principles,
Applications,
Experimental Techniques*

This is Volume 19 in
PHYSICAL CHEMISTRY
A series of monographs
Edited by ERNEST M. LOEBL, *Polytechnic Institute of Brooklyn*

A complete list of the books in this series appears at the end of the volume.

Phase Equilibria

*Basic Principles,
Applications,
Experimental Techniques*

Arnold Reisman
*IBM Watson Research Center
Yorktown Heights, New York*

QD501
R377p
1970

Academic Press
New York and London *1970*

Copyright © 1970, by Academic Press, Inc.
ALL RIGHTS RESERVED.
NO PART OF THIS BOOK MAY BE REPRODUCED IN ANY FORM,
BY PHOTOSTAT, MICROFILM, RETRIEVAL SYSTEM, OR ANY
OTHER MEANS, WITHOUT WRITTEN PERMISSION FROM
THE PUBLISHERS.

ACADEMIC PRESS, INC.
111 Fifth Avenue, New York, New York 10003

United Kingdom Edition published by
ACADEMIC PRESS, INC. (LONDON) LTD.
Berkeley Square House, London W1X 6BA

LIBRARY OF CONGRESS CATALOG CARD NUMBER: 70-107575

PRINTED IN THE UNITED STATES OF AMERICA

CONTENTS

Preface, xiii
Acknowledgments, xv

1. Some Preliminary Remarks and Observations, 1

2. Thermodynamics and the Phase Rule, 4

3. Definition of Terms and Concepts

    A. Introduction, 9
    B. Systems, 9
    C. Phases, 10
    D. The Limits Imposed by an Experimental Situation, 11
    E. The Component, Species, and Mole Terms, 19
    F. Defining the Number of Components Present, 24
    G. Chemical, Compound, Substance, and Constituent, 25

4. The Thermodynamic Basis of the Phase Rule

    A. Introduction, 27
    B. The Phase Rule and Its Basis, 28
    C. Choice of Systems and the Reduced Phase Rule, 32
    D. The Vapor Pressure, the Evaporation Rate, and the Reduced Phase Rule, 33
    E. The Time Factor in a Non-Time-Dependent Science, 36

5. Systems of One Component—Temperature Effects

    A. Introduction, 38
    B. The Analytical Description of Univariance, 40
    C. Univariant Equilibria Involving Solids and Gases, 41

D. Univariant Equilibria Involving Liquids and Gases, 43
E. Univariant Equilibria Involving Solids and Liquids, 45
F. Further Aspects of Univariant Systems, 47
G. Metastability in One-Component Systems, 47
H. Structure Changes in Unary Systems, 52

6. **Systems of One Component—Pressure Effects and the Continuous Nature of Metastability**

    A. Metastable-to-Stable Transformations with Increasing Pressure, 61
    B. Solid Phases That Do Not Exhibit Stable Solid–Vapor Equilibria at Any Temperature: Monotropes, 63

7. **Complex Metastability in One-Component Systems**

    A. Systems with Solid–Vapor Univariance, 71
    B. Stable High-Pressure Systems, 74

8. **Other Aspects of One-Component Behavior**

    A. The Effects of Inert Gas Pressures on the Vapor Pressure of Single-Component Solids or Liquids, 79
    B. The Effect of Surface Tension on the Vapor Pressure of a Single Component, 82

9. **Enthalpy and Entropy Diagrams of State for Unary Systems**

    A. Introduction, 85
    B. Enthalpic Relationships and the Enthalpy Diagrams of State, 86
    C. Entropic Relationships and Entropy Diagrams of State, 89

10. **Multicomponent Systems, Homogeneous Systems, and the Equilibrium Constant**

    A. Introduction, 92
    B. The Species and the Homogeneous Equilibrium Constant, 93

C. Standard States and Their Significance, 101
D. The Homogeneous Equilibrium Constant in Terms of Mole Fractions, 103
E. The Variation of Equilibrium Constant with Temperature, 105

11. **Multicomponent Systems, the Equilibrium Constant, and Heterogeneous Systems**

    A. Introduction, 108
    B. The Chemical Potential for Components and Species, 109
    C. Equilibrium Constants in Terms of Partial Molar Quantities, 112

12. **The Thermodynamic Parameters—Fugacity and Activity**

    A. Introduction, 117
    B. Rationale for a Thermodynamic Approach to the Treatment of Real Systems, 118
    C. The Fugacity, Standard States, and Free Energy Relations, 122
    D. Thermodynamic Concentrations— The Activity, 123

13. **Two-Component Systems—Systems in Which One Phase Is Pure and in Which Species and Component Mole Terms Are Equivalent— Simple Eutectic Interactions**

    A. Introduction, 125
    B. The Analytical Expression of Univariance in a Simple Eutectic Interaction, 127
    C. An Alternative Derivation of Univariant Equations for Eutectic Systems, 131

14. **Graphical Representations of Simple Eutectic Interactions**

    A. Introduction, 139
    B. The Phase Diagram—A Qualitative View, 140
    C. The Lever Arm Principle, 145
    D. Specifics of Application of the Lever Arm Principle, 149

15. **Eutectic Systems Continued—The Parameter $\Delta H_{\text{fusion}}/T_{\text{melting}}$ and Its Influence on the Contour of Eutectic Liquidus Curves**

    A. Introduction, 152
    B. Theoretical Analysis, 156
    C. Application of the $\Delta H_A/T_A$ Principle to Real Systems, 158
    D. Pseudobinary Systems, 164

16. **Eutectic Interactions Continued—The Effects of Homogeneous Equilibria on Liquidus Contours**

    A. Introduction, 166
    B. Dissociation of One of the Liquid-Phase Species, 168
    C. Solution for the AB Liquidus Curve, 170
    D. The Liquidus for the Nondissociating Component, 173

17. **Common Species Effects in Simple Eutectic Systems**

    A. Introduction, 177
    B. The Liquidus Curve for AB, 181
    C. The Liquidus for B, 182

18. **Eutectic Interactions Continued—The Effects of Association on Liquidus Contours**

    A. Introduction, 185
    B. The A Liquidus, 186
    C. The Liquidus for the Component B, 189

19. **Eutectic Interactions Continued—Effects of End Member Species Interactions on Liquidus Contours**

    A. Introduction, 192
    B. Solution of the A Liquidus, 195
    C. Solution of the B Liquidus, 196

20. **Eutectic Interactions Continued—Effects of Complete Dissociation on Liquidus Contours**

    A. Introduction, 197
    B. The AB Liquidus, 201
    C. The B Liquidus, 203

21. Eutectic Interactions Continued—Graphical Description of the Results of Chapters 16–20

   A. Introduction, 204
   B. Test Cases, 205
   C. Dissociation of a Pseudounary Component in the System AB–C, 206
   D. Dissociation with a Common Species Effect, 210
   E. Association of One of the Liquid-Phase Species, 211
   F. Reaction between Liquid-Phase Species, 213
   G. Effects of Complete Dissociation on Liquidus Contours, 215

22. Eutectic Interactions Continued—Single Compound Formation in Binary Systems and Coordinate Transformations

   A. Introduction, 217
   B. A Single Compound Is Generated, 217
   C. Scaling Factors and Coordinate Transformations, 222
   D. Examination of Solubility Curves in Systems Exhibiting Compound Formation, 227

23. Eutectic Interactions Continued—Multiple Compound Formation in Binary Systems and Coordinate Transformations in Multiple Compound Systems

   A. Introduction, 229
   B. Two Compounds Are Generated, 230

24. Eutectic Interactions Continued—p-T-M_X Representations and Phase Changes

   A. Introduction, 234
   B. The T–p–M_X Diagram of State, 236
   C. Phase Changes—A Second Quadruple Point, 255

25. Eutectic Interactions Continued—Intermediate Compound Instability—Incongruently Melting Compounds

   A. Introduction, 257
   B. The Intermediate Compound A_xB_y Melts Incongruently, 258
   C. The Three-Dimensional Model of Incongruently Melting Systems, 264

D. The Effect of Pressure When the S–L Slope of the Intermediate Incongruently Melting Compound Is Positive, 267

26. **Liquid Immiscibility in Eutectic Systems**

   A. Introduction, 271
   B. The Miscibility Gap Is Confined to the Liquid Regions, 272
   C. The $T-p-M_X$ Representation of the Eutectic System Showing Liquid Phase Immiscibility, 273
   D. The Intersection of a Miscibility Gap with the Liquidus Regions, 276
   E. The $T-p-M_X$ Representation of the MG Intersecting a Liquidus, 278

27. **Systems Exhibiting Solid Solubility—Ascending Solid Solutions**

   A. Introduction, 284
   B. The Analytical Description of Univariance for Ascending Solid Solutions, 286
   C. The Graphical Representation of Ascending Solid Solutions, 289
   D. The Purification of Materials, 290
   E. The General Experimental Approach to the Resolution of Solid Solution Diagrams, 294
   F. The Space Model for Solid Solution Systems, 295

28. **Solid Solutions Continued—Graphical Representation of Idealized Systems and Some Comment on Real Systems**

   A. Introduction, 301
   B. Temperature–Composition Diagrams for Cases I–XIV, 302
   C. Real Systems, 312

29. **Solid Solutions Continued—Complete Liquid Phase Dissociation and Minimum Type Solid Solutions**

   A. Introduction, 313
   B. The Qualitative Aspects of Fig. 1, 315
   C. Complete Dissociation in Solid Solution Systems, 315
   D. Case I—Dimer–Monomer Behavior, 317
   E. Case II, 321

F. The Case Where Both Components Are Dimers in the Solid, 324
G. Space Models, 326

30. **Hypothetical Examples Using the Equations of Chapter 29**

    A. Introduction, 330
    B. Some Hypothetical Examples, 331
    C. The System Na_2CO_3–K_2CO_3—A Comparison between Experimental and Theoretical Solidus and Liquidus Curves, 335

31. **Condensed-Vapor Phase Binary Diagrams**

    A. Introduction, 340
    B. Boiling and Sublimation Point Diagrams (Constant Pressure Diagrams), 341
    C. Constant Temperature Condensed-Vapor Phase Diagrams, 344
    D. Some Hypothetical Examples of Maximum Type S–V Equilibria in Isothermal Diagrams, 352
    E. Isothermal Distillations, 359
    F. Isobaric Distillations, 362

32. **Condensed-Vapor Phase Binary Diagrams Continued—Incongruently Vaporizing Systems**

    A. Introduction, 367
    B. Vaporous Curves in Isothermal Diagrams, 368
    C. Immiscible Solid–Vapor Systems Exhibiting Congruently Vaporizing Compounds, 370
    D. Immiscible Solid–Vapor Systems Containing Incongruently Vaporizing Compounds, 373
    E. Hydrate Systems, 377
    F. Further Remarks on the Regions of Variable Composition, 382
    G. Deliquescence and Efflorescence of Hydrate Systems, 385
    H. Efflorescence and Deliquescence in Aqueous Systems Not Generating Hydrates, 386
    I. The Variation of p with T of Isothermally Invariant Incongruently Vaporizing Systems, 387

33. **Limited Solid Solubility**

    A. Introduction, 389
    B. Limited Solubility Systems Exhibiting a Eutectic, 390

C. The Case Where Limited Solubility Occurs below the Solidus, 397
D. Three-Dimensional Representations of Limited Solid Solubility Systems, 401

34. Three-Component Systems

A. Introduction, 407
B. Representation of Three-Component Condensed-Phase Equilibria, 408
C. Simple Ternary Eutectic Interactions, 415
D. Ternary Diagrams for Systems in Which Binary Congruently Melting Compounds Are Formed, 432
E. Ternary Diagrams for Systems in Which Binary Congruently Melting Compounds Occur, One of Whose Eutectic Extensions Generates a Ternary Peritectic, 436
F. Ternary Systems in Which Incongruently Melting Binary Compounds Are Present, 447
G. Ternary Compound Formation—Congruently Melting Ternary Compounds, 448
H. Ternary Compound Formation—An Incongruently Melting Ternary Compound Is Present, 448
I. Isothermal Sections for Selected Ternary Systems, 451
J. Systems Exhibiting Solid Solubility, 459

35. Three-Component Solid–Vapor Equilibria—Chemical Vapor Transport Reactions

A. Introduction, 477
B. Disproportionation Processes, 480

36. Experimental Techniques

A. Introduction, 487
B. Pressure Measurements, 489
C. Measurement of Thermal Anomalies, 499
D. Other Useful Tools, 514

Appendix: References and Additional Remarks for Chapters 1–36, General Bibliography, 517

Subject Index, 536

PREFACE

Until the mid to late 1950's, Phase Equilibria was a topic of concern primarily to the classical metallurgist. The emphasis, naturally, was then aimed at understanding the behavior of metal alloy phenomena and, indeed, was heavily oriented to the steel industry. With the advent of the electronics, nuclear, and space ages, however, we have witnessed an increasing involvement in this field by the physicist, the physical chemist, the inorganic and organic chemist, the device oriented electrical engineer, and the interdisciplinary product of these ages, the materials scientist. Accordingly, the contributions to the literature by this grouping have increased enormously, with the emphasis shifted to nonmetallic-compound and solid-solution systems. In addition, where previously the domain of interest of the classical metallurgist was confined mainly to interactions among condensed phases, the recent entrants in the field, including the modern metallurgist, have been confronted by a need to better understand condensed-vapor phase behavior. This need exists because many systems of interest to them are volatile, or employ the vapor as a vehicle for the transport, synthesis, purification, and single crystal growth of materials.

It is because of this change in emphasis and need, as well as the fact that existing treatments are basically descriptive in nature that the present work was undertaken. Analytical treatments, for example, have in the past been provided in texts on chemical thermodynamics. These have in general focused on the phenomenological approaches involving activities and fugacities, and therefore, on the largely defensive concepts of ideal and nonideal behavior. In addition, except for advanced thermodynamic treatments, heterogeneous equilibria has been treated generally in a cursory fashion. Books devoted to phase equilibria exclusively have, for the most part, minimized analytical concepts. Instead, the subject has been developed graphically. Even a topic as widely employed by the experimentalist as the lever arm principle is almost always presented without rationalization.

The present treatment has been formulated with these and related considerations in mind. It has been organized in a fashion that affords the theoretical elements sufficient emphasis for the casual participant to understand at least the principles involved, and the serious worker may be sufficiently motivated to delve further. The method of species model systems is employed broadly following introduction of pertinent thermodynamic concepts. In a number of chapters, computer analysis of derived relationships is presented to provide graphical visualization of the consequences of species changes within coexistent phases. The chapter arrangements have been ordered so that in a course of formal instruction the presentation may include or omit the more mathematically oriented material depending upon the needs of the audience. This feature has been utilized successfully in formal presentations by the author within the IBM Research Professional Education Program.

The subject content includes the substance of classical treatments of the subjects and also more modern aspects such as concern with condensed phase–vapor phase equilibria and vapor transport reactions, zone refining techniques, nonstoichiometry, among others. Further, a fairly extensive treatment of experimental techniques is provided in the text and appendix. In addition, the appendix presents additional commentary on specialized topics,* as well as selected general and specific references to major areas treated within the body of the text. It was the intent to provide a reference work for the involved experimentalist, a step by step introductory graduate text for departments of chemistry, metallurgy, and materials science. While it would be presumptuous to assume that all these needs have been met, it is hoped that together with existing treatises, obvious voids have been reasonably filled.

---

* For example, a fairly detailed discussion of the effects of vacancy concentration on the dissociation pressure of binary compounds (defect structures) is given in the appendix on p. 524 *et seq*. Other topics, e.g., phase transformations, Vegard's law, crystal growth, vapor pressure, evaporation rate, and accommodation coefficient, are also discussed in some detail with appropriate references.

ACKNOWLEDGMENTS

During the several years that this manuscript was being formulated, a number of the author's associates at IBM provided invaluable assistance without which the complicated undertaking would probably have never seen fruition. During the early stages, the late Dr. G. Mandel acted as a sounding board for a number of the proposed approaches and it was through these interactions that the current format began to take shape. Mr. G. Cheroff devoted many hours to a critical evaluation of the mathematical approaches, and in addition offered numerous suggestions which were subsequently incorporated. Mrs. S. A. Papazian provided much needed help in validating derivations and her contributions as a co-worker in the evaluation of the theoretical approach employed in Chapter 34 for the description of multicomponent systems were key ones. The application of the model equations developed in Chapters 16–20 and 27–31 to the generation of hypothetical phase diagrams using computer techniques is due in large measure to Mrs. J. E. Landstein. Her patience in the face of innumerable obstacles, both those of a mathematical nature and those due to the author's inadequacies, has been commendable. Appreciation is due also to the late Dr. H. R. Leonhardt for many stimulating discussions and to Dr. T. O. Sedgwick for many excellent suggestions about improvement of the presentations in the introductory chapters. The author is particularly indebted to Mrs. M. H. Lindquist for her tireless efforts in bringing the manuscript through its several drafts, then in putting it into final form; proofreading, correcting, typing, and what have you. Finally, the author is grateful to the IBM Corporation for providing the extensive time and necessary resources.

I

Some Preliminary Remarks and Observations

The starting point for all that ensues is the body of accumulated knowledge concerned with interrelationships among work, heat, and energy for macroscopic systems existing in a state of chemical and physical equilibrium. From this body of knowledge, known as thermodynamics, a mathematical generalization called the "phase rule" was first deduced by Willard Gibbs. This equation functions usefully as a constraint on the final state of almost any conceivable macroscopic chemical or physical process. In terms of the parameters, temperature, pressure, and the number of independent chemical variables, none of which is dependent on the total quantity of material under scrutiny, one is able to predict without exception the total number of variables that must be specified in order to define uniquely the state of the system. This total number of variables is termed the degrees of freedom possessed by the system. Specifically, for a system possessing a specified number of independently variable chemical quantities (system components) present in one or more distinctly identifiable phases, all at a defined temperature and pressure, one is able to predict precisely the total number of variables that may be varied independently of the others, namely, how many may be varied simultaneously without changing the physical state of the system. Once this number of required variables is specified, all other intensive properties of the system, i.e., density, thermal conductivity, etc., are uniquely, if not explicitly, defined. The equation known as the phase rule then, is not an equation of state that, for example in the case of an ideal gas, relates what change in pressure results from a specified change in temperature at constant composition. It functions in more general terms, and defines whether the temperature may or may not be varied in the first place without causing a change in the number of phases present. It tells us that by varying only the pressure, or only the temperature, or only the composition, or any two of these, or all three of them (if such is an allowable condition), the remaining parameters will dependently acquire unique values.

Based on the implications of universality inherent in the phase rule, an extensive field of study has developed whose impact is felt but frequently not realized in many outwardly unrelated areas. The designation given an equation of thermodynamic origin has thus become synonymous with a broad field of intellectual and experimental endeavor having considerable practical significance. Since the practical implications are of greatest import in situations where the properties of nonhomogeneous (heterogeneous) agglomerations of matter are under scrutiny, alternative names such as phase equilibria and heterogeneous equilibria have been invoked frequently to characterize the subject. Furthermore, as the culmination of an effort in the field is in the quantitative description of the variations in temperature, pressure, and composition of a given system, such description being amenable to graphical capsulization in the form of a phase diagram or diagram of state, the name phase diagrams has also been employed as a description of the field.

In any event, all of these perfunctory designations are somewhat misleading since they imply a concern with a restricted and self contained area of activity which presumably is divorced from other areas of scientific endeavor. Coincidentally, then, one finds that phase equilibria is treated as a separate, unrelated topic in existing treatises on thermodynamics and physical chemistry. The role of the phase rule in treatments of analytical, organic, and inorganic chemistry is almost never mentioned, and workers in these fields frequently entertain only vague notions of its existence or implications concerning their particular activity. This state of affairs is certainly surprising when it is realized that there does not exist a single area of materials studies, that comes to mind, that does not depend on a sound knowledge of the phase rule and its ramifications. For example, it is generally the case that the practical and theoretical bases for analytical gravimetry, inorganic and organic synthesis, physical measurements of systems in equilibrium, and, for that matter, in transit toward equilibrium, are bounded by the constraints of the phase rule.

1. SOME PRELIMINARY REMARKS AND OBSERVATIONS

To be more specific, consider, for example, the precipitation of an insoluble sulfate in an analytical procedure, the melting of snow by deposition of calcium chloride, the fractional distillation or crystallization of a mixture of organic or inorganic materials, the zone refining and the single crystal growth of an ingot of germanium, silicon, or other semiconductor from the melt, the synthesis of a compound of defined stoichiometry from two or more starting materials, the manufacture of synthetic gems, the dehydration of a hydrate, the measurement of the vapor pressure of a compound formed from one or more elements, or the fabrication of a transistor via an alloying process. The above and hundreds of other seemingly unrelated phenomena, e.g., the making of potable water from sea water, are all describable using only a very few generalized principles concerned with phase equilibria. The limiting states of such processes are graphically depictable in a phase diagram and in principal at least, these diagrams can be described analytically by equations whose parameters include many of the useful variables of physics and chemistry, e.g., temperature, pressure, composition, energy, electrical conductivity, thermal conductivity, compressibility, solubility, etc.

While heretofore, the classical metallurgist, almost uniquely, has been exposed to an extensive formalism in the tenets of phase equilibria, and has then alone contributed to the available literature, such is no longer the case. In recent years the hybrid known as the materials scientist has contributed increasingly to the literature on phase equilibria. Unfortunately, because of his generally deficient formalism in the subject, too often one encounters implausible conclusions based on imperfectly defined experimental procedures.

The enhancement of activity in this field with the advent of the electronics and space ages has indeed provided the activation energy necessary for the writing of this monograph. Hopefully, it will serve the function of filling the void that exists between general texts concerned primarily with the thermodynamics of homogeneous systems and those concerned with purely descriptive introductions to the field of phase equilibria. Hopefully also, it will provide, together with the extant formal treatments of the subject, a sufficiently useful tool to allow for wider treatment of this technologically significant field in departments of chemistry, physics, ceramics, materials science, etc.

2

Thermodynamics and the Phase Rule

The science of thermodynamics represents an attempt to summarize analytically the sum total of man's experiences with respect to processes involving energy changes attending the interplay of heat and work. Classical thermodynamics is concerned solely with the initial states from which such processes originated and the final state in which they terminate. It is not concerned at all with any transient aspects of the processes involved. Based on the underlying principle that an energy change in a given process confined to a given system is the same, irrespective of the pathway followed in achieving the final state, a network of relationships has evolved which enables one to predict whether processes can occur spontaneously and what energy changes are associated with such processes. Thermodynamics does not have the time element as one of its parameters. Consequently, the knowledge that a process can take place does not imply that the process will take place in a practical time interval. The transient aspects of processes occurring within the boundary or equilibrium states considered in thermodynamics lie within the realm of the science of kinetics, and one may bridge the gap between thermodynamics and kinetics by invoking certain assumptions between rate and equilibrium parameters as the transient occurrences approach the limiting boundaries. These assumptions

2. THERMODYNAMICS AND THE PHASE RULE

require the introduction of concepts alien to thermodynamics, because the latter in its rigorous form makes no attempt to dissect systems which are being focused upon. For example, in order to define the kinetics of a reaction it is essential to identify both the number and the nature of the species which participate in the reaction. In thermodynamics, the concept of the species does not exist. Rather, one is concerned with an energy possessed by an entire system, or a phase in the system, or in a quantity called a component mole. This latter quantity, in fact, serves as the bridge between the two sciences, since it may be related to a quantity termed the species mole via the use of conservation equations. The latter conserve the component quantity in terms of sufficient species quantities. The approach to the analytical description of the behavior of heterogeneous systems may be in the form of a rigorous adherence to the thermodynamics in which equilibrium constants remain either undissected or are delineated in terms of the phenomenological concepts of activity and fugacity. These delineations provide little insight since any process which does not adhere to very restricted concepts of ideality generated within such regimes is termed nonideal. On the other hand, one may invoke quasi-thermodynamic approaches which, for the most part, trace back to kinetic origins. Such approaches, being based on a physical model, tend to provide greater insight. In fact, these quasi-thermodynamic approaches have been incorporated into the thermodynamics to the degree that their nonthermodynamic origins are lost. For example, the perfect gas equation is kinetic in origin, depending on an assumption as to what constitutes a mole of gaseous molecules, and is used extensively in otherwise rigorous thermodynamic treatments.

In any event, whether the approach is purely or only quasi-thermodynamic, the variables of thermodynamics remain the same. For a system under constant constraints of gravity, electrical and mechanical fields, and what have you, one need use as tools only the pressure, temperature, and the composition of a system. Should, for example, our studies be conducted in an environment where the system is subjected to varying electrical, magnetic, or gravitational fields, such fields being capable of perturbing the equilibrium states of the system, possible effects must be taken into account by considering these fields as potential variables.

Assuming for the purposes of our discussion that these fields are constant in the earth's environment, to a good first approximation, the phase rule may be stated as

$$V = C - P + 2. \tag{1}$$

The term V is the variance or degrees of freedom possessed by the system. It is the number of parameters that must be specified simultaneously in order to establish unambiguously the state of a system. The term C represents the number of independent chemical variables, components, possessed by the system, e.g., if we add Cu to a system initially consisting of Pb, the quantity of Pb is not altered by the quantity of Cu added and vice versa. Obviously then, copper and lead are independent chemical variables, and the system contains two such variables. The term C in Eq. (1) for the case mentioned would then be equal to 2. The term C may have any integral value consistent with such a concept. The term P represents the number of phases within which the components of the system are distributed. Since the degrees of freedom may not be less than zero, it is evident that P may not exceed C by more than 2. The number 2 in Eq. (1) shows the fixed number of variables, T and p. Since, in a system at equilibrium, the temperature and pressure must be uniform everywhere, the maximum number of variables these intensive parameters may contribute to a system is two.

Equation (1) does not say that a number V greater than the value determined may not be specified. It simply states what the minimum number is that is necessary to specify the state of the system completely. Implicitly, it precludes the larger number from being comprised only of independent parameters. Nor, as we have stated, does it define how all other possible variables change as this minimum number is varied. It may, for example, lead to the conclusion that the system possesses one degree of freedom from among the possible variables T, p, and C. Let us assume that we chose T as the variable. If we now desire a knowledge of how the pressure varies with T we require an equation which specifically relates T and p for a defined system. When only one degree of freedom is possessed by a system, the system is univariant. An equation for such a univariant situation would describe the variation of a chosen dependent variable as a function of the variation of a chosen independent variable.

Consider, for example, the case of water in equilibrium with its vapor. Such a system, as we shall see, is univariant, that is, it possesses only one degree of freedom. Suppose we choose as this degree of freedom, the temperature, and desire to know the variation of the pressure as a function of the temperature. An equation providing this relationship is

$$dp/dT = \Delta H/T \Delta V. \tag{2}$$

This equation, known as the Clapeyron equation, specifies the pressure as a single valued function of the temperature in terms of the molar heat

2. THERMODYNAMICS AND THE PHASE RULE

of vaporization of water and the difference in molar volumes of gaseous and liquid water. The phase rule, while not providing us with the form of this relationship, demands that the system in question be univariant. Thus, if in Eq. (1), we insert the number 1 for the number of components (pure water behaves as only a single chemical variable in many situations as we shall see), and the number 2 for the number of phases (liquid and gas), the number V is seen to be 1.

As implied previously, Eq. (1) in general form may be written as

$$V = C - P + X, \qquad (3)$$

where X may have any integral value depending upon the experiment in question and represents the number of possible variables that affect all parts of the system simultaneously. Where X has some arbitrary value, it is seen from Eq. (3) that one component present in two phases possesses $X - 1$ degrees of freedom. Alternately it is seen that if the system consists of only one phase,

$$V = (C + X) - 1. \qquad (4)$$

Since the maximum possible number of variables is the sum $C + X$, the degrees of freedom are always the maximum possible number less the number of coexisting phases.

The effectiveness of the phase rule lies in its utter simplicity and complete generality, the nature of the system under scrutiny being of little consequence. Thus, it may be stated, where Eq. (1) is applicable, that the maximum number of phases possible in a two-component system is four, in a one-component system, three, in a system of twenty components, twenty-two and in a system of m components, $m + 2$. Similarly, it may be stated that for a system of one component existing in two phases there will only be a single value of the pressure for each single value of the temperature and that for a system of n components in $n + 1$ phases there will also be a single value of the pressure for each single value of the temperature.

These statements will assume greater significance as we proceed further with a systematic discussion of the phase rule. For the present, the intention is meant to convey some feeling for the possibility of consolidating the treatment of outwardly different situations. In this context, a precise and thorough development of thermodynamics with particular emphasis on those aspects relating to heterogeneous systems would be of considerable value. Such a development is outside the planned scope of this treatise. Instead, only those treatments which make for continuity in the development of the topic will be presented, such treatments being in the form of thermodynamic models for the most part. This thrust obviously represents a bias on the part of the author with which the reader is at liberty to agree or to reject. It is expected, however, that the general approach is palatable on the whole and provides enhanced insight as to the origins of the diagrams of state even though one might disagree with specific arguments.

Finally, it is to be emphasized that while the phase rule enables certain predictions, and equations of state enable certain other predictions and data fitting to existing curves, neither enables the prediction of the specific type of phase interaction that might result as an outcome of an unknown chemical process. Thus, in the reaction of A with B we are unable to predict with any degree of certainty, in the absence of prior knowledge, whether these react at all in a given time interval, whether if they do react they form AB, A_2B, A_xB_y, etc., whether the different constituents are soluble in one another or not, or whether the products are soluble, partially soluble, or immiscible in one another. Such capability of prediction, while not realized as yet, falls within the ultimate encompassing domain of quantum mechanics.

What follows, attempts to introduce systematically the types of relationships that are known to occur in one, two and, to a lesser extent, three and higher component number systems, to explain such behavior where possible on the basis of models, and then to describe experimental techniques useful in the resolution of equilibrium diagrams. Furthermore, as this work is not intended to be a compendium of reported phase diagrams there will be few references to "natural" diagrams since these have been known to vary with time, a distinctly nonthermodynamic parameter. More to the point, however, the use of hypothetical diagrams, is intended for the most part to focus on the general applicability of the subject to a diverse number of fields.

3
Definition of Terms and Concepts

A. INTRODUCTION

Because much of what follows depends on a clear understanding of certain concepts basic to problems in polyphase equilibria, it will serve us well to discuss these concepts at the present time. It will become apparent that often one must consider a definition in the context of a given experimental situation. It will become equally evident that one strives to delineate a situation such that the minimum number of parameters necessary to define completely a given state of a body under investigation are invoked. Since experiments are generally performed in a restricted area whose boundaries, real or otherwise, are defined, we must first consider the meaning of the term system.

B. SYSTEMS

The variables employed in classical thermodynamic treatments are the temperature, pressure, and concentrations of independent chemical variables. One attempts to express analytically the changes in the extensive parameters—E, the internal energy; H, the enthalpy; S, the entropy; F the Gibbs' free energy; and A, the Helmholtz free energy—

that occur when a delineated portion of space, more or less filled with matter, is subjected to certain changes in pressure, temperature, or concentrations of the components of the system.

This delineated portion of space whose properties are being scrutinized is termed a system. While matter may distribute within the system in different ways, a prime prerequisite is that the mass of the system must be conserved for the system as a whole in the process of moving from an initial to a final state. While an experiment may be conducted in an open vessel, the system in question then being defined as consisting of condensed phases alone, the total mass of these condensed phases must remain constant. Furthermore, if an experiment is conducted in an open vessel, the pressure cannot be treated as variable. Experiments conducted in systems open to the environment are therefore constant pressure experiments. On the other hand, experiments conducted in systems isolated from the environment may or may not represent constant pressure situations, depending upon whether or not the volume of the containing chamber is maintained constant.

C. Phases

In general, the state of matter may be described as solid, liquid, or gas. Without resorting to philosophical arguments such as whether glasses are really solid or highly supercooled liquids, the distinction between the states of matter is not, in general, obscure. In the broad sense therefore, we have little problem deciding on the number of states of matter present. Each of these general states of matter is termed a phase. In the thermodynamic sense, however, our distinctions must be somewhat more precise. If, for example, a system under investigation consists of a mechanical mixture of solid table salt and sugar, it is clear that all of the material is present in the solid phase. More to the point, however, is the fact that two distinctly different solid phases are present, i.e., sodium chloride and sucrose. A phase then is an agglomeration of matter having distinctly identifiable properties, e.g., a distinct X-ray pattern, viscosity, density, electrical conductivity, index of refraction, etc.

The state of subdivision of this agglomeration of matter is inconsequential so long as the state of subdivision does not affect properties within the context of the experiment under consideration. The precise composition of a phase is also inconsequential in defining the number of phases present. For example, assume that a container holds water in equilibrium with vapor. Clearly, only two phases are present. If we dissolve table salt in the water two results accrue. First, the vapor pressure

D. EXPERIMENTAL LIMITS

of the water will decrease. Second, the composition of the liquid will be altered. Attendant with these changes, other properties will change. For example, the density of the liquid will be altered as will its electrical conductivity, etc. What will not change is the number of phases. These will remain as two. Each of the original phases will have had something about it perturbed, but will coincide effectively with the initial phases. If we continue to add salt, a point will be reached at which the maximum solubility of the salt is reached and then exceeded. At this point a new phase, solid salt, will become part of the system, and the number of phases will increase to three.

Consider now an experiment in which we have pure salt alone in a container. Certain impurities may be present in the salt. Each of these impurities, depending upon its atomic size, when incorporated in the salt crystalline lattice, will perturb the lattice dimensions to a lesser or greater extent. The number of phases present, however, will still be the same. Equally important is the fact that whether we look at pure or impure NaCl we are looking at essentially the same phase, one which might be characterized as an NaCl phase.

D. The Limits Imposed by an Experimental Situation

In any real situation, the ability to differentiate it from any other situation depends on the type of experiment one performs, and the limits of sensitivity of the measuring equipment employed. In the measurement of the pressure of a given amount of gas as a function of temperature, the accuracy of the results will depend on the ability to measure the volume of the container, the temperature, and the pressure. If impurities are present in the gas, their effect is apparent only when such effect is greater than the normal "noise" or sensitivity present in the detection equipment. Specifically, when one speaks of the purity of a material, one is defining this purity within the context of a given noise level. Thus, wet chemical techniques may be able to detect variations in stoichiometry to parts per thousand while electrical measurements may be useful down to one part in 10^{14} or more.

Aside from the question of impurities causing a variation in data depending upon the sensitivity of measuring equipment, a boundary condition that exists relates to the size of the sample that must be looked at in order to have confidence that the result is representative of the bulk material. In addition to sample size, the sensitivity of the measuring equipment plays a part here also. Another factor that influences the results of an experiment relates to ratios such as surface area per unit volume, particularly as surface properties of matter are frequently different than bulk properties.

By bulk properties we mean specifically properties that may be attributed to a material having large enough (infinite) dimensions such that all local fluctuations average out. Local phenomena may be termed microscopic while bulk phenomena are termed macroscopic. Since the field of thermodynamics deals with macroscopic events, it is important that the sample size be sufficiently large to ensure adherence to macroscopicity. A rule of thumb is that the expected deviation in the value of a measured property is proportional to the $\sqrt{n}$ where n represents the number of particles involved.

To amplify further on the difference between microscopic and macroscopic behavior let us consider the following: Assume that we have a quantity of material containing n discrete particles whose total energy we wish to measure. Hypothetically, this chore could be accomplished in two general ways. We could measure the energy of all n particles simultaneously, or alternately we could measure the energy of a single particle, and multiply by the total number of particles to obtain the energy of the system. In the extreme cases described, the distinction between microscopic and macroscopic is clear. It is also clear that if the energy of all the particles is not the same, but only that their average energy is the same, the second method could yield a result completely different from the first. An intermediate approach might involve measuring the total energy of any y of these n particles and multiplying this value by n/y. If, within the limits of our ability to measure, this latter result was the same as the direct measurement of the total energy of all n particles simultaneously, we might be inclined to say that the y particles were representative of bulk properties. If different answers were obtained, then we might suspect that the degree of statistical fluctuation among the y particles (a deviation $\propto \sqrt{y}$) is sufficient to make their behavior unrepresentative of bulk properties. The point qualitatively, at which the number of particles studied becomes a factor in a measurement of a specific property may be visualized from the following: Suppose we have 10 distinguishable particles which can be distributed randomly between two regions. This distribution of the 10 particles between the two regions may be accomplished in a number of ways as seen from Table I.

Each of these means of randomly separating the 10 particles into two groups is termed a distribution D_i, of which eleven are seen to be possible in the case cited.

Let us now examine distribution 2 (D_2): that in which region 1 contains nine particles and region 2 contains one particle. It is evident that before

D. EXPERIMENTAL LIMITS

TABLE I
Distributions of Distinguishable Particles in Two Regions

Distribution D_i	Particles in region 1	Particles in region 2
1	10	0
2	9	1
3	8	2
4	7	3
5	6	4
6	5	5
7	4	6
8	3	7
9	2	8
10	1	9
11	0	10

we begin distributing the particles so as to make up D_2, there are 10 ways to choose this first particle since we have 10 distinguishable particles to choose from. Having chosen the first particle there now are nine ways to chose the second particle, eight ways to choose the third particle, etc. If we call our choices starting at the first choice 1, 2, 3, 4,..., r we see that D_2 can be accomplished with the following freedom of choices

Choice	1,	2,	3,	4,	...,	r
No. of ways a choice can be made	n,	$n-1$,	$n-2$,	$n-3$,	...,	$n-r+1$

where r is the number of particles in region 1 (nine in our case).

The product of the numbers n, $n-1$, etc. represents the total number of ways r particles may be chosen from among n particles. (Note that for each of the 10 ways of choosing the first particle, there are nine ways of choosing the second. Thus the first and second particles may be chosen a total of 90 different ways, i.e., 10×9.) This may be expressed by

$$D_2 = n(n-1) \cdots (n-r+1). \tag{1}$$

If we multiply Eq. (1) by the identity

$$\frac{(n-r+1-1) \cdots n-n+1}{(n-r+1-1) \cdots n-n+1},$$

we have

$$D_2 = n(n-1)(n-2)(n-3) \cdots (n-r+1) \frac{(n-r+1-1) \cdots n-n+1}{(n-r+1-1) \cdots n-n+1}, \tag{2}$$

we see that the numerator drops incrementally in units of 1 from n to 1 which is simply $n!$. The denominator ranges from $n-r$ to 1 which is $(n-r)!$. Thus,

$$D_2 = n!/(n-r)!. \tag{3}$$

Since we are unconcerned, in a given region, about how the r_i particles in it are arranged relative to one another, this value being $r_i!$, our derived Eq. (3) must be devided by the value $r!$, where $r!$ is the number in region 1. Thus

$$D_2 = n!/(n-r)!\, r!. \tag{4}$$

Note, however, that if r is the number of particles in one of the regions, $n-r$ is the number in the other region. In general then, for n distinguishable particles distributed among X boxes containing n_i particles in the first box, n_j particles in the second box, etc., the number of ways of arriving at the desired distribution is

$$D_i = \frac{n!}{n_i!\, n_j! \cdots n_z!} = \frac{n!}{\pi n_a!}. \tag{5}$$

Returning now to our 10-particle problem and applying Eq. (5), we can determine the number of ways of achieving each of the eleven distributions shown in Table I. These ways are given in Table II.

For the case given, in fact, the total possible ways of arriving at the

TABLE II

Ways of Achieving the Distributions Shown in Table I

Distribution no.	Particles in region 1	Particles in region 2	Ways of achieving D_i	
1	10	0	10!/10! · 0! =	1
2	9	1	10!/9! · 1! =	10
3	8	2	10!/8! · 2! =	45
4	7	3	10!/7! · 3! =	120
5	6	4	10!/6! · 4! =	210
6	5	5	10!/5! · 5! =	252
7	4	6		210
8	3	7		120
9	2	8		45
10	1	9		10
11	0	10		1
			Total	1024

D. EXPERIMENTAL LIMITS

different distributions is $2^n = 2^{10} = 1024$. By definition, probability is the number of favorable events divided by the total number of events possible. The number of ways of achieving a given distribution in Table II represents the number of favorable events for that distribution, the total of all events favorable or otherwise being 1024. The probability of achieving a 10–0 distribution is 2/1024. On the other hand a 5–5 distribution has a probability of occurrence of 252/1024, some 126 times as great. We are of course simplifying our problem by assuming that no weighting factors influence a given distribution. From the preceding we see then, that the most probable single distribution is D_6, namely equidistribution. However, the sums D_5 and D_7 have a greater probability of one or the other occurring than does D_6. As the number n increases, the probability of one distribution (equidistribution) becomes overwhelmingly greater than that of any other distribution, or for that matter the sum of all other distributions. If the most probable distribution referred to is that for a given values of a property, it provides us with the macroscopic property of the grouping of n particles. If in the example given, which involves only 10 particles, space 1 refers to a temperature of 100°C while space 2 refers to a temperature of 200°C, we see that while the distribution with the greatest probability would yield an experimental result of 150°C, the probability of measuring 150°C is only 252/1024. The probability of measuring 140°C is 210/1024, a not very different probability.

Thus, we see that when only a small number of particles is involved, the probability of a given number other than the equidistribution number influencing our observations is large, large enough to confuse the issue as to what the average value should be if a large number of particles were involved. It is an interesting exercise now to provide semirigorously a better feeling as to how the number of particles being distributed affects the probability of a particular distribution occurring. It is also pertinent to provide some evidence as to why the expected deviation from the most probable distribution can be expected to be of the order of $\sqrt{n}$. In other words, we want to demonstrate that if the most probable distribution of particles between two boxes is $n/2$, the root mean square deviation, $\overline{m}$, from this most probable distribution will have a value on the average proportional to $\sqrt{n}$.

A species mole of a chemical contains of the order of 10^{23} particles, a large number indeed. An impurity present to only one part in a billion contains some 10^{14} particles, again a very large number, but nonetheless a small number when compared to 10^{23}. For cases of real interest to us then, the numbers we will concern ourselves with are always large on an absolute scale.

The total number of ways of achieving all possible distributions of n particles in two boxes is given by

$$\#_{\text{total}} = \sum_n n!/(n-r)!\,r! = 2^n. \tag{6}$$

Using Stirling's approximation for $n!$ where n is large, and neglecting terms which contribute less than 1 % or so to the answer, any particular distribution of n particles between two boxes can be achieved in $\#D_i$ ways, this number of ways being given by Eq. (7) after invoking the approximation mentioned.

$$\ln \#D_i = n \ln n - (n-r)\ln(n-r) - r \ln r. \tag{7}$$

Differentiating Eq. (7) to find the value of r that provides us with the maximum value of $\#D_i$ we obtain

$$\ln r - \ln(n-r) = 0; \tag{8}$$

therefore

$$r = n/2. \tag{9}$$

Equation (9) states that the distribution which has the most ways of being created and which therefore has the greatest probability of occurrence is the equidistribution case. Remembering that the number of ways of arriving at all distributions is 2^n, it is interesting to evaluate the number of ways at arriving at equidistribution. We can then estimate the probability of equidistribution. If we insert the conclusion, Eq. (9) into Eq. (7), we obtain

$$\ln \#D_{\text{eqi}} \sim n \ln n - (n/2)\ln(n/2) - (n/2)\ln(n/2) \sim n \ln 2. \tag{10}$$

Therefore,

$$D_{\text{eqi}} \sim 2^n; \tag{11}$$

therefore

$$P_{\text{eqi}} \sim 2^n/2^n \sim 1. \tag{12}$$

Thus, we have shown that to a first approximation the probability of equidistribution is approximately unity, the number of ways of achieving equidistribution approximating the number of ways of achieving all possible distributions.

Next we ask the question, "Assuming a deviation m from equidistribution, what is the probability of such a deviation occurring?" Thus, in

D. EXPERIMENTAL LIMITS

Eq. (7) we want to find $\ln \#D_i$ when $(n - r) = (n/2) - m$, and $r = (n/2) + m$. Substituting in Eq. (7) we obtain

$$\ln \#D_m = n \ln n - \{[(n/2) - m][\ln[(n/2) - m]]\}$$
$$- \{[(n/2) + m][\ln[(n/2) + m]]\}. \quad (13)$$

Recalling that

$$\ln 1 - \alpha \text{ (with } \alpha \text{ small)} \sim -\alpha, \quad (14)$$

and

$$\ln 1 + \alpha \text{ (with } \alpha \text{ small)} \sim +\alpha, \quad (15)$$

we can rewrite Eq. (13) as follows

$$\ln \#D_m = n \ln n - \left\{\left(\frac{n}{2} - m\right)\left[\ln \frac{n}{2}\left(1 - \frac{m}{n/2}\right)\right]\right\} - \left\{\left(\frac{n}{2} + m\right) \ln \frac{n}{2}\left(1 + \frac{m}{n/2}\right)\right\}$$

$$= n \ln n - \left\{\left(\frac{n}{2} - m\right)\left[\ln \frac{n}{2} - \frac{m}{n/2}\right]\right\} - \left\{\left(\frac{n}{2} + m\right)\left[\ln \frac{n}{2} + \frac{m}{n/2}\right]\right\}$$

$$= n \ln n - \left\{\frac{n}{2} \ln \frac{n}{2} - m \ln \frac{n}{2} - m + \frac{m^2}{n/2}\right\}$$

$$- \left\{\frac{n}{2} \ln \frac{n}{2} + m \ln \frac{n}{2} + m + \frac{m^2}{n/2}\right\}$$

$$= n \ln 2 - \frac{4m^2}{n}. \quad (16)$$

Thus,

$$\ln \#D_m - \ln \#_{\text{total}} = \ln(\#D_m/\#_{\text{total}}) = \ln P_{D_m}$$
$$= n \ln 2 - (4m^2/n) - n \ln 2 = -(4m^2/n) \quad (17)$$

and

$$P_{D_m} = \exp(-4m^2/n). \quad (18)$$

Equation (18) gives the probability of any particular distribution occurring. Since m and n are both positive numbers it is clear that P_{D_m} will vary between 0 and 1. When m is small, that is, when the deviation from equidistribution approaches zero we see that

$$P_{D_m} \to 1. \quad (19)$$

When m is large, i.e., of the order of n, $(m^2/n) \to 10^{23}$ and $e^{-10^{23}} \to 0$. Thus, Eq. (18) is general for all possible distributions including equidistribution. It is clear that the probability of a large deviation from equidistribution is highly improbable. Specifically, we might ask what is the average root mean square deviation of all possible deviations from equidistribution likely to be. In other words, for each possible deviation m, there is associated with it a probability given by Eq. (18). For all such probabilities we want to know the root mean square value of m, i.e., Eq. (20)

$$\bar{m}_{\text{rms}} = \{[m_1^2 + m_2^2 + m_3^2 + \cdots + m_n^2]/n\}^{1/2}. \tag{20}$$

The Gaussian error function equation has the form

$$P(m) = (h/\sqrt{\pi}) \exp(-h^2 m^2), \tag{21}$$

where $P(m)$ is the probability associated with a given deviation m, and h is a constant.

The mean square deviation, $\bar{m}^2$ is given by

$$\bar{m}^2 = \int_{-\infty}^{\infty} m^2 P(m) \, dm = 1/2h^2, \tag{22}$$

and

$$(\bar{m}^2)^{1/2} = 1/(\sqrt{2} \cdot h). \tag{23}$$

If we note the similarity between Eqs. (18) and (21), we see that

$$h^2 = 4/n. \tag{24}$$

From Eqs. (23) and (24) then we see that

$$(\bar{m}^2)^{1/2} = \bar{m}_{\text{rms}} = 1/(\sqrt{2} \cdot h) = \sqrt{n}. \tag{25}$$

Equation (25) is the conclusion previously offered without any demonstrated argument. To reiterate then, Eq. (25) states that if there are n particles being scrutinized, on the average m of these n particles will be deviating from the average behavior of all of them. If the number of particles is $\sim 10^{22}$, then 10^{11} of these will, on the average, be possessed of a behavior different than the average of all of them. While the number 10^{11} is indeed large it only accounts for approximately one in each 10^{11} particles and this indeed is a small fraction. Being a small fraction, this deviating number will perturb the average value only slightly. More to the point, even if n is only a small part of a species mole, for example, a number like 10^{10}, we see that m will still only account for about one particle in 100,000 deviating from the average behavior of all 10^{10} particles. This number again is not likely to perturb the average value by much. Since the average behavior of the 10^{10} particles is what we would like to call the thermodynamic or macroscopic behavior, we see then, that even minute fractions of a species mole might be expected to exhibit behavior representative of the behavior exhibited by a species mole.

As the number of particles under scrutiny continues to decrease, the expected number deviating from the average behavior becomes a significantly larger part of the total number participating. The instantaneous perturbation, on the average, by these deviating particles is now appreciable, and the average value of a property being measured can become significantly different than the average value of a large number of particles, particularly if the property is measured only a few times in order to obtain its "average" value.

E. THE COMPONENT, SPECIES, AND MOLE TERMS

Of all the concepts associated with the phase rule, none is more fundamental than that of the component. Unfortunately, confusion as to its significance may arise, not due to any inherent complexity in the concept, but rather to an ambiguous use of terms in elementary as well as more advanced works in inorganic and physical chemistry and thermodynamics. Thus, one encounters frequently interchangeable usages of words such as compound, chemical, species, constituent, and component even though not all properties are necessarily common to each of these terms as normally defined.

Basically, the problem arises because of the attempt to relate the properties of matter to the number n of discrete particles present as atoms or molecules. As implied earlier, this gives rise to the essentially nonthermodynamic concept of the gram atom or the gram mole which necessitates the introduction of the Avogadro constant N. This number has physical meaning only when the precise disposition of matter is known. If we designate a chemical substance by the symbol AB, all that is really intended, is to conserve the stoichiometry (overall composition) of the substance. In other words, we are stating simply that for each A atom we have one B atom. We should not assume *a priori* that each individual molecule comprising the substance has a molecularity AB in addition to a stoichiometry AB. In fact, the substance may be primarily made up of molecules having the composition A_2B_2 or A_3B_3 or A_xB_x. Nonetheless, for each A atom there will be one B atom and the stoichimetry of the substance will still be conserved and given by AB. Obviously, if the substance is made up of A_3B_3 molecules, a molecular weight based only on the molecularity AB will contain only $\frac{1}{3}$ of the

Avogadro number. Any property sensitive to the actual number of particles present per unit volume, i.e., the pressure, cannot be described physically if only the stoichiometry is known, and it will be difficult to gain insight into unless the molecularity of actual particles is known. Any unique molecular aggregation is termed a species. Thus, if we wish to know the partial pressure of one constituent of a gaseous mixture, it is imperative that we know the fraction of the total number of particles present that this constituent accounts for. In other words, the mole fraction of this specific type of particle must be known. For the most part, such information is not available and this leads us to the expedient of inserting equation-satisfying terms such as fugacity and activity in place of partial pressure and composition. Fugacities and activities are based on overall stoichiometric rather than particle molecularity (species) knowledge.

A component of a system has been partially defined as an independent chemical variable, by which we mean that introduction of this component into a system does not alter the actual quantity of any other component already present. Lest this lead to confusion, the significance of the term "independent chemical variable" must be unambiguously understood. An independent chemical variable is not something we have a choice in designating. Rather it is a variable having thermodynamic significance. As we shall see, the phase rule specifically sets the number of independent chemical variables present (chemical with reference to the effect on composition). Thus, if we add a pure substance to a system we cannot, without additional knowledge, state that we have added only a single component to the system.

The component is a hypothetical quantity in the sense that its designation can be symbolized in many ways so long as we retain the proper overall chemical stoichiometry. Once having arbitrarily defined the molecularity of a component, a molecular weight based on this designation is automatically established, and for the system in question the number of moles of this component must be conserved. Note, however, that this number of moles does not necessarily relate to the number of moles of a species that may be present; it relates to a reference base for counting purposes. For example, let us consider a substance whose chemical symbol is AB. This substance may behave as a single component in accordance with restrictions imposed by the phase rule which are yet to be discussed. We may choose to define the component molecularity by the symbol A_3B_3. A mole of this component would then possess $6.023 \cdots \times 10^{23}$ hypothetical particles each having the molecularity A_3B_3. Assume now that when present in the gas phase, this component is actually present as discrete atoms of A and atoms of B as

E. THE COMPONENT, SPECIES, AND MOLE TERMS

well as AB molecules. The actual states of aggregation of the component AB, i.e., discrete A and B atoms and AB molecules, as we have mentioned are known as species. The quantities of specific species may vary with temperature or composition of the system. What will not change are the quantities of the components from which the different species are generated.

Thus, if we redefine the component molecularity as AB and designate the number of moles of this component by the symbol m_{AB}, we know that independent of any changes undergone by the system

$$m_{AB} = \text{constant}. \tag{26}$$

Let us designate the moles of species present by the symbols n_A, n_B, and n_{AB}. In order for m_{AB} to remain a constant it follows that

$$m_{AB} = \text{constant} = n_A + n_{AB} = n_B + n_{AB}. \tag{27}$$

Since the complete dissociation of one mole of AB species gives rise to one mole each of A and B species, it is obvious that both cannot be counted simultaneously to determine how many moles of the component AB, m_{AB}, they came from. As can be seen from Eq. (27), if we know either n_A or n_B and add this quantity to the quantity n_{AB}, it will tell us the value of the constant quantity m_{AB}.

Let us assume that we had originally designated the component by the formula A_2B_2, and that the species actually present are still A, B, and AB. The conservation equation for the component A_2B_2 would now be given by

$$m_{A_2B_2} = (n_A/2) + (n_{AB}/2) = (n_A + n_{AB})/2 = \text{constant}, \tag{28}$$

or

$$m_{A_2B_2} = (n_B + n_{AB})/2 = \text{constant}. \tag{29}$$

When a system of given mass is being specified, it is evident that we have wide latitude in designating component molecularity and therefore component composition. For example, in the cases just discussed let us assume that the system contains the two components AB or A_2B_2 and the component C. The composition of the system is conveniently expressed in terms of a quantity termed the mole fraction. This is simply the ratio of the moles of the component whose composition is being specified divided by the total number of moles of all components present.

If we have 100 grams of the component AB (whose gram molecular weight we will assume arbitrarily to be 50), then we have two moles of the component AB present. Let us assume that the component C has a molecular weight, based on this designation of its molecularity, of 200 and that 200 gm of C are added.

The mole fraction of the component AB designated M_{AB} is given then by

$$M_{AB} = m_{AB}/(m_{AB} + m_C) = 2/(2 + 1) = 0.067. \tag{30}$$

The mole fraction of C present is given by

$$M_C = m_C/(m_{AB} + m_C) = 1/(2 + 3) = 0.333. \tag{31}$$

Note now that the sum $M_{AB} + M_C$ is given by

$$M_{AB} + M_C = [m_{AB}/(m_{AB} + m_C)] + [m_C/(m_{AB} + m_C)] = 1. \tag{32}$$

The sum of the mole fractions of the components comprising a system is equal to unity.

Suppose, starting with the same mass of each component but choosing the molecularity A_2B_2 for the component we previously termed AB, we recalculate the mole fractions of the components. The moles of $A_2B_2 = 1$ since its molecular weight is based now on the molecularity A_2B_2

$$M_{A_2B_2} = m_{A_2B_2}/(m_{A_2B_2} + m_C) = 1/(1 + 1) = 0.5, \tag{33}$$

$$M_C = m_C/(m_{A_2B_2} + m_C) = 1/(1 + 1) = 0.5. \tag{34}$$

Thus for a given mass of each component we see that the composition in terms of mole fractions depends on how we choose to designate the molecularity of a given component. On the other hand, if we choose to define the composition of our system in terms of mass fractions, these latter simply being the mass of the component in question divided by the total mass, we would find that independent of how we choose to designate the components their mass fractions would remain the same.

In the first example given above where we designated one of the components by the symbol AB we would obtain the following: Using the symbols g_{AB} and g_C for the masses of the components AB and C, respectively, and the symbols f_{AB} and f_C for their respective mass fractions we obtain

$$f_{AB} = g_{AB}/(g_{AB} + g_C) = 100/(100 + 200) = 0.333, \tag{35}$$

$$f_C = g_C/(g_{AB} + g_C) = 200/(100 + 200) = 0.667. \tag{36}$$

E. THE COMPONENT, SPECIES, AND MOLE TERMS

Assuming A_2B_2 is the molecularity of one of the components we obtain

$$f_{A_2B_2} = g_{A_2B_2}/(g_{A_2B_2} + g_C) = 100/(100 + 200) = 0.333,$$
$$f_C = g_C/(g_{A_2B_2} + g_C) = 200/(100 + 200) = 0.667. \quad (37)$$

We see therefore that when component concentrations are expressed in mass fraction terms, the designation of the component by symbols implying different molecularities does not change the composition of the component in the mass system. When expressing the composition in the mole system, the composition depends completely on the assumed molecular formula.

Intuitively, it can be seen that thermodynamic treatments might be best handled in the mass system since this system is unaffected by how we choose to define our components. Thus, we might wish to discuss the variation of the total system enthalpy caused by addition of a quantity of a component, i.e., $\partial H/\partial X_y$ where H is the enthalpy and X is a quantity term of the component y. If we operate in the mass system and specify g_y, no confusion can occur even if we define the molecularity of the component by y_2 or y_3 or y_n. On the other hand, if we operate in a mole regime, it is seen that the quantity that need be added (mass quantity, that is) depends upon whether we use as our component designation y, y_2, y_3, etc.

Most equations of thermodynamics are, however, specified with reference to mole quantities and consequently it is more convenient to use these, particularly as the relationships of greatest utility are quasi thermodynamic and are dependent on species mole quantities.

In order to apply the criterion of conservation of component quantities it is imperative that not only must we be able to conserve the component quantity for the system as a whole, but also for each of the phases in which it is present. For example, if we start with a solid chemical whose stoichiometry is AB and find that upon heating it, it vaporizes in such a way that only particles having the stoichiometry A go into the vapor phase while all particles having the stoichiometry AB and B remain in the solid phase, it would be inconvenient to do our component counting for each phase in terms of the component AB or any multiple of this designation. It would be more convenient to choose A and B as components of the system, since the component composition of each phase may then be designated always in terms of positive numbers. This requirement is not purely one of convenience as we shall see in the following section where thermodynamically via the phase rule the minimum number of components is predicted.

F. Defining the Number of Components Present

In theory, two approaches might be applied in determining the minimum number of independent chemical terms possessed by a system. The first is concerned with the maximum observable number of phases. The second approach depends on the conservation of component stoichiometry in coexistent phases. In the limit, the total number of components present in a system must coincide with the total number of elements present. In practice, however, over relatively wide ranges of interest, a substance consisting of two or more elements may behave as a component. This behavior traces back to a restriction imposed by the phase rule which we reiterate here for the case where the parameters affecting the entire system simultaneously are the pressure and temperature

$$V = C - P + 2. \tag{38}$$

If the number of components, C, is one, the number of phases present when $V = 0$ is given by

$$P = C + 2 - 0 = 3. \tag{39}$$

Since the system cannot possess less than zero degrees of freedom, Eq. (39) immediately sets an experimental criterion for the number of components present in a system. Namely, a one-component system may not exhibit more than three phases coexisting simultaneously. A two component system may not exhibit more than four phases coexisting simultaneously. Thus, if we study a system over a wide range of conditions and do not observe more than three phases coexisting in equilibrium situations, the system is behaving as a one-component system. A chemical such as H_2O behaves in this manner over wide ranges of temperature and pressure conditions and may be treated as a component. Other substances, such as hydrates, e.g., $CuSO_4 \cdot 5H_2O$ do not behave as single components, i.e., more than three phases may be found to exist simultaneously. When four phases are observed, e.g., saturated solution, solid $CuSO_4 \cdot 5H_2O$, solid $CuSO_4 \cdot 3H_2O$, and vapor containing only H_2O, the system is binary in nature with respect to its component number. The second distinguishing attribute relative to component number as mentioned is the conservation of the stoichiometry of the component in each of the phases in which it is present. Frequently, the phase rule criterion of total number of phases may not be easily evaluated. However, the consequence of this limitation, i.e., phase conservation of stoichiometry of a component, may be readily determined.

If, for example, in the case of solid $CuSO_4 \cdot 5H_2O$ it is found that upon formation of a vapor phase this latter has the stoichiometry H_2O, it is clear that this stoichiometry cannot be reconciled with the component stoichiometry $CuSO_4 \cdot 5H_2O$. Having partially dissociated, if the solid phase is now separated from the other (or others), it also cannot be reconciled with a component stoichiometry $CuSO_4 \cdot 5H_2O$ there now being a deficiency of H_2O in this solid phase.

Thus, while the designation of a component in terms of its molecular formula has wide latitude, there is limited latitude in choosing the *number* of components present. The freedom of designation arises only because it is immaterial how the component quantity is conserved, i.e., what we use as our basis for counting. The number of components present, however, is directly concerned with the degrees of freedom possessed by the system. Here the thermodynamic behavior, rather than a convenience, is the determining factor.

In a real case, any latitude attendant with the specification of the number of components arises as a consequence of the sensitivity of the experimental monitoring equipment. As mentioned, in the limit, the number of components present must equal the number of elements present. If, for example, our equipment is unable to detect the dissociation of H_2O, and component phase stoichiometry is maintained within the limits of experimental detectability, a system containing H_2O will always appear as if water adds only a single component.

G. Chemical, Compound, Substance, and Constituent

Up to this point we have used these terms assuming *a priori* knowledge on the part of the reader as to their meaning. We will now attempt brief definitions of these terms as employed throughout the text.

The term *chemical* refers to a material having a known stoichiometry and stability in the condensed or vapor phase over at least some range of conditions. A chemical has some of the attributes of a component in that its stoichiometry is definable. A chemical need not, however, behave as a single component, e.g., if its stoichiometry is not conserved in coexistent phases. Furthermore, whereas a chemical once given a designation is generally always referred to in the same way, the method by which it is counted when treated as the source of a component (or components) need not be the same each time. Thus HCl is a chemical. For purposes of counting in systems containing HCl we might choose to use the designation H_2Cl_2. Thus while HCl is a chemical present in this system we may, if we choose, prefer to count it as the component H_2Cl_2.

$CuSO_4 \cdot 5H_2O$ is a solid chemical but, as pointed out, the designation $CuSO_4 \cdot 5H_2O$ cannot, in general, be used for a single component. Thus while the term chemical refers to a physically definable state of something, the designation of a chemical, e.g., $CuSO_4 \cdot 5H_2O$ need not necessarily represent a useful designation for the hypothetical counting quantity we call a component. The chemical then may represent the physical entity from which the components are subsequently defined.

The term *compound* has a somewhat more restrictive connotation than the term chemical. Compounds are chemicals containing two or more elements, and possessing distinct chemical and physical properties such as crystallographic lattice constants, indices of refraction, etc. More restrictively, however, we shall employ the term to denote a chemical of fixed stoichiometry that is capable, hypothetically at least, of exhibiting a maximum in a solubility curve and existing in a completely ordered lattice state.

A *substance* is any physically viable material and is more general in connation than either the term chemical or compound. Thus, specification of the term chemical or compound implies a knowledge of its stoichiometry or composition. Specification of a material as a substance need not imply a knowledge of stoichiometry, e.g., a vaporous substance as opposed to the chemical HCl in the vapor state.

The term *constituent* is synonymous with the term chemical and assumes a knowledge of stoichiometry, e.g., HCl is a constituent of the vapor phase. It is further employed to denote species molecularity, e.g., the specie, HCl, is a constituent of the vapor phase.

4

The Thermodynamic Basis of the Phase Rule

A. INTRODUCTION

In the preceding chapter, general concepts to be used in subsequent discussions were presented. In addition, the phase rule has been discussed without derivation, in order to point out the fact that a given experimental situation has associated with it a unique, albeit an unknown set of restrictions. Thus, while a given system possesses a set number of degrees of freedom ranging from 0 to n, it is not always obvious how many the system actually possesses. This is not surprising since thermodynamics is a tabulation of the sum total of man's experiences with energy, heat, and work, namely an ordering of experimental encounters. Each new experimental encounter then, is a new thermodynamic experience and it is the problem of the scientist to determine how the results of this new encounter fit into the heretofore established order.

In this chapter, we will present a derivation of the phase rule and then, with a semiquantitative discussion of the process of evaporation of a pure liquid, point out the difficulty of the task of defining equilibrium situations.

B. The Phase Rule and Its Basis

Consider the addition of a known quantity of the chemical AB to a gas composed entirely of the chemical A. We might ask the question, "What is the change of volume accompanying this addition?" If, based on the molecular formulas A and AB, we determine a molar volume for A and AB, then to a first approximation the final volume of the resulting system, V_s, is given by

$$V_s = \text{moles AB} \cdot V_{AB} + \text{moles A} \cdot V_A, \tag{1}$$

where V_{AB} is the molar volume of AB based on this assumed molecularity. Similarly for A. Such an approximation depends on two things: First, for Eq. (1) to hold, it is necessary that upon adding a mole of the chemical designated by the symbol AB, the vapor be enriched by 6.023×10^{23} particles. Second, it is necessary that each of the molecules of AB in the vicinity of A atoms be unable to differentiate these A-type particles from AB particles and vice versa for the A particles.

Suppose, however, that the chemical denoted by the symbol AB, which, it will be recalled, refers to overall stoichiometry and not necessarily to species molecularity, is present as a mixture of A, B, and AB species, the relationship among them being given by an equilibrium constant, K, as shown in Eq. (2)

$$K = \frac{[A][B]}{[AB]} \tag{2}$$

where each of the bracketed terms is a composition.

Obviously, if the chemical AB is added to the gaseous A (let us assume this latter gives rise to species having only the molecularity A), the composition of A in Eq. (2) will be greater than when only pure AB is present alone. Since K is a constant at a particular temperature, it is evident that if the quantity of A is increased, the quantities of B and AB species must change accordingly in order that K does not change. In fact, [B] will decrease while [AB] will increase.

In general, the densities of the chemicals A, B, and AB, when in the pure state, are not the same. Consequently, the effective molar volumes of the two chemicals A and AB will not be the same once they are mixed. Depending, then, on how extensively the dissociation of AB is in the pure state at the temperature in question, Eq. (1) will be a good or poor approximation. An example of an interaction in which an equation like (1) is a very poor approximation is that between ethyl alcohol and water where a large volume decrease attends mixing. In the absence of information on what is occurring in such a mixture, or what causes the

B. THE PHASE RULE AND ITS BASIS

shrinkage, we generally refer to phenomena which are not additive with respect to certain extensive properties as nonideal.

Having no information concerning the molecular aggregation of a chemical present in one or more phases of a system, it would be a difficult, if not impossible, task to attempt representation of changes in the extensive properties such as E, H, F, S, and A (internal energy, enthalpy, Gibbs free energy, entropy, and Helmholtz free energy, respectively) in terms of species properties. Our approach, however, may be made completely general if we simply consider the effect on these properties caused by the addition of a component mole to a system of one or more phases. This leads to a restriction which lies at the heart of the derivation of the phase rule.

In considering a system of one or more phases, properties such as the ones listed above may be considered in two different contexts. For example, if the system is of constant composition and consists of a single phase, then as a function of T and p, the variation of the property may be defined generally by

$$\partial X = (\partial X/\partial T)_p \, dT + (\partial X/\partial p)_T \, dp \tag{3}$$

More explicitly the Gibbs free energy would be written

$$dF = -S \, dT + V \, dp \tag{4}$$

Equations (3) and (4) refer specifically to closed systems, closed, that is, with respect to compositional variation. In the most general sense these equations are not restricted to single phase systems although it is most convenient to do so.

When we consider the effect on the free energy of a phase caused by the addition of a quantity of a component A to the system, Eqs. (3) and (4) must be modified to include the effect of composition on the property in question. Systems to which material may be added are generally termed open systems. If a system is composed of several phases among which the components may distribute, it is termed internally open.

Rewriting Eqs. (3) and (4) for single phase systems so as to allow for compositional variation due to changes in component quantities in addition to changes in T and p we obtain

$$\partial X^{(1)} = (\partial X/\partial T)_{p,m_A,m_B\ldots} \, dT + (\partial X/\partial p)_{T,m_A,m_B\ldots} \, dp + (\partial X/\partial m_A)_{p,T,m_B\ldots} \, dm_A^{(1)}$$
$$+ (\partial X/\partial m_B)_{p,T,m_A} \, dm_B^{(1)} \cdots. \tag{5}$$

$$dF^{(1)} = -S^{(1)} \, dT + V^{(1)} \, dp + (\partial F^{(1)}/\partial m_A)_{T,p,m_B\ldots} \, dm_A^{(1)}$$
$$+ (\partial F^{(1)}/\partial m_B)_{T,p,m_A\ldots} \, dm_B^{(1)} + \cdots. \tag{6}$$

The superscripts refer to the phase in question and the terms m_A, m_B represent a mole of the components A, B, etc.

For each coexisting phase, an equation of type (6) would be specified.

The partials $(\partial F^{(1)}/\partial m_i)_{T,p,m_j,\ldots}$ represent the variation of the free energy of the phase per mole of added component at constant T, p and constant quantities of the other components. It is indeed the slope of the curve on a graph of free energy versus composition of component i. This variation of free energy with component quantity is termed a partial molar free energy, or chemical potential, and is generally designated $\mu_i^{(a)}$. Using this notation we may rewrite Eq. (6) as

$$dF^{(1)} = -S^{(1)}dT + V^{(1)}dp + \mu_A^{(1)} dm_A^{(1)} + \mu_B^{(1)} dm_B^{(1)} + \cdots . \quad (7)$$

At constant T and p this reduces to

$$dF^{(1)} = \mu_A^{(1)} dm_A^{(1)} + \mu_B^{(1)} dm_B^{(1)} + \cdots = \sum_i \mu_i^{(a)} dm_i^{(a)} . \quad (8)$$

In a system at equilibrium made up of several phases containing i components, dF system $= 0$. In other words,

$$dF_{\text{system}} = dF^{(1)} + dF^{(2)} + \cdots = \sum_a dF^{(a)} = 0. \quad (9)$$

Substituting for the $dF^{(a)}$ terms the expansion of $dF^{(a)}$ provided in Eq. (8), we obtain

$$0 = \sum_a \sum_i \mu_i^{(a)} dm_i^{(a)}. \quad (10)$$

If now the system is externally closed (i.e., its composition is fixed) but internally open (i.e., transfer of components between phases is possible), it is obvious that for each component the sum of its variations in the several phases must be given by

$$dm_i^{(a)} + dm_i^{(b)} + dm_i^{(c)} + \cdots = 0. \quad (11)$$

In order for Eq. (10) to be valid within the constraint imposed by Eq. (11), it follows that

$$\mu_i^{(a)} = \mu_i^{(b)} = \mu_i^{(c)} \cdots . \quad (12)$$

In other words, it is a consequence of the criteria for chemical and physical equilibrium that the chemical potential of a component must be equal in all coexisting phases.

B. THE PHASE RULE AND ITS BASIS

The conclusion arrived at in Eq. (12) is the basis for all arguments relating to heterogeneous systems (those containing more than one phase) in a state of equilibrium. It immediately opens the door to defining the variance of a system of one or more phases in terms of the variables T, p, and number of components C.

If there are P coexisting phases and C components are distributed among them, then for each phase there are only $C - 1$ composition terms that need be specified to show the compositional state of the system, the remaining composition being determinable by subtracting the sum of the $C - 1$ known values from one. For P phases then, the maximum number of composition terms that need ever be specified is $P(C - 1)$. At equilibrium, both thermal and mechanical equilibrium must obtain, that is, the temperature and pressure must be uniform through the system. Thus, in addition to the $P(C - 1)$ composition terms that might be specified, the temperature and pressure should be specified. Therefore, the maximum number of specifications that might be made to define the state of the system is

$$\text{number of possible specifications} = P(C - 1) + 2 \tag{13}$$

This number of specifications is potentially very large, e.g., in a system of many components distributed among many phases. Note, however, that the chemical potential defines the variation in free energy as a function of m_i at a specific composition, and that if the value of the composition of a component in a phase is specified, its chemical potential in that phase and all coexisting phases is fixed. Not only is it fixed but it is the same in all coexisting phases. In other words, if the composition of a component in one phase is known, then its composition in all coexisting phases becomes fixed, and not capable of independent variation. Its composition in all other phases must be such that its chemical potential is the same in each phase. Thus for each component i present in P phases all but one of the composition values for that component are dependent values. For each component present in P phases then, $P - 1$ composition terms are not independently variable. For C components in P phases then, $C(P - 1)$ of the total possible number of composition terms are not independently variable. This number subtracted from the total number of composition specifications that might be made provides us with the total number of concentration specifications that may be varied independently. This number represents the degrees of freedom or variance V (freedom of independent variation) possessed by the system. Performing this operation we obtain

$$V = P(C - 1) + 2 - C(P - 1) = C - P + 2. \tag{14}$$

Our derivation of the phase rule assumed that each of the components was present in each of the coexisting phases. It is a simple matter to demonstrate that Eq. (14) is the result even if all of the components are not distributed among all of the phases.

Assume phase 1 contains $C - X$ of the number of components C present in the total system where X has any value from 0 to $C - 1$ (a phase must consist of at least one component). To define completely the composition of phase 1, the composition of all but one of the components present must be known. There are at most, therefore, $C - X - 1$ composition terms necessary to specify the composition of phase 1. Assume phase 2 contains $C - X'$ components where X' may have any value from 0 to $C - 1$. We need then $C - X' - 1$ composition terms at most to define the composition of phase 2.

For P phases therefore, the maximum number of composition terms that need be specified in addition to the temperature and pressure is given by Eq. (15)

$$\text{specifications} = C - X - 1 + C - X' - 1 + C - X'' - 1 + \cdots$$
$$= PC - \sum_P X - (P \cdot 1) + 2. \qquad (15)$$

Phase 1 is defined by $C - X$ chemical potential terms, phase 2 by $C - X'$, etc., the total number being $PC - \sum_P X$. Of these, all but C of the chemical potential terms are superfluous because of the criterion of chemical potential equality for a component in coexisting phases.

Thus the excess number of potential terms is

$$PC - \sum_P X - C = \text{excess number of compositions}. \qquad (16)$$

Subtracting Eq. (16) from Eq. (15) provides us with the independent number of variables

$$V = PC - \sum_P X - P + 2 - \left[PC - \sum_P X - C \right] = C - P + 2, \qquad (17)$$

which is the same result arrived at in Eq. (14).

C. Choice of Systems and the Reduced Phase Rule

Under any set of experimental conditions where one or more of the parameters, T, p, or one or more of the compositions is maintained constant, it is obvious that the variance of a system is reduced appropriately. In many experiments it is convenient to consider the state of a

D. VAPOR PRESSURE AND EVAPORATION RATE

system under conditions of constant pressure. Since the pressure is not varied, the number 2 in the phase rule is reduced to the number 1 and

$$V = C - P + 1. \tag{18}$$

To apply the "reduced" phase rule properly to constant pressure experiments, i.e., an experiment conducted open to our ambient atmosphere, it must be ascertained that the mass of the system is constant during the course of the experiment. This requires that the evaporation rate of the components be small. It is also necessary that the constituents of the atmosphere be neither soluble in, nor reactive with the constituents of the system. Otherwise, the number of components present will be different than expected.

Having ascertained that these criteria are satisfied, studies conducted on the system obviously relate only to condensed phases, no vapor phase being assumed present. If the identical system is studied in an environment in which the pressure is that of the system itself, the variance of the system remains the same since, concomitant with an increase in the number 1 to the number 2, the number of phases also increases to take into account the presence of a vapor phase.

In general, the applicability of a reduced phase rule to a system is based on the assumption that the vapor pressures rather than the evaporation rates are low. While for many substances vapor pressure and evaporation rates track one another, this need not be the case.

D. THE VAPOR PRESSURE, THE EVAPORATION RATE, AND THE REDUCED PHASE RULE

Since it is often convenient to examine heterogeneous systems in containers open to the ambient atmosphere, or an otherwise inert environment at 1 atm pressure, it is of interest to consider the implications of terms such as low evaporation rate and low vapor pressure.

The gaseous pressure measured in a system of condensed phases in equilibrium with vapor is termed the vapor pressure of the system. The rate at which this equilibrium vapor pressure is achieved when the condensed phases are allowed to evaporate into an evacuated chamber of appropriate dimensions (one in which only sufficient free space is available to allow for a vapor phase, but not large enough to cause complete evaporation of the condensed phases) is termed the evaporation rate.

If a solid A is placed in an evacuated chamber of volume V at a temperature T it will tend to evaporate. The rate of such evaporation r_e, in number of evaporating particles per unit time, will be proportional to the total number of particles exposed to the environment at the surface, n_o, at a given time, and a probability factor k_e defining the number per unit time per unit area that will leave the surface. k_e may be termed the evaporation rate coefficient or constant; r_e may be termed the evaporation or volatility rate of the material. This rate, assuming that n_e is a constant (the surface area is assumed constant), is the steady-state rate at which particles leave the surface of a condensed phase and enter a vapor available region. It is not the net rate of departure, which varies as a function of time, since a reverse process must begin after the first particle has evaporated. This reverse process is termed condensation and may be defined in terms of a condensation rate r_c.

The evaporation and condensation rates may be expressed as

$$r_e = n_e k_e \tag{19}$$

and

$$r_c = n_c k_c, \tag{20}$$

respectively, where r_c is the rate of condensation, n_c is the number of particles striking the surface per unit time and k_c is the probability that a strike results in condensation. k_c is a condensation rate constant. The rates of evaporation and condensation, r_e and r_c, respectively, may be expressed as a number per unit time per unit area. If we then express n_e and n_c as the number at the surface per unit area, then k_e and k_c are given by (unit time)$^{-1}$.

Since condensed phases are essentially unchanged by a buildup of pressure over them (they are essentially incompressible), n_e, and therefore r_e, may be considered constant at constant temperature.

The number n_c may not, however, be considered constant, except when equilibrium obtains; its magnitude depends on the gaseous pressure of A at any given time during the evaporation process. n_c may be written as

$$n_c = cp, \tag{21}$$

where c is a constant and p is the pressure at the instant of measurement. If we assume ideal gas behavior [Eq. (22)]

$$pV = nRT, \tag{22}$$

then

$$n = pV/RT, \tag{23}$$

where n is the number of moles of species present in the vapor, p is the pressure, V is the volume, R is the gas constant, and T is the temperature.

D. VAPOR PRESSURE AND EVAPORATION RATE

Designating Avogadro's number as N, we see that

$$n_g = N \cdot pV/RT, \tag{24}$$

where n_g is the number of particles present in the vapor.

The variable n_c, however, represents only those particles present at the surface. n_c, therefore, is proportional to n_g and represents a fraction of it.

$$n_c = fn_g = fN \cdot pV/RT, \tag{25}$$

from Eqs. (25) and (21) we see that

$$c = fNV/RT. \tag{26}$$

When our hypothetical system achieves a state of equilibrium, the pressure of the system acquires a constant value p_{eq}, the equilibrium vapor pressure. This means of course that r_e and r_c are macroscopically equal. Up to this point r_e has always been greater than r_c, as r_c was increasing.

At equilibrium, then,

$$r_e = r_c \tag{27}$$

or

$$n_e k_e = c p_{eq} k_c. \tag{28}$$

From Eq. (28) we see that

$$p_{eq} = n_e k_e / c k_c = (n_e RT/fNV) \cdot k_e/k_c. \tag{29}$$

Note that whereas r_e and r_c are proportional to k_e or k_c, respectively, p_{eq} is proportional to the ratio of these rate constants. The physical implications are significant. The constants k_e and k_c may both be small and similar in magnitude, in which event p_{eq} is small, and either p_{eq} or k_e enables us to assess the applicability of the reduced phase rule in a given situation since they track each other. On the other hand, k_e and k_c may both be large and of similar magnitude. Here, while p_{eq} will again be small, the vapor pressure and evaporation rates will not track. A system made up of such a material (being studied in a system open to the environment) will be changing composition rapidly as a function of time, and the reduced phase rule will not be applicable. A similar inapplicability arises when k_e is large and k_c is much larger than k_e. p_{eq} will, in this instance, be a very small number but the system will change composition rapidly when exposed to the ambient atmosphere.

The saving feature that enables a certain level of casualness on the part of the investigator is the fact that the term n_e/c is about the same for most substances (the solid densities are similar as is the ideality of behavior of the gas phase at low pressures), and the condensation rate constant k_c (sometimes called a sticking probability coefficient) probably does not vary much for many materials. With this the case, for the most part, we may then, with trepidation, assume that

$$k_e \propto p_{eq}. \tag{30}$$

E. The Time Factor in a Non-Time-Dependent Science

While thermodynamics does not contain time as a parameter, the fact that experiments pertaining to equilibrium must be conducted within a reasonable time period inadvertently makes time a part of any experimental result. The question that must be answered is whether, within the time allotted to attain equilibrium and within experimental limits of detectability, the system has come to some state of rest, or whether it is in a transient state. This introduces a concept known as a time constant τ for an experiment. For exponential relationships (which for many natural processes is the case), this constant is the time it takes to achieve a value $1/e$ of the final value of the property being measured.

Consider, for example, the experiment discussed in the last section. The net buildup of vapor pressure is directly proportional to the change in the number of particles present in the vapor available space dn_g, this number being different from the number n_c at the surface. This buildup at any chosen time is related to the instantaneous difference in evaporation and condensation rate as shown in Eq. (31):

$$dn_g/dt = r_e - r_c = n_e k_e - cpk_c. \tag{31}$$

Assuming ideal gas behavior, we may substitute for n_g the term $c'p$, where c' is simply NV/RT from Eq. (24). For a system of constant vapor available volume, c' is a constant. Therefore, from Eqs. (24), (26), and (31) we see that

$$dp/dt = (n_e k_e/c') - (c/c')pk_c = (n_e k_e/c') - fpk_c. \tag{32}$$

As noted previously r_e may be taken as constant due to the structural stability of a condensed phase with nominal variation in pressure. For ideal gases at constant volume c and c' are also constants, as is k_c. We may therefore rewrite Eq. (32) more simply as

$$dp/dt = c_1 - c_2 k_c p. \tag{33}$$

E. THE TIME FACTOR IN A NON-TIME-DEPENDENT SCIENCE

Integration of this equation ($du/dx = A - BY$) between the limits $p = 0$ at $t = 0$ and $p = p_{eq}$ at $t = t_{eq}$ yields

$$p = p_{eq}[1 - \exp(-t_{eq}c_2 k_c)]. \qquad (34)$$

The time it takes for a value $(1/e)\, p_{eq}$ to be achieved is, as mentioned, termed the time constant τ which in an exponential represented by

$$x = x_0\, e^{-\alpha t}, \qquad (35)$$

is

$$\tau = 1/\alpha.$$

From Eq. (34) we see that τ is proportional to k_c^{-1}. If the condensation-rate coefficient is large, then equilibrium is approached quickly while if k_c is small (the probability of sticking in a collision is small), a long time is involved in attaining equilibrium.

The discussion above points out some of the difficulty encountered in ensuring that equilibrium for a system under study has been approximated.

5

Systems of One Component—Temperature Effects

A. INTRODUCTION

In the limit, there does not exist a real system populated by only a single component. Thus, any single element substance contains impurities and isotopes, and must be contained in something in which some level of mutual solubility exists. Consequently, discussions of single or multiple component behavior must be considered in the context of a given experimental frame of reference, this reference being bounded by limits of experimental sensitivity. A poly-element material such as water behaves, to a first approximation, as if it consisted of a nondivisible unit over a relatively wide range of conditions, and may be conveniently treated as a single component in most cases. On the other hand, a substance such as copper sulfate pentahydrate exhibits behavior characteristic of unary systems over only a very narrow range of conditions, and is more conveniently treated as binary in nature. Even this designation deteriorates with relatively minor environmental changes, necessitating treatment in terms of ternary component behavior. Our discussions will be predicated, therefore, on the assumption that the component number designation of a system is experimentally viable, it being realized that under different conditions than those being considered, the designation may be insufficient.

A. INTRODUCTION

Of the variables (temperature T, pressure p, and concentration M_x) available to describe the state of a system, only the first two have significance in a single component system, the value M_x always being unity. Without recourse to the phase rule, therefore, it is immediately evident that the maximum variance of a unary system is two. Utilizing the phase rule, it is seen that this maximum variance occurs when only a single phase is present, namely in a homogeneous system. When two phases coexist, the independent choice of parameters decreases to one and when three phases coexist, no degrees of freedom are available. Thus, the number of phases that may be observed in a one-component system ranges between one and three. The latter boundary is a unique attribute of one-component systems, and may be employed as a criterion of unary behavior. Thus, if over the entire experimental range of interest no more than three phases are found to coexist simultaneously, the behavior of the system is unary, i.e., only a single component is present. Another criterion of perhaps greater experimental value is the equality of composition of coexisting phases, i.e., if the stoichiometry of each of the coexisting phases is observed to be invariant over the entire range of conditions being examined, the system is unary in behavior.

As noted above, a unary system possesses one degree of freedom when two phases coexist. Consequently, one predicts that if a single component solid is in equilibrium with its vapor, the vapor pressure will be determined uniquely once the temperature is specified. Conversely, given a vapor pressure for such a system, and providing that such a situation is physically possible, there exists only a single temperature at which the single-component solid may be in equilibrium with the vapor at this pressure. An exactly analogous set of restrictions apply to liquid–vapor, solid–solid, or solid–liquid coexistences in a single component system. In the last two cases, the pressure considerations refer to hydrostatic pressures rather than to vapor pressures since no vapor-available space exists.

Certain situations obtain where the behavior of a unary component is examined in the presence of an inert gas, this inert gas exerting a pressure on a condensed phase beyond that exerted by the vapor of the material itself. Such a system is only pseudounary in nature, it being assumed that the "inert" gas neither reacts with nor dissolves in the condensed phase, and does not interact with the vapor derived from the condensed phase. One other situation merits comment in introduction, namely that relating to the effects of surface area. The behavior of bulk systems containing condensed phases depends, for reproducibility, on the assumption that the geometries of the individual aggregates comprising a phase be such as to obviate effects due to surface tension.

We will, at the conclusion of our discussions, present brief treatments describing the effects of inert gas pressures on vapor pressures and of area variations of idealized geometrical shapes.

B. The Analytical Description of Univariance

In a system at equilibrium, it is a thermodynamic conclusion that the molar free energies of a given component in coexisting phases are equal. We may write for the total derivative of the Gibbs' free energy of the component in a condensed phase in equilibrium with its vapor

$$dF^{(1)} = -S^{(1)}\, dT + V^{(1)}\, dp. \tag{1}$$

The superscript (1) refers to the phase of lower heat content, in this instance the condensed phase, F to the molar Gibbs free energy, S to the molar entropy, V to the molar volume, and p to the pressure in the system (in this instance to the vapor pressure of the condensed phase). The total derivative of the gas phase Gibbs free energy is given by Eq. (2) where the superscript (2) refers to the phase of higher heat content among the two coexisting phases necessary to define a condition of univariance.

$$dF^{(2)} = -S^{(2)}\, dT + V^{(2)}\, dp. \tag{2}$$

The equivalency of the molar free energies in each of the coexisting phases permits us to equate Eqs. (1) and (2) to give

$$-S^{(2)}\, dT + V^{(2)}\, dp = -S^{(1)}\, dT + V^{(1)}\, dp. \tag{3}$$

Collecting terms yields

$$\frac{S^{(2)} - S^{(1)}}{V^{(2)} - V^{(1)}} = \frac{dp}{dT}. \tag{4}$$

The entropy difference $S^{(2)} - S^{(1)}$, written ΔS, is by definition related to the enthalpy change accompanying the process of isothermally vaporizing one mole of condensed phase by the relation

$$\Delta S_\text{v} = \Delta H_\text{v}/T, \tag{5}$$

where the subscript v refers to a vapor-condensed phase process. Consequently, the exact Eq. (4) may be rewritten in the equally exact form

$$\Delta H_\text{v}/T\, \Delta V = dp/dT. \tag{6}$$

Equation (6) represents the fundamental relationship describing the behavior of a univariant equilibrium in a one-component system. Although we have derived it in terms of a condensed-phase–gas-phase equilibrium, it is evident that an identical relationship results from a consideration of solid–solid or solid–liquid coexistences.

C. Univariant Equilibria Involving Solids and Gases

When treating a solid–gas equilibrium, the basic relationship (6) may be modified conveniently to yield an approximate equation more amenable to use and requiring less factual information. Since the molar volume of a solid is many times smaller than the molar volume of its coexistent vapor, the term ΔV may be set equal to the gas molar volume. This volume V_g, by dint of the perfect gas equation may be set equal to RT/p assuming the gas phase behavior is perfect. Substituting these approximations into Eq. (6) leads to

$$\Delta H_v \, dT/RT^2 = dp/p. \tag{7}$$

Integrating between the limits p_2 at T_2 and p_1 at T_1, assuming constancy of ΔH_v with temperature, leads to

$$\ln(p_2/p_1) = (-\Delta H_v/R)[(1/T_2) - (1/T_1)]. \tag{8}$$

It is seen that a plot of the logarithm of the vapor pressure versus the reciprocal of the absolute temperature provides us with a slope equal to $-\Delta H_v/R$. Integration of Eq. (7) to yield the indefinite integral provides the analytical relationship of vapor pressure versus temperature (Eq. 9):

$$\log_{10} p = (-\Delta H/2.303 \, RT) + C. \tag{9}$$

The constant of integration may be evaluated once ΔH_v has been determined either by the use of Eq. (8) or directly from a semilogarithmic plot. Dimensionally, if p is in atmospheres and ΔH is in calories, R must be given in units of calories per mole degree.[1]

---

[1] When one plots the logarithm of the pressure versus the reciprocal of the absolute pressure for a given set of data representing a real system whose behavior is nearly ideal, it is difficult to tell whether the slope is indeed linear. It is apropos, in general, to refer to the derived slope as the temperature coefficient rather than as the heat of the process. The determination of the ΔH for the process as well as comparison of the derived temperature coefficient with the ΔH can be made by invoking the so called "third law" test which makes use of free energy functions in the following manner. The basic relation-

A plot of vapor pressure versus temperature on linear coordinates is exponential and has the appearance shown in Fig. 1.

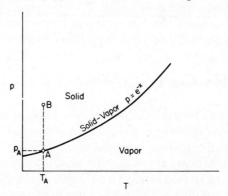

FIG. 1. The vapor pressure curve of a solid.

At any point along the univariant curve, a solid is in equilibrium with a vapor. Below the line, at any temperature in question, no solid can exist and above the line no vapor can exist. Experimentally, the different states of the system are achievable as follows. If into a thermostated, evacuated cylinder fitted with a piston, all at temperature T_A, we introduce 1 mole of vapor and then reduce the volume, the pressure will rise along T_A–A.

If the gas is perfect, the pressure at any point along the line T_A–A will be given by

$$p = RT_A/V, \qquad (10)$$

---

ship between the standard free energy, enthalpy, and entropy of a process is given by

$$\Delta F^0 = \Delta H^0 - T\,\Delta S^0. \qquad (a)$$

The standard free energy for the process is given further by

$$\Delta F^0 = -RT \ln K = -RT \ln p, \qquad (b)$$

for a unary solid in equilibrium with its vapor where K is the equilibrium constant and p the vapor pressure.

From Eq. (a) we see that

$$\Delta S^0 = (\Delta H^0 - \Delta F^0)/T = -(\Delta F^0 - \Delta H^0)/T. \qquad (c)$$

Values of the quantities $[(F^0 - H^0)/T]$ are obtainable for each of the phases from free energy function tables, enabling the calculation of ΔS^0. Once this quantity is known, ΔH^0 can be computed since ΔF^0 is derivable via Eq. (b). If the computed value of ΔH^0 is equal to the temperature derivative of the pressure, then the slope is indeed representative of the heat change accompanying the process and the plot one has obtained is linear.

D. UNIVARIANT LIQUID–GAS EQUILIBRIA

which describes the bivariant behavior of the vapor phase. Since T_A is assumed constant we have removed one degree of freedom from our gas phase and the isotherm T_A–A is effectively a line of gas phase univariance. Thus, the pressure varies inversely with the chamber volume for the special case defined. When the gas pressure reaches a value p_A, where conditions are correct for solid to coexist with its vapor, sufficient solid will precipitate out such that the pressure remains constant. Note, that along the isotherm T_A–A this precipitation or sublimation set of conditions is unique, and that at the point A the system becomes isothermally invariant. Upon continuing to compress the vapor further, no pressure change is possible as long as two phases coexist. In order to maintain this state, solid must continue to precipitate. Since both p and T remain constant during the precipitation process, and recalling that the experiment was begun with 1 mole of the component in the vapor state, it is evident that the number of moles of the component in the vapor decreases with continuing decrease in the volume of the system such that the vapor phase continues to obey Eq. (10).

Continued reduction of the system volume ultimately results in condensation of all of the vapor, a phenomenon which is coincident with the disappearance of all vapor available space. Further compression results in the pressure increasing along the isotherm A–B. Here, as in the region in which only vapor was present, the system is bivariant (or *isothermally univariant*). Thus, in either single-phase region one must specify the pressure and temperature in order to define the system completely.

An equation of state describing the isothermal univariance of a solid phase may be derived from the compressibility relationship

$$K = -(1/V)\,dV/dp, \tag{11}$$

where K is the compressibility, V the volume, and p the pressure. Integration of (11) yields (12) and (13) for the indefinite and definite integrals assuming K to be independent of p.

$$\ln V = -Kp + C, \tag{12}$$

$$\ln V_2/V_1 = -K(p_2 - p_1). \tag{13}$$

As K is generally expressed in units of cm^2/dyn, it is obvious that p should be defined in units of dyn/cm^2 (1 atm = 1.01325×10^6 dyn/cm^2).

D. Univariant Equilibria Involving Liquids and Gases

The relationships (7)–(9) describing the solid–gas univariant behavior are equally applicable to single component liquid–gas phenomena.

Together with the diagram of state for the solid–gas interaction, the liquid–gas diagram is depicted in Fig. 2. The same experiment described in Section C of this chapter may be envisioned for the liquid–gas region

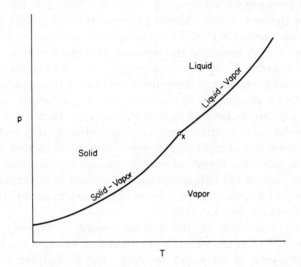

FIG. 2. The vapor pressure curves of a solid and a liquid.

of the diagram, with the regions described for the solid–gas case having exact analogs in the liquid–gas temperature intervals of existence, the first appearance of liquid, however, being termed an evaporation point.

In general, each temperature along the solid–gas curve is termed a sublimation temperature with the ΔH for the process being termed the heat of sublimation ΔH_s. Along the liquid–gas curve each temperature is termed a vaporization temperature, the latent heat of the process being termed the heat of vaporization ΔH_v.

Providing that it is physically possible to achieve a vapor pressure of 1 atm along the sublimation curve, the temperature at which this condition obtains is called the "normal" sublimation point, normal that is for our environment on earth. A similar occurrence along the liquid–gas curve is termed the "normal" boiling point. For either case the implication is that, upon heating in air, the substance cannot develop a vapor pressure greater than 1 atm.

Upon tracing the liquid–vapor curve to higher temperatures, the quantity of vapor molecules per unit volume increases as the temperature increases. This results, therefore, in an increase in the gas phase density. Simultaneously, the increase in temperature causes a diminution in the density of the coexisting liquid phase. At a point called the critical

E. UNIVARIANT SOLID–LIQUID EQUILIBRIA

point, the densities of liquid and vapor become equal and the interface between the two phases vanishes. Above this critical temperature, therefore, only a single phase exists, and univariance is no longer possible. We may define the critical point temperature as that temperature above which a gas cannot be liquified, or more precisely, that temperature above which univariance cannot be realized. Since both the temperature and pressure must be specified simultaneously to define the state of the system above the critical point, it is evident that if in a defined volume we place different quantities of the component and heat to temperatures above the critical point, the pressures generated will be greater at a given super critical temperature for a system containing a larger quantity of the component. This procedure has great practical utility in a technique for synthesis of materials and single crystal growth, referred to as hydrothermal synthesis and growth. Because the solubility of one material in another is a function of pressure, it has been useful to dissolve relatively insoluble materials in aqueous solutions above the critical point of water. The desired pressure values have been achieved by loading the bomb volumes to different levels.

E. Univariant Equilibria Involving Solids and Liquids

Whereas the molar volumes of solids or liquids are small, as compared to those of vapors, the molar volumes of solids and liquids are similar. Consequently, the simplifying assumption employed in Section C of this chapter cannot be invoked in treating condensed phase equilibria. One type of condensed phase equilibrium of interest is that involving a solid and liquid. If, in our previously employed experimental system, a temperature is achieved in a solid–vapor equilibrium at which the solid begins to melt, such as the point x in Fig. 2, a situation exists in which three phases coexist simultaneously, and the system is invariant. If the piston is now lowered so as to compress and thereby condense the gas phase, a point is finally reached at which only solid and liquid occupy the entire volume of the system. At this point the system becomes univariant and further force applied to the piston, with resultant increase in pressure on the condensed phases, in general, demands a change in the temperature simultaneously in order that both solid and liquid continue to coexist. Equation (6) rewritten in the form given by

$$dT/dp = T\Delta V/\Delta H_f, \qquad (14)$$

describes the change in the melting point of a solid with pressure. Since ΔH_f, the heat of fusion, is positive, the sign of the slope of this melting point curve will depend on whether the term ΔV is positive or negative. Noting that

$$\Delta V = V_2 - V_1, \tag{15}$$

where V_2 represents the molar volume of the phase of higher heat content, the liquid in this instance, and where V_1 represents the molar volume of the solid, it is clear that if the solid is denser than the liquid, as is normally the case, ΔV is positive, and the slope of the melting point curve is positive. If on the other hand, the reverse is the case (a typical example being coexisting solid and liquid water), ΔV is negative, and the slope of the melting point curve is negative. If the densities of liquid and solid are identical, the slope becomes infinite since dT/dp must $= 0$. The three cases are depicted in Fig. 3. If, in utilizing Eq. (14), molar

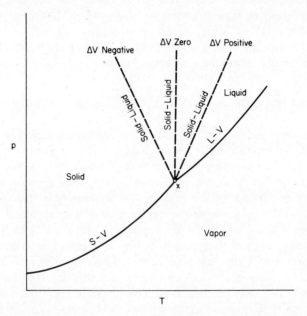

FIG. 3. Variation in the slope of the melting point curve depending on the molar volumes of solid and liquid.

volumes are expressed in liters per mole, the pressure in atmospheres, ΔH_f in calories per mole, and T in degrees absolute, we obtain an equation that, given in units, has the appearance

$$\frac{\text{liters/mole} \cdot (\text{degrees} \cdot \text{atmospheres})}{\text{calories/mole}} = \text{degrees}. \tag{16}$$

For the left-hand side to yield an answer in degrees, the equivalency must exist that

$$x \text{ liter atmospheres} = y \text{ calories.} \tag{17}$$

This equivalency is 0.04 liter atm = 1 calorie. Consequently, Eq. (14) may be written

$$dT = \Delta V \, dp / 0.04 \, \Delta H_f \tag{18}$$

where dT represents the change in melting point that must accompany a change in pressure, dp, in a single component solid–liquid equilibrium and ΔV, dp, and ΔH have the units mentioned. For water at 0°C, the change in melting point with an increase in pressure of 1 atm at low pressures is approximately -0.008°C. Thus, in air one might expect water to melt at -0.008°C rather than at 0°C, its value under its equilibrium vapor pressure.

In Fig. 3 it will be noted that the curves S–V, L–V, and S–L intersect in a point. Such a construction is made necessary by the restriction that a three phase coexistence be invariant. The point intersection of the three univariant curves is termed a triple point, further insight into its occurrence being given in the following section.

F. Further Aspects of Univariant Systems

Since the origin of the analytical expressions describing the univariant behavior of two-phase single-component systems lies in equating free energies of a component in coexisting phases, it is evident that each point along a univariant curve depicts a situation of equiphase chemical potential. Consequently, the p–T diagrams presented represent projections of equipotential states in an F–p–T diagram onto the p–T plane. Since the Gibbs free energy F is a function of the intensive parameters p and T, the three-dimensional representation consists of free energy surfaces. When two surfaces intersect, the lines of phase equipotential (univariance) are generated. This is shown schematically in Fig. 4 in which the intersections of the bivariant surfaces representing solid or liquid or vapor intersect to form the univariant two-phase lines. The point of intersection of the S, L, and V surfaces to form the triple point T_t and the projections of the lines of intersection on the p–T plane are also depicted schematically.

G. Metastability in One-Component Systems

At the lines of intersection of the free energy surfaces discussed in the preceding section, the transitions from regions of bivariance to those

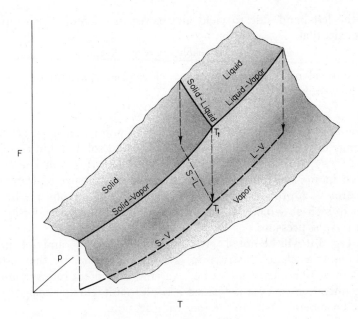

FIG. 4. Schematic representation of free energy surfaces and projections on the p–T plane.

of univariance, or in the special case of the triple point, to invariance, are presumed to proceed smoothly. It is frequently the case in real systems that transitions of liquid to solid, liquid–gas to solid–gas, or solid–gas to solid–gas do not take place under the thermodynamically prescribed set of conditions. Excepting for the solid–gas to solid–gas transition, the nonadherence to the diagram of state generally involves the failure of a phase of lower heat content to nucleate as the phase of higher heat content is either cooled or compressed. The net result is that the higher heat content phase persists to lower temperatures or higher pressures. In these lower temperature or higher pressure states, the higher heat content phase is metastable relative to the thermodynamically stable phase. Nonetheless, the metastable phase continues to obey the relationships describing its behavior in its range of stability, thus the designation *metastable equilibrium*. In the region of metastability, the molar free energy of the component must be greater than it would have been in the stable situation, consistent with the requirement of a negative free energy change accompanying a spontaneous process. An interesting aspect of the free energy effect relates to the relative pressures of stable and metastable states at constant temperature in condensed-gas phase equilibria. This can be seen from the following.

G. METASTABILITY IN ONE-COMPONENT SYSTEMS

Consider the experiment in which the system liquid–vapor smoothly transforms at the triple point T_t to the system solid–vapor, after which it is cooled to the temperature $T_t - \Delta T$ with ΔT not too large. At the conclusion of this process, the vapor phase in equilibrium with the solid is separated from the solid by sliding a shutter across the container as shown in Fig. 5. Next, let us perform the identical experiment excepting

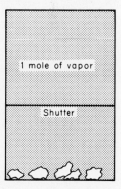

FIG. 5. A hypothetical experiment to determine relative vapor pressures of stable and metastable phases at the same temperature.

that now the liquid–vapor system is supercooled below the triple point to the same temperature $T_t - \Delta T$. The shutter is positioned so that the contained, isolated vapor is comprised of the same number of moles of vapor as in the preceding experiment. Since the systems solid–vapor and liquid–vapor individually fulfill the equilibrium requirement that the molar free energies of coexisting phases are equal we need only consider the free energies of the gas phase. For simplicity let us assume that we have isolated precisely 1 mole of each of the gas phases and then proceed to compare them.

At constant temperature, the derivative of the Gibbs free energy is given by

$$dF = V\,dp. \tag{19}$$

Assuming ideal behavior of the vapors, the relation (19) may be rewritten as

$$dF = RT(dp/p). \tag{20}$$

Integrating between the limits F_2 at p_2 and F_1 at p_1 we obtain

$$\Delta F = RT \ln p_2/p_1. \tag{21}$$

From Eq. (21) we see that in order for a spontaneous change to be possible, that is where $F_2 - F_1$ is negative, F_2 representing the final, more stable state and F_1 representing the initial, less stable state p_1, the pressure in the initial state, must be greater than p_2, the pressure in the final state. Thus, where p_1 represents the pressure in the metastable liquid–vapor system and p_2 the pressure in the stable solid–vapor system at the same temperature, it is necessary that p_1 be greater than p_2 in order for the molar free energy to show a decrease in the process of the system transforming spontaneously from a metastable to a stable state. This restriction on relative vapor pressures in stable and metastable states is depicted in Fig. 6.

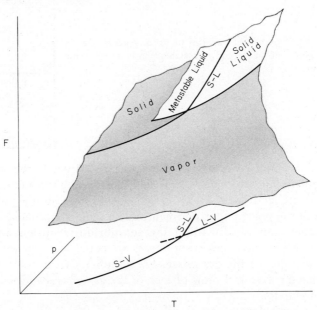

FIG. 6. Metastable extension of the liquid free energy surface.

A similar conclusion can be arrived at by inspection of the p–T projection of the free energy surface intersections. Such a projection is shown in Fig. 7.

If we focus our attention in the vicinity of the triple point, we observe from Eq. (22) that the slope of either the S–V or L–V curve is solely a function of the term $\Delta H_p/RT$, where ΔH_p refers to either the process of vaporization or sublimation.

$$\ln p = (-\Delta H_p/RT) + C. \tag{22}$$

Since the heat of sublimation ΔH_s bears the relationship to the heat of vaporization and fusion

$$\Delta H_s = \Delta H_f + \Delta H_v, \tag{23}$$

G. METASTABILITY IN ONE-COMPONENT SYSTEMS

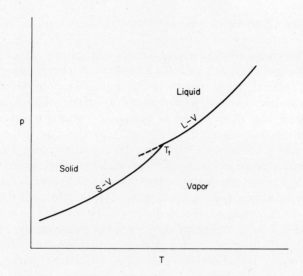

FIG. 7. Projection of the S–V and L–V curves of Fig. 6 on the p–T plane.

it is concluded that the slope of the solid–vapor curve must be greater than the slope of the liquid–vapor curve. Consequently, the continuation of each of these curves beyond the triple point places them above the curve for the stable phase in that temperature interval. Thus, it is seen again that the vapor pressure of a metastable unary system is greater than that of the stable system at the same temperature.

While observations of metastable liquid–vapor extensions have been made during cooling cycles in which the systems have been maintained as unperturbed as possible, many systems are known in which metastability occurs even when efforts are made to induce nucleation of the stable phase via the use of seeding techniques. Such metastable-prone systems are frequently those in which a significant degree of polymerization occurs in the liquid state, e.g., in the freezing of sulphur or selenium. These materials may be supercooled below their solid–liquid–vapor triple points and maintained in metastable glassy states indefinitely. Substances such as water may be supercooled to an extent, but are more or less easily perturbed with a resultant spontaneous evolution of heat and simultaneous change to the stable structure. Other liquid substances may be supercooled somewhat with a subsequent spontaneous conversion to a metastable solid structure which is then converted to the stable solid structure only with great difficulty.

H. Structure Changes in Unary Systems

Many unary or pseudounary materials are known which undergo structural changes in the solid state as a function of temperature, pressure, or both. In general, such structural changes are accompanied by discontinuities in heat content, vapor pressure, molar volumes, and other properties and are referred to as first-order phase transformations. Where such discontinuities are observed, it is the case that the transformation takes place isothermally, namely, at a triple point. Frequently, the isothermal change is kinetically impeded, in which event the ΔH_p for the change is not readily detected because of the slow rate of heat evolution or absorption. In such instances, it is difficult to determine whether or not the change is first order or second order. The term second order has been used to designate other types of phase changes in which, for example, a disordered solid structure becomes more ordered, or in which a primitive liquid structure acquires a different primitive liquid structure, liquid He being the classic example. It has not been demonstrated unequivocally, however, that lambda point transitions (as the second-order transformations are often termed because of the shape of the heat content curves) really do occur. It is possible that such transformations are kinetically limited and also that the ΔH_p's are very small but still finite.

We will concern ourselves solely with temperature-induced transformations of the first order variety, including in our considerations problems associated with kinetically limited transformations which may be termed sluggish or reconstructive. The latter designation has been used frequently because kinetic limitations appear most often when involved reconstitution of lattice spatial arrangements takes place in the transformation. Polyoxide pseudounary systems frequently exhibit so-called reconstructive phase changes.

At the temperature of transition where the three phases — solid 1, solid 2, and vapor coexist, we have a unique situation exactly analogous to the invariant solid–liquid–vapor triple point previously discussed. Again, as with the solid–liquid transitions, the analogous equations to those already derived may be similarly developed. Again also, the curve solid 1–solid 2 may exhibit negative, infinite, or positive slope depending on the molar volumes of the coexisting solids. Figure 8 depicts schematically the case of a solid 1 undergoing a phase transformation to yield the solid 2. It will be noted that the mestastable extensions of the S_1–V and S_2–V curves obey the same relationships developed in Section G of this chapter. The solids 1 and 2 each exhibit stable fields of existence and are alternately termed polymorphs, allotropes, or enantio-

H. STRUCTURE CHANGES IN UNARY SYSTEMS

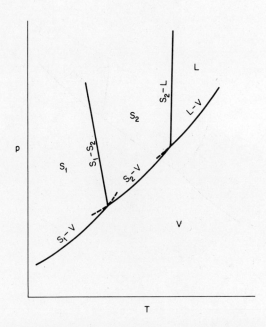

FIG. 8. A one-component system exhibiting a solid–solid phase transformation.

morphs, the first and last terms enjoying greatest usage. As depicted, the S_1–S_2 transformation line exhibits a negative slope indicating that the molar volume of solid 1 is greater than the molar volume of solid 2. Alternately, the density of solid 1 is less than the density of solid 2. The slope of the solid-2–L transformation curve, on the other hand, is positive. This implies that the density of the liquid is less than that of solid 2. It is again a consequence of the additive nature of the enthalpy that the slope of the S_1–V curve is greater than that of the S_2–V curve which in turn is greater than that of the L–V curve, the relationships (24) and (25) being appropriate in the neighborhood of the triple points; i.e.,

$$\Delta H_s \text{ (solid 1)} = \Delta H_{\text{trans}} \text{ (solid 1–solid 2)} + \Delta H_s \text{ (solid 2)}, \quad (24)$$

$$\Delta H_s \text{ (solid 2)} = \Delta H_f \text{ (solid 2)} + \Delta H_v, \quad (25)$$

where ΔH_{trans} is the heat of transformation.

A three-dimensional schematic representation of the stable relationships in the system depicted in the p–T projection of Fig. 8 is shown in Fig. 9.

Since the system solid 1 has its free energy surface interrupted by the solid 2 surface, a stable solid-1–liquid intersection cannot occur, under

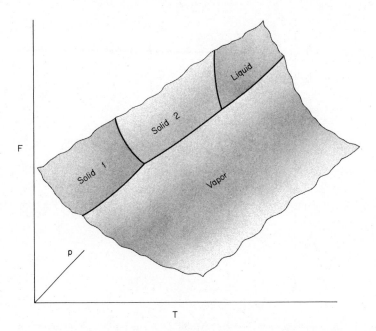

Fig. 9. Free energy surface representation of Fig. 8.

the equilibrium vapor pressure of the system. However, if the transformation solid 1 → solid 2 does not occur because of kinetic limitations, it is possible to realize a metastable solid-1–vapor equilibrium and a melting of solid 1 to form the metastable solid-1–liquid–vapor triple point. This triple point is as unique as any other and occurs at the temperature at which the metastable extension of the liquid–vapor curve downward in temperature from the S_2–L–V triple point, T_t, intersects the upward extension of the S_1–V curve from the triple point T_t'. This occurrence is depicted in Fig. 10.

In Fig. 10, the dashed lines represent univariant equilibria which are metastable with respect to other univariant equilibria at the same temperature. The stable equilibria are shown as solid lines, and it should be recognized that when two alternatives are depicted in a given temperature interval, one or the other of these will, in general, prevail experimentally.[2] Along the line X–T_t', the stable univariant system is S_1–V. When the triple point temperature T_t' is attained with increasing

---

[2] In instances where solid–solid phase transformations are sluggish, it is possible to have both metastable and stable equilibria occur together. Clearly, such a situation is a nonequilibrium one.

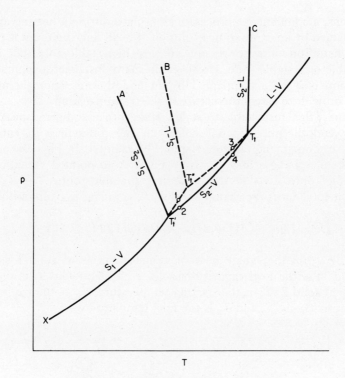

Fig. 10. The melting of a metastable solid.

temperature, solid 1 should isothermally and isobarically transform to yield solid 2 in equilibrium with vapor. Upon continued heating to the triple point T_t, the stable solid 2 is expected to melt yielding a liquid in equilibrium with its vapor.

In the event that, at the transformation point, T_t' the solid-1–vapor equilibrium persists, the vapor pressure of this system follows the path $T_t'-T_t''$ in accordance with the analytical expression describing the behavior of the S_1-V equilibrium. If this metastable equilibrium persists to high enough temperatures, the triple point T_t'' is reached, at which point the solid-1 melting point is achieved. If, in cooling a liquid to the transition point T_t, it fails to crystallize to form the stable solid-2 phase, continued cooling will follow the path T_t-T_t''. At the triple point T_t'', metastable solid 1 will crystallize. In the temperature interval $T_t''-T_t'$ solid 1 will be metastable and when the temperature drops to T_t' and lower, solid 1 becomes the stable phase. During any of the heating or cooling cycles described, a spontaneous transformation from the metastable to stable states might have been observed. Under certain conditions, a spontaneous process taking place during a heating cycle is accompanied by an exothermic evolution of heat. Such an event is clearly a demonstration of a conversion from a metastable to either a less metastable or a stable state. However, in many instances a spontaneous conversion may be accompanied by an endothermic anomaly making it more difficult to assess the nature of the transformation.

Whether the metastable-to-stable state process during heating is exo- or endothermic depends solely on the differences in enthalpic changes accompanying the conversion. For example, in Fig. 10, consider the enthalpic changes in going from point X to point 2 via either the metastable route $X \rightarrow T_t' \rightarrow 1 \rightarrow 2$ or the stable route $X \rightarrow T_t' \rightarrow 2$. ΔH_p, the total enthalpic change via the last sequence is given in Eq. (26):

$$\Delta H_p = C_p(\text{solid 1})[T_t' - T_x] + \Delta H_{T_t'} + C_p(\text{solid 2})(T_2 - T_t'), \qquad (26)$$

where $\Delta H_{T_t'}$ is the latent heat of transformation of solid 1–solid 2, C_p (solid 1) is the heat capacity of solid 1 and C_p (solid 2) is the heat capacity of solid 2, T_x is the starting temperature, T_t' is the triple point temperature S_1–S_2–V, and T_2 is the final temperature.

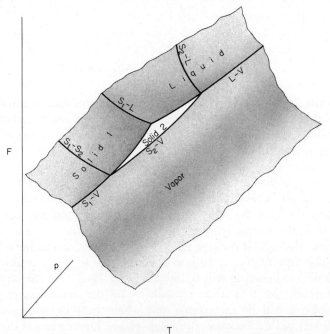

FIG. 11. Free energy surfaces showing the generation of the stable triple points T_t' and T_t as well as the metastable triple point T_t''.

ΔH_p, the total enthalpic change via the metastable sequence must be identical to that for the stable sequence and is given by

$$\Delta H_p = C_p(\text{solid 1})(T_1 - T_x) - C_p(\text{solid 1})[T_1 - T_t'] \\ + \Delta H_{T_t'} + C_p(\text{solid 2})[T_2 - T_t']. \tag{27}$$

After being heated metastably to T_1, the solid must lose the excess heat absorbed in the heating process $T_t' \to T_1$. The solid must, however, absorb the latent heat of transformation, and the enthalpy associated with solid 2 being heated from T_t' to T_2. Since the $C_p \Delta T$ terms for solid 1 and solid 2 are probably similar, the enthalpic endo- and exothermic phenomena due to temperature changes probably cancel to a great extent. On the other hand, the $\Delta H_{T_t'}$ latent heat anomaly is not balanced by an exothermic effect, and the total metastable process in moving from the temperature T_1 to the same temperature T_2 with a structural change is most likely to be endothermic.

Consider now the process starting at point X (Fig. 10) and terminating at the point 4 by either the stable sequence $X \to T_t' \to 4$ or the

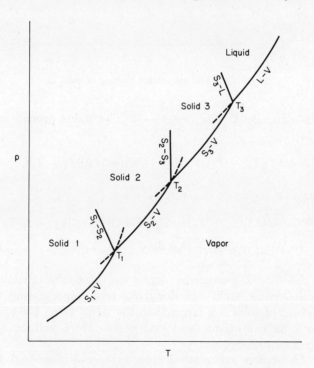

FIG. 12. A system exhibiting three stable solid phases.

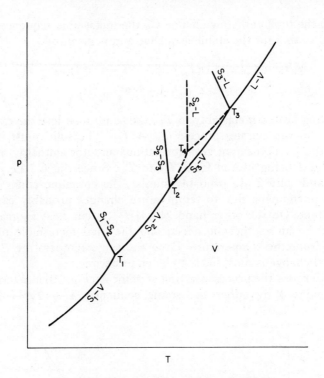

Fig. 13. The metastable melting of solid 2.

metastable sequence $X \to T_t'' \to 3 \to 4$. The stable process is depicted by [Eq. (28)]

$$\Delta H_p = C_p(\text{solid 1})[T_t' - T_x] + \Delta H_{T_t'} + C_p(\text{solid 2})[T_4 - T_t']. \tag{28}$$

The metastable sequence is given by

$$\Delta H_p = C_p(\text{solid 1})[T_t'' - T_x] - C_p(\text{solid 1})[T_t'' - T_t'] - C_p[T_3 - T_t'']$$
$$- \Delta H_{T_t''} + \Delta H_t' + C_p(\text{solid 2})[T_t - T_t']. \tag{29}$$

Again, the $C_p \Delta T$ exothermic terms are probably cancelled by the $C_p \Delta T$ endothermic terms. If the $\Delta H_{T_t''}$ term representing the latent heat of melting of solid 1 is larger than the $\Delta H_{T_t'}$ term, the spontaneous process will be exothermic and vise versa. More will be said about similar processes subsequently.

Figure 11 depicts schematically on a three-dimensional graph the stable and metastable free energy surfaces for the system shown in Fig. 10.

H. STRUCTURE CHANGES IN UNARY SYSTEMS

In Fig. 11, the metastable extension of the vapor surface to join with the metastable solid 1 and metastable liquid is left unconstructed. It is evident, however, that the vaporous surface must intersect the S_1–V and L–V surfaces in their regions of metastability. It is also evident that the melting point of the metastable phase must lie below that of the stable phase.

Up to this point we have considered systems in which only a single structural change has been involved, and in which the two solid phases exhibited stable intervals in which solid–vapor univariant equilibria obtained. Many systems are known in which more than one structural change is observed as a function of increasing temperature, with each of the generated solid phases being stable and, consequently, with each of the solid phases exhibiting a solid–vapor range of coexistence. Such a system is depicted in Fig. 12 where three stable solid structural entities are possible.

Metastable extensions of each of the univariant equilibria are indicated, and it is evident from the preceding discussions that one could depict metastable melting of the lower temperature phases, solid 1 and solid 2. It is to be expected, however, that realization of the metastable melting of solid 1 is, in general, less probable than that of solid 2 since the liquid

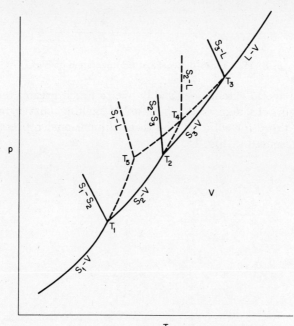

FIG. 14. The metastable melting of solid 1.

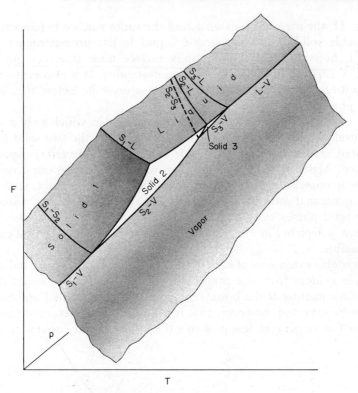

Fig. 15. Three-dimensional representation of Fig. 14.

field would have to extend metastably over a much greater temperature interval. Figure 13 depicts the metastable melting curves for solid 2, and Fig. 14 those for solid 1. The three-dimensional representation of Fig. 14 is presented in Fig. 15.

6

Systems of One Component—Pressure Effects and the Continuous Nature of Metastability

A. Metastable-to-Stable Transformations with Increasing Pressure

In the solid–solid transformation phenomena considered in the preceding chapter, the slopes of the univariant solid–solid and solid–liquid curves were arbitarily chosen to be of opposite sign. When the lower temperature condensed-phase–vapor-phase curve exhibits a positive pressure slope, and the higher temperature one exhibits a negative slope, the situation depicted in Fig. 1 may arise. It is to be noted that the intersection of the S_1–S_2 with the S_2–L curve generates the all-condensed-phase triple point T_c. The slope of the univariant line T_c–A is of interest to us since its origin is not immediately evident. Clearly the intersection T_c–A cannot represent the equilibrium S_2–L since it separates the S_1 and liquid fields. Similarly, it cannot represent a continuation of the S_1–S_2 equilibrium for the same reason. Logically, since this line is bounded by the solid 1 and liquid fields, it must represent the stable extension of the metastable curve S_1–L into pressure regions where S_1 can coexist with liquid and provide a system of minimal free energy. This situation is depicted two dimensionally in Fig. 2 and three

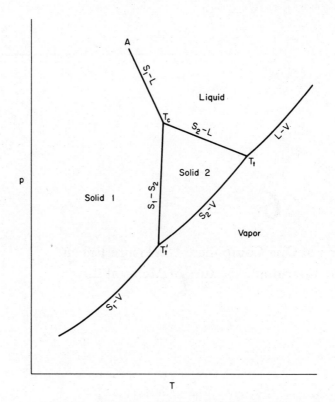

FIG. 1. Formation of the triple point S_1–S_2–L.

dimensionally in Fig. 3. In Fig. 2, the region bounded by T_t', T_c, T_t represents stability conditions for solid 2. Within this region solid 1 may coexist metastably with liquid, or vapor, or both. In the bivariant regions surrounding the triangle T_t'–T_c–T_t, the only stable bivariant equilibria are those involving S_1 or liquid. Consequently at pressures greater than that at T_c the univariant equilibrium S_1–L emerges as the stable one.

Assuming that the slopes of the curves originating at condensed-phase–vapor triple points remain constant with increasing pressure it is a consequence of the preceding that:

(1) When the higher temperature transformation exhibits a positive slope and the lower one a negative slope, the lower temperature solid phase can never exhibit a stable solid–liquid equilibrium.

(2) When the sign of the slopes is reversed, both condensed phases may exhibit a stable solid–liquid equilibrium, but only the higher temperature phase will exhibit a stable liquid–vapor equilibrium.

B. MONOTROPY

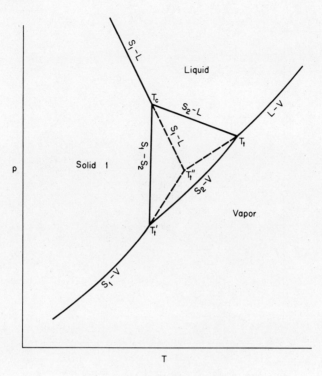

FIG. 2. Origin of the S_1–L curve.

B. Solid Phases That Do Not Exhibit Stable Solid–Vapor Equilibria at Any Temperature: Monotropes

Up to this point we have purposely oriented our discussion in such a way that we would think of a stable phase as something normal, and a metastable phase as something that occurs accidentally. This orientation serves to add confusion to the process of understanding the behavior of unary systems, but has been employed to maintain consistency with previously published treatises on heterogeneous equilibria. Such an approach is, however, purely semantic and may lead one to setting up a catalog of unique phenomena, each describing a given case of heterogeneous behavior. This quantizing of heterogeneous phenomena contradicts macroscopic physical events since it is evident, with some thought, that heterogeneous behavior must indeed represent orderly variation as would be predicted by the laws that govern it. In this and the following sections we will present a somewhat alternate approach

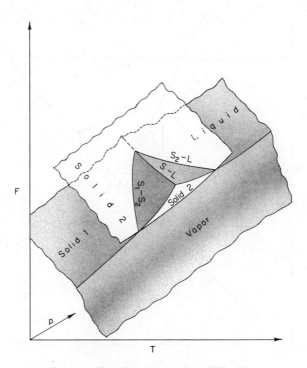

Fig. 3. F–p–T representation of Fig. 2.

to the understanding of unary systems which is based solely on an examination of what variations in the thermodynamic properties of macroscopic systems would demand.

Consider a unary system which in the solid state may exhibit more than one structure at a given temperature with changing pressure. If each of these structures is capable of coexisting with a vapor, it is clear that thermodynamically all of these structures are metastable with respect to one of them. For our purposes, we ignore the unique case where the free energies of two possible structures are precisely the same. Two boundary situations may be visualized for the system under scrutiny with all other possibilities being intermediate to them. The first boundary is that in which the metastable structure at our arbitrary temperature exhibits a free energy surface whose temperature slope is greater than that of the stable structure. The second boundary is that in which it exhibits a smaller temperature slope. In order not to prejudice the potential stable–metastable existences afforded both phases, let us designate one of these phases the A phase and the other the V phase.

Three dimensionally, the two cases may be depicted in Figs. 4 and 5.

B. MONOTROPY

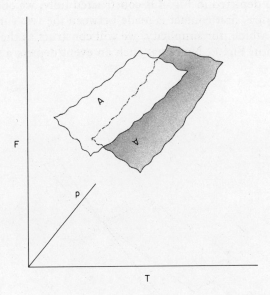

FIG. 4. The A phase surface exhibits a smaller temperature coefficient than the ∀ surface.

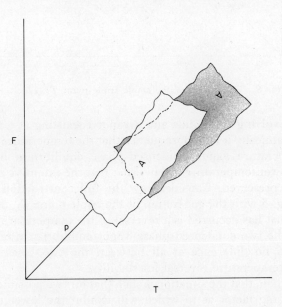

FIG. 5. The A phase exhibits a larger temperature coefficient than the ∀ phase.

If the situation depicted in Fig. 4 is constructed fully, we observe that at some temperature, first contact is made between the two surfaces giving rise to a point which, for simplicity, we will construct as the triple point T_{A-V-V} shown in Fig. 6. Note that such an event depicts a phase trans-

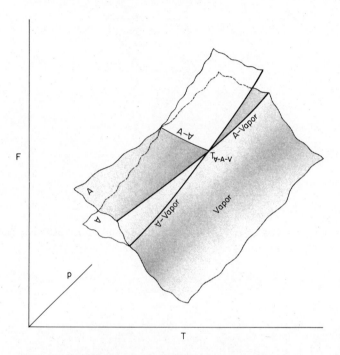

FIG. 6. Generation of the stable triple point T_{V-A-V}.

formation involving two solids and a vapor coexisting at a triple point. Thermodynamically it is a consequence that the temperature slope of the higher temperature vapor–condensed-phase equilibrium be less than that of the lower temperature one in order that the extensive nature of the enthalpy be preserved. Consider now the full construction of the case shown in Fig. 5 with the generation of the triple point T_{A-V-V} shown in Fig. 7. All that has occurred is a reversal of the temperature intervals of stability of the two condensed-phase–vapor equilibria. In effect, we see that there is no difference at all between the two cases considered, excepting a temperature reversal for stability.

It is clear then, that the situation is such that for two possible structures, the determining factor as to which will exhibit the lower temperature field of stability is the magnitude of its heat of sublimation. Depending on

B. MONOTROPY

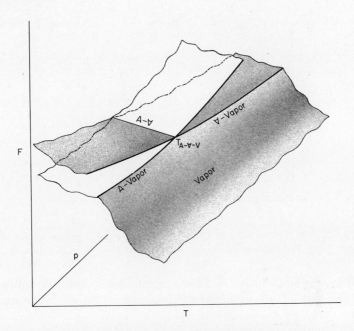

FIG. 7. Generation of the stable triple point T_{A-V-V}.

this magnitude, an infinite number of situations can develop ranging from an intersection of the free energy surfaces below the absolute zero of temperature to above the melting point of both of the structures. A series of such variations is depicted in Figs. 8–11. For the sake of

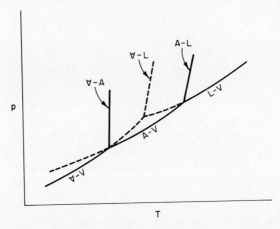

FIG. 8. The V–V curve exhibits a larger temperature coefficient than the A–V curve and intersects the latter above the absolute zero generating a stable V–V equilibrium.

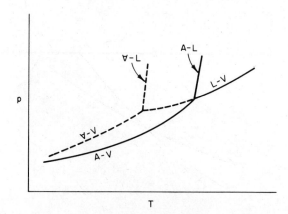

Fig. 9. The V̄–V curve exhibits a larger temperature coefficient than the A–V curve, but intersects the latter below the absolute zero. No stable V̄–V range is possible.

simplicity, the slope of the A–V univariant equilibrium (the sublimation curve of Solid A) is kept constant while that of the V̄–V equilibrium is varied.

The situations represented by Figs. 8 and 11 have been discussed previously so that we may focus our attention on the implications of Figs. 9 and 10. In Fig. 9 we have the case where, although the V̄–V curve exhibits a greater temperature coefficient than that of the A–V curve, the difference is not too great and intersection of the two curves might occur hypothetically below the absolute zero. Experimentally,

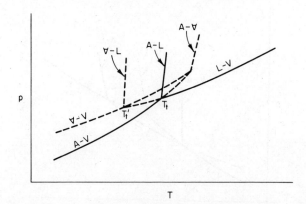

Fig. 10. The V̄–V curve exhibits a smaller temperature coefficient than the A–V curve and intersects the latter above the melting points of both V̄ and A. No stable V̄–V range is possible.

B. MONOTROPY

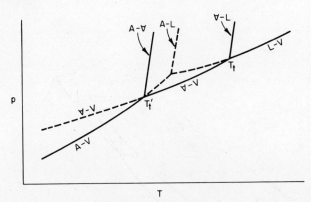

FIG. 11. The V–V curve exhibits a smaller temperature coefficient than the A–V curve and intersects the latter below the melting point of A.

however, the V–V system may form if the L–V curve is supercooled to low enough temperatures. It is possible therefore to observe heating of the V phase to its melting point and beyond, and regeneration of this phase upon cooling. The V phase, while exhibiting no stable field, may be readily formed in cyclic heating and cooling cycles due to kinetic phenomena.

In Fig. 10 the V phase has a smaller temperature coefficient than the A phase but not much smaller. In this instance the A–V and V–V curves do not intersect hypothetically until past the melting point of both structures. Since solid metastability beyond the melting point has not been observed, it is evident that the metastable extensions A–V and V–V resulting in the intersection A–V–V will not be observed experimentally.

A structural modification which cannot form a triple point with another more stable structural modification excepting above the melting of both phases is termed a monotrope.

In Fig. 11 we have simply reversed the case defined by Fig. 8, and the A and V phases have simply changed positions relative to their temperature intervals of stability and metastability.

Figure 12 depicts three dimensionally the monotropy of Fig. 10 in order to point out an important feature of one-component diagrams that is not perhaps evident from Fig. 10. Note that whereas in the p–T projection of Fig. 10, the curve V–V intersects the A–L curve, this intersection is an artifact introduced in projection. In fact the V surface does not meet the A–L boundary at any place. In general, when considering curve intersections and the possibility of whether they may lead to metastable extensions, the criterion to be applied is whether the

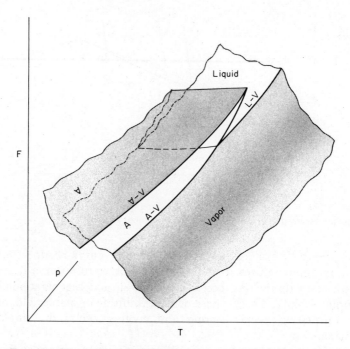

Fig. 12. Three-dimensional representation of a monotropic system.

latent heats of the intersecting curves are related. Thus the A–L curve does not bear a direct relationship to the V–V curve whereas the A–V and L–V curves are related, the ΔH_s of the solid A being equal to the sum $\Delta H_f + \Delta H_v$. Below the triple point A–L–V, T_t, in Fig. 10, the triple point V–L–V is possible, and here the situation prevails that

$$\Delta H_s(V) = \Delta H_f(V) + \Delta H_v(V).$$

7

Complex Metastability in One-Component Systems

A. Systems with Solid–Vapor Univariance

While the general behavior of one-component systems follows the pathways discussed in the preceding two chapters, a significant number of unary and pseudounary systems exhibit related behavior of more complex character. In addition, many systems are known in which high pressure complexities arise. In the following sections we will present discussions of the nature of more complex equilibria, primarily of the kind where the free energy surfaces of the complex equilibria lie at higher values than those existing stably under the vapor pressure of the system.

Figure 1 shows the case where only a single stable phase is present but where a monotrope forms part of series of metastable enantiomorphic structures. In addition to the stable α structure, two other structures, β and γ, may be formed metastably. The relationship of the α and β structures is such that β is monotropic. However, the relationship between the γ and β structures is such that they are enantiomorphic. In the stable–metastable sequence, the melting point of the β phase may be realized if the liquid is cooled below the melting point of the α phase. An even higher order metastability may be realized in the crystallization

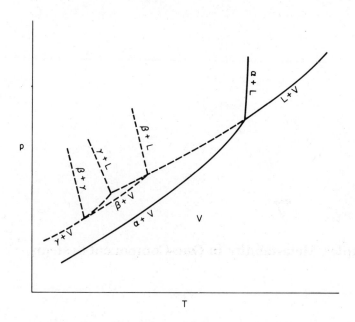

FIG. 1. Complex metastability exhibiting polymorphism.

of the γ phase if the liquid phase is further supercooled below the melting point of the metastable β phase. The probable occurrence of the first part of this sequence has been observed in the case of the compound Nb_2O_5. This material, whose behavior in oxygen or air is pseudounary, exhibits only one stable phase which melts at ~1490°C. Upon cooling a melt from above this temperature, the stable phase α very rarely crystallizes without continuous seeding through the melting point. Upon further cooling without seeding, a metastable phase ϵ crystallizes at approximately 1435°C. This ϵ phase is however quite unstable and if it is cooled to approximately 1370°C, it will spontaneously revert to the α form with great rapidity. That the ϵ phase is metastable relative to the α phase can be demonstrated readily in the following manner. If, once the ϵ phase has been crystallized, it is remelted and heated further and a seed of the α phase is dropped into the melt during the heating cycle, the liquid will solidify spontaneously forming the α phase with the liberation of heat. Subsequent heating to 1490°C results in the observed melting of the α phase. A qualitative analysis of the type offered in Section H of Chapter 5 serves to provide insight as to the cause of the exotherm during the heating cycle, a direct evidence of metastability.

In heating the ϵ phase from 10°C below its melting point to higher

A. SYSTEMS WITH SOLID–VAPOR UNIVARIANCE

temperatures, the enthalpic change in the metastable and stable systems, respectively, would be as follows:

$$\Delta H_\epsilon = C_{p_\epsilon}(T_m^\epsilon - T_{\text{start}}) + \Delta H_f^\epsilon + C_{p_L}(T_c - T_m), \quad (1)$$

where T_m^ϵ is the melting point of ϵ, T_c is the highest temperature to which we heat the ϵ phase, and ΔH_f^ϵ is the latent heat of fusion of ϵ, and

$$\Delta H_\alpha = C_{p_\alpha}(T_c - T_{\text{start}}). \quad (2)$$

Since we might not expect large differences between the specific heats of the ϵ, α, and liquid phases it is clear that the absorption of heat during the melting of the ϵ phase is responsible for the subsequent exotherm attending the spontaneous solidification of the liquid during heating. In general, the enthalpic increase due to $C_p\, dT$ terms over small temperature intervals is small compared to latent heat absorptions. When the metastable liquid phase solidifies to form the α phase it is simply releasing the heat it acquired during the melting of ϵ. Looking at it in another way, all crystallizations have been observed to be exothermic and such a phenomenon during the heating of a liquid is definite evidence that the liquid was metastable relative to the phase that formed.

In general, therefore, we would expect that anytime a metastable phase exhibits a latent heat phenomenon in a temperature range in which the stable phase would not exhibit a latent heat anomaly, an exothermic evolution will subsequently accompany the irreversible process metastable phase → stable phase.

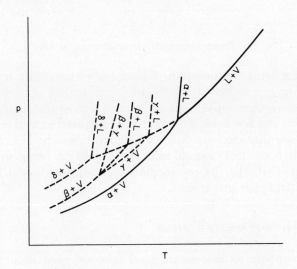

FIG. 2. Complex metastability involving multiple monotropy.

A slightly more complex situation than the one just described is that in which metastable phases present are monotropic, one with respect to the other. This situation is depicted in Fig. 2. Here, the γ phase is monotropic relative to the α phase, but enantiomorphic relative to the β phase. In addition, the δ phase is monotropic relative to both the β and γ phases. The three-dimensional representation of Fig. 2 is given in Fig. 3.

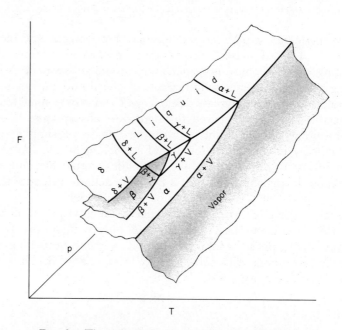

FIG. 3. Three-dimensional representation of Fig. 2.

Without belaboring the question of complex metastability it should be clear that a large number of hypothetical interactions can be generated. Whether real representatives can be found is a moot question since even the resolution of a few complex systems has required extensive investigations over long time-periods. More important than whether real analogs of possible hypothetical cases do exist is the ability to test a given system and determine whether its representation is reasonable and how it fits into sequential behavior patterns.

B. Stable High-Pressure Systems

Up to this point, all discussions have been oriented about univariant equilibria involving vapor phases, with all-condensed phase equilibria

B. STABLE HIGH-PRESSURE SYSTEMS

arising from triple points at which a vapor was present. Thus, questions of metastability were treated in the context of a vapor available space being present, excepting for the condensed phase extension from a triple point. In this section, we will consider the effects of hydrostatic pressure on systems in which vapor-available space is not present. Many unary systems are known in which the application of large pressures generates phases stable at those pressures. Two classic examples are those for the systems carbon and water. Whereas, the graphical appearance of univariant equilibria on conventionally constructed p–T graphs is generally as already shown, high temperature curves frequently exhibit two phase lines of unconventional shape. A little thought will convince the reader that these curves are not really different than the exponentials previously treated if it is remembered that ΔV terms determine the sign of the slope but not its magnitude, while ΔH terms determine the magnitude. Also the presentation requires less reorientation if it is realized

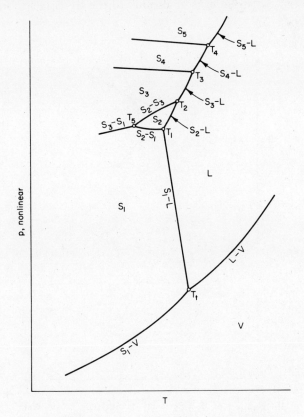

FIG. 4. High pressure polymorphism in a unary system.

that solid–solid curves will generally be analogous to the solid–vapor, or liquid–vapor curves of the earlier systems.

Figure 4 shows a schematic representation of a hypothetical high pressure system containing features present in reported real systems. The pressure axis is presented nonlinearly in the high pressure range, i.e., the scale in the high pressure range is greatly compressed. In addition, the univariant lines are not completed below a certain temperature, this because little work has been performed in low temperature–high pressure systems making it difficult to anticipate the nature of the curves in such temperature ranges. Five structures are depicted, four of which are stable only at high pressures.

Figure 5b shows the nature of the S_1, S_2, and S_3 surfaces three dimensionally. It should be noted that the S_1 surface now fills the role of the vapor phase in the interaction S_3–S_1 and S_2–S_1. Furthermore, the transition line S_2–S_3, representing the projection of the intersection of the S_2 and S_3 surfaces, is analogous to the S_1–L projection at lower

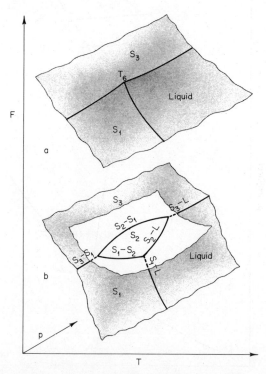

FIG. 5. (a) Intersection of S_1, S_3, and L surfaces to generate the metastable triple point T_6. (b) Intersection of S_1, S_2, S_3, and L surfaces.

B. STABLE HIGH-PRESSURE SYSTEMS

pressures. As with the lower pressure behavior, the enthalpic change accompanying processes of different path between starting and concluding states must be identical. This is not apparent from the high pressure graphs since the curves do not necessarily exhibit ascending univariant behavior, this because the solid molar volumes may be very nearly the same, unlike the S–V or L–V cases. If we reconstruct the S_1, S_2, and S_3 relationships of Fig. 4 somewhat differently, this point may become more evident (see Fig. 6).

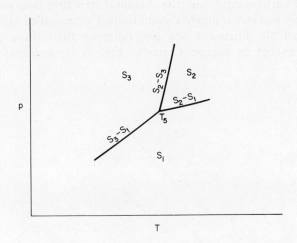

FIG. 6. A modified construction of the S_1, S_2, and S_3 relationships.

The S_3–S_1 curve exhibits positive slope demanding that the molar volume of S_1 be greater than that of S_3. On the other hand this situation is reversed for the S_2–S_1 interaction. Thus $V_{S_1} > V_{S_3}$, $V_{S_1} < V_{S_2}$, and $V_{S_3} < V_{S_2}$. For such to be the case the curve S_3–S_2 must exhibit positive slope. If both the S_3–S_1 and S_2–S_1 curves exhibited positive slope, one may conclude that the S_3–S_2 curve is capable of exhibiting either positive or negative slope.

The heats of transition are such that

$$\Delta H_{(S_3 \to S_1)} = \Delta H_{(S_3 \to S_2)} + \Delta H_{(S_2 \to S_1)}. \tag{3}$$

Consequently, the slope of the S_3–S_1 curve must be greater in absolute value than the slope of the S_2–S_1 curve, even though both are of opposite sign.

The S_2–L, S_3–L, S_4–L, and S_5–L curves of Fig. 4 are all ascending so that no immediate restriction is placed on the sign of the S_3–S_4 and S_4–S_5 curves. As constructed they are all negative indicating that the phase of higher heat content is more dense than the phase of lower heat content.

The final feature of interest is the closed nature of the S_2 field, a characteristic of many high pressure systems. It simply indicates that the S_2 surface cuts across the surfaces S_1, S_3, and L as shown in the sequence Figs. 5a and 5b. Whereas Figs. 4, 5, and 6 depict cases of high pressure stability, it is evident that situations of metastability involving vapor–solid equilibrium with these high pressure phases are possible. For example, the element carbon exhibiting the diamond structure does indeed exist under our normal environmental conditions. Our so-called high pressure phase metastable situations are not different than those previously considered except in degree. Consider Fig. 7, for example. If the S_3

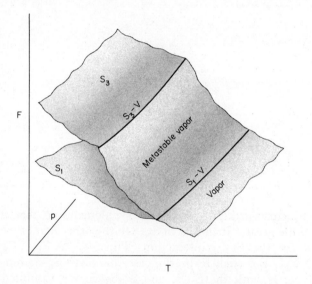

Fig. 7. Generation of the metastable system S_3–V.

surface of Fig. 5a is extended beyond its intersection with the S_1 surface while the vapor surface is extended beyond its intersection with the S_1 surface, the S_3 and vapor surfaces will intersect to form the S_3–vapor metastable curve depicted in Fig. 7. In similar fashion other solid–vapor curves may be generated.

8

Other Aspects of One-Component Behavior

A. The Effects of Inert Gas Pressures on the Vapor Pressure of Single-Component Solids or Liquids

Generally speaking, if we add an inert gas pressure to a system containing a single-component solid or liquid in equilibrium with its vapor, the component number rises to two. If however, this inert gas is insoluble in the condensed phase and interacts ideally with the vapor present, the effect of the second component is simply that of increasing the pressure on the condensed phase. In essence, the inert gas system as described can be thought of as involving a semipermeable membrane separating two phases, the surface of the condensed phase functioning as the membrane. Since the inert gas is insoluble in the condensed phase, it cannot penetrate the "surface membrane." On the other hand, the molecules of the component under examination are free to penetrate the surface in either direction.

At constant temperature, the total derivatives of the Gibbs free energies of the component in each of the phases is given by:

$$dF_c = V_c \, dp_c, \tag{1}$$

$$dF_g = V_g \, dp_g, \tag{2}$$

where the subscripts "c" and "g" refer to the condensed and gaseous phases, respectively, dF and dp are the changes in molar free energy accompanying a change in pressure, and V is the molar volume of the component in the phase in question. When equilibrium obtains in the pure single-component system, it is a consequence of the criterion of equality of molar free energies that

$$V_c \, dp_c = V_g \, dp_g. \tag{3}$$

From Eq. 3 we see that if the pressure on the condensed phase dp_c is increased, the vapor pressure of this solid dp_g will have to increase in order for the equilibrium to be reestablished, viz for dF_g to precisely balance the change dF_c. That this change is inversely proportional to the molar volumes of the component in the coexisting phases is seen from Eq. (4), which is obtained by rearrangement of Eq. (3):

$$V_c/V_g = dp_g/dp_c. \tag{4}$$

If for the molar volume of the gas V_g, we insert the ideal gas equivalency $V_g = RT/p_g$, then

$$V_c p_g/RT = dp_g/dp_c, \tag{5}$$

where p_g is the partial pressure of the component, and dp_g is the change in partial pressure due to a change in the total pressure on the condensed phase dp_c.

From Eq. (4) it is evident that the change dp_g with a change dp_c is small at nominal total pressures because the ratio V_c/V_g is small. We may rewrite Eq. (5) to give

$$\Delta p_g/p_g = V_c \, \Delta p_c/RT. \tag{6}$$

From Eq. (6) it is evident that the units for Δp_g and p_g must be the same and the unit for R must be consistent with those for V_c and Δp_c. Thus with Δp_c in atmosphere and V_c in liters/mole, R must be specified in liter atmospheres/mole degree. For water at approximately room temperature, the increase in p_g (namely, Δp_g) with an increase in total pressure to 1 atm is ~0.02 torr. At higher inert gas pressures, the assumption of ideal gas behavior becomes less viable and the approximation Eq. (5) breaks down.

By series of steps similar to those described, but making use of the total derivative of the free energy, i.e., $dF = -S \, dT + V \, dp$, one can deduce an equation relating the change in vapor pressure of a condensed phase at *constant total pressure* with temperature. The general form of this equation is

$$(\partial p/\partial T)_{p_c} = \Delta S/V_g. \tag{7}$$

A. INERT GAS PRESSURES

Substituting for V_g the ideal equivalency and the ΔS equivalency we obtain

$$dp_g/dT = \Delta H p_g/RT^2, \qquad (8a)$$

and

$$\ln(p_g^{(2)}/p_g^{(1)}) = -(\Delta H/R)(T_2^{-1} - T_1^{-1}), \qquad (8b)$$

which are seen to be identical to the approximate Clausius–Clapeyron equation. However, it should be realized that the value $p_g^{(2)}$ arrived at, at any temperature, will not be the same as that obtained from the Clausius–Clapeyron relationship since the value $p_g^{(1)}$ at some other temperature will be slightly higher due to the pressure effect previously considered.

The net effect of what we have just discussed is that the vapor pressure curves of a pseudounary system, that is one under an inert gas pressure will be displaced upwards from the unary curves as shown in Figs. 1 and 2.

Of interest to us is the effect on the triple point due to the pressure effect under consideration. Since the projected S–L curve is the result

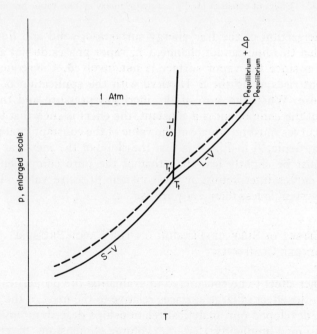

FIG. 1. Effect of a constant inert gas pressure on unary system vapor pressures—total pressure not constant.

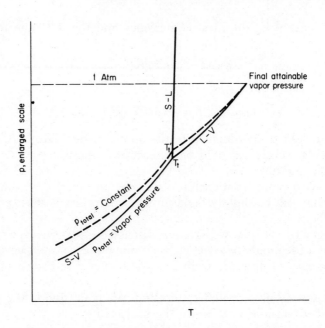

FIG. 2. Effect of constant total pressure on unary system vapor pressures.

of the intersection of the free energy surfaces of solid and liquid, it is obvious that this line cannot change, i.e., vapor phases do not affect this intersection since the vapor surface is not involved. Consequently, the triple point rises along the S–L curve with the application of an inert gas pressure. When the sum of the inert gas pressure and the partial pressure of the component is a constant, the effect is such that the vapor pressure curves must terminate at the value of the constant total pressure. The effect, depicted in Fig. 2 is that the slope of the inert gas pressure system must be slightly less than that of the pure component system with the curves intersecting at the constant pressure value. Thus ΔH inert gas system is less than ΔH pure system.

B. The Effect of Surface Tension on the Vapor Pressure of a Single Component

One other effect to be considered in evaluating the properties of unary systems is the effect of large surfaces relative to the mass of the material. When we developed our analytical relationships describing univariance, it was assumed implicitly that the volume dimensions of condensed phases are large in comparison to surface areas. When this condition

B. SURFACE TENSION EFFECTS

does not prevail, the values of the vapor pressure increase somewhat, the added contribution being due to surface energy effects. The magnitude of such effects can be estimated as follows:

Consider a given number of moles n of material present as small liquid spherical droplets. Consider this same number of moles in a single aggregate in a beaker. It is evident that in the first instance the surface area of this number of moles is much greater than in the second instance.

For a change in surface area ds, the free energy change of the condensed phase is given by

$$dF_s = \gamma \, ds, \tag{9}$$

where γ is the surface tension.

The volume of a sphere of radius r is $\frac{4}{3}\pi r^3$. If each sphere contains n moles of the substance of molar volume V, then the volume per sphere is given by

$$nV = \tfrac{4}{3}\pi r^3. \tag{10}$$

The change in the number of moles per drop with radius is

$$dn = (4\pi r^2/V) \, dr. \tag{11}$$

The surface area per sphere is $4\pi r^2 = s$, and the variation of surface area with sphere radius is

$$ds = 8\pi r \, dr. \tag{12}$$

From Eqs. (11) and (12) we see that

$$ds = 8\pi r V \, dn/4\pi r^2 = 2V \, dn/r, \tag{13}$$

and

$$dF = 2\gamma V \, dn/r. \tag{14}$$

For equilibrium to obtain, the free energy change of the condensed phase must be equal to that of the vapor. For each mole in the vapor phase undergoing a pressure change $p_1 \to p_2$ the gaseous free energy change is

$$dF = RT \ln(p_2/p_1), \tag{15}$$

where p_1 is the bulk vapor pressure, and for a change of dn moles

$$dF = dn \, RT \ln(p_2/p_1). \tag{16}$$

Equating Eqs. (14) and (16) and solving for $\ln(p_2/p_1)$ we obtain

$$\ln(p_2/p_1) = 2\gamma V/rRT. \tag{17}$$

It is seen that as r increases, the ratio p_2/p_1 must decrease. In other words, for a material of molar volume V, the effect of increasing the surface area (decrease in r) is to increase the vapor pressure. As r becomes large the value $p_2 \rightarrow p_1$ and we realize the bulk vapor pressure of the material. Furthermore, for materials of larger molar volume, surface per unit of bulk is obviously larger than for more dense materials of similar dispersed size. Consequently, we might expect greater deviations from bulk properties in low density materials.

9

Enthalpy and Entropy Diagrams of State for Unary Systems

A. Introduction

We have previously explored the origin of $p-T$ diagrams of state in terms of projections of equipotential lines generated at the intersections of free energy surfaces. Since the Gibbs free energy represents a mathematical conclusion resulting from the interaction of other thermodynamic parameters, it should come as no surprise that the planar $p-T$ projection of any $X-T-p$ representation, X representing an extensive thermodynamic parameter, will again provide the now familiar $p-T$ diagrams of state. The construction of three dimensional $H-T-p$ or $S-T-p$ relations for example is not as obvious, however, as are their $F-T-p$ counterparts since, in general, latent heat anomalies are associated with phase changes. Thus, enthalpic and entropic surfaces do not intersect as do the free energy surfaces, and points of equiphase enthalpy and entropy do not, in general, occur. It is the case, however, that when a heterogeneous system exists in a state of dynamic equilibrium, the coexistent phases are at the same temperature and total pressure. It is via the routes of isobaric and isothermal relationships that we may deduce the three-dimensional enthalpy and entropy diagrams. The following discussion will briefly explore enthalpic and entropic relationships in order to provide further insight into the thermodynamic origins of phase diagrams.

B. Enthalpic Relationships and the Enthalpy Diagrams of State

The processes involved in heating a homogeneous system to different temperatures and pressures result in changes in the heat content of the system. The nonexplicit expression of the total derivative of the enthalpy of the homogeneous system may be stated as

$$dH(T, p) = (\partial H/\partial T)_p \, dT + (\partial H/\partial p)_T \, dp \qquad (1)$$

Depending on the temperature and pressure in a two phase coexistence, the latent heat anomaly associated with this change may also vary and the heterogeneous counterpart of Eq. 1 may be expressed as

$$d\Delta H(T, p) = (\partial \Delta H/\partial T)_p \, dT + (\partial \Delta H/\partial p)_T \, dp. \qquad (2)$$

In explicit form, the total derivative of the enthalpy of 1 mole of a single component present in a single phase may be defined as

$$dH = T \, dS + V \, dp, \qquad (3)$$

where S is the molar entropy and V the molar volume.

The partial derivative of H, Eq. (3), with respect to the pressure at constant temperature is given by

$$(\partial H/\partial p)_T = T(\partial S/\partial p)_T + V(\partial p/\partial p)_T. \qquad (4)$$

From Eq. (3) one may deduce the Maxwell relationship

$$(\partial S/\partial p)_T = -(\partial V/\partial T)_p, \qquad (5)$$

by utilizing the reciprocity equivalency, which is applicable to complete differentials. Substituting Eq. (5) into Eq. (4) yields

$$(\partial H/\partial p)_T = -T(\partial V/\partial T)_p + V. \qquad (6)$$

This is the exact relationship describing the variation of the enthalpy with pressure at constant temperature.

An approximate equation useful for condensed phases [see Eq. (8)] may be arrived at by employing the definition of the cubical coefficient of expansion

$$(\partial V/\partial T)_p/V = \beta \qquad (7)$$

B. ENTHALPY DIAGRAMS OF STATE

to obtain

$$(\partial H/\partial p)_T \cong -T\beta V + V \cong V(1 - \beta T), \tag{8}$$

for condensed phases. For ideal gases, the approximate Eq. (11) may be arrived at by making use of the ideal gas equation reiterated in Eq. (9) and the derivative of the volume with respect to the temperature at constant pressure, Eq. (10), i.e.,

$$pV = RT, \tag{9}$$

$$(\partial V/\partial T)_p = -R/p, \tag{10}$$

$$(\partial H/\partial p)_T = -(RT/p) + V = 0, \quad \text{ideal gas phase.} \tag{11}$$

From Eq. (11) we see that at constant temperature, the enthalpy of an ideal gas does not vary with pressure. From Eq. (8) we see that, in general, the enthalpy of a condensed phase does vary somewhat with pressure, the extent of variation depending on the difference $1 - \beta T$. If this difference tends toward zero, the variation tends toward zero since the molar volume V is not a large number. For example, if we assume that the molar volume of most condensed phases is approximately 10 cm³ and $\beta = 0.3 \times 10^{-3}$ cm³/deg cm³, and use units of dyn/cm² for p and dyn cm for H, viz dyn cm/(dyn/cm²) = cm³, we can approximate enthalpic changes per atmosphere applied pressure at room temperature. To obtain answers in calories let us employ the following unity conversions:

$$1 \text{ cm}^3 = \frac{1 \text{ dyn cm}}{1 \text{ dyn/cm}^2} \times \frac{1 \text{ erg}}{1 \text{ dyn cm}} \times \frac{10^7 \text{ dyn/cm}^2}{\text{atm}} \times \frac{1 \text{ cal}}{4.2 \times 10^7 \text{ erg}}$$

$$\cong 0.25 \text{ cal/atm}. \tag{12}$$

At 300°K

$$(\partial H/\partial p)_T = 10 \text{ cm}^3 \left(\frac{1 - 0.3 \times 10^{-3} \text{ cm}^3}{\text{deg cm}^3} \times 0.03 \times 10^4 \text{ deg} \right)$$

$$\cong 10 \text{ cm}^3 \cong 2.5 \text{ cal/atm} \quad \text{at} \quad 300°C. \tag{13}$$

From Eq. (13) we see that, in general, Eq. (8) may be approximated even further to give

$$(\partial H/\partial p)_T \text{ condensed phases} \cong V_c, \tag{14}$$

where V is the molar volume.

The change in enthalpy with pressure is not very large and, for purposes of our construction of condensed surfaces, we will assume that β does not vary with T and that $(\partial H/\partial p)_T$, for condensed phases, tends toward zero. Since the equilibrium pressure for the two-phase equilibrium, condensed-vapor phase is a single valued function of the temperature, it is evident that the above conclusions provide us with the characteristics of the homogeneous regions along constant pressure lines until these lines intersect two phase regions. Thus, starting at low temperatures and moving to higher temperatures, the terminus of an isobar is the point of appearance of a phase of higher total heat content. Alternately, starting at higher temperatures and moving to lower temperatures, the terminus is the appearance of a phase of lower heat content. The univariant ascending nature of condensed-vapor phase equilibria demands that the temperature value of termination of successively increasing isobars also increase. Since the slope of condensed–condensed phase equilibria need not be positive, successively greater condensed-phase isobars need not of course terminate at higher temperatures.

The variation of the enthalpy of a single phase with temperature at constant pressure is [from Eq. (3)] given by

$$(\partial H/\partial T)_p = T(\partial S/\partial T)_p = T((C_p/T)\,\partial T/\partial T)_p = C_p, \tag{15}$$

where C_p is the molar specific heat. For vapors we may assume a value of the order of 5–7 cal/mole deg and for condensed phases a value of 6–10 cal/mole deg, which provides a not insignificant temperature coefficient of the enthalpy.

Having evaluated the individual parts of Eq. (1), i.e., Eqs. (8), (11), and (15), which, as has been stated, sums the pressure and temperature effects on the heat content of a homogeneous phase, we may now attempt a similar set of approximations to permit of the summation specified in Eq. (2).

Consider a condensed phase in equilibrium with its vapor at a temperature T_1. For the heat content of the condensed phase we may write $H^1_{T_1}$ and for the vapor phase $H^2_{T_1}$. The difference in heat contents of the two coexistent phases, viz. $-H^2_{T_1} - H^1_{T_1}$ is of course the latent heat of sublimation ΔH_s. If we raise the temperature of the condensed phase to some higher value T_2 at the same total pressure p_1, the heat content of that phase will, according to Eq. (15), assume a new value, i.e.,

$$H^1_{T_2 p_1} = H^1_{T_1 p_1} + C_{p_c}(T_2 - T_1), \tag{16}$$

where C_{p_c} is the molar specific heat of the condensed phase. Similarly, the heat content of the vapor phase in equilibrium with the condensed

C. ENTROPY DIAGRAMS OF STATE

phase at the new temperature T_2 and constant total pressure will assume a value given by

$$H^2_{T_2 p_1} = H^2_{T_1 p_1} + C p_g (T_2 - T_1), \tag{17}$$

where C_{p_g} is the molar specific heat of the gas phase. The difference between the heat contents of the phases, the latent heat of sublimation will vary, therefore, as

$$(\partial \Delta H_s / \partial T)_p = \Delta C_p. \tag{18}$$

Obviously then, ΔH_s at constant pressure is not a constant as we have previously implied, but depends on the difference ΔC_p. As ΔC_p approaches zero, ΔH_s is more nearly constant with increasing temperature.

As the two-phase equilibrium, condensed-phase–vapor is not isobaric, it is evident also that a pressure effect must be considered when deciding on the shape of the enthalpic surface of transformation. For ideal vapors, we have seen from Eq. (11) that the effect here is zero. For condensed phases, we have seen from Eq. (14) that the effect depends on the molar volume. Utilizing this information we may write for Eq. (2) the approximate relationship

$$d\Delta H(T, p) = \Delta C_p \, dT + V_c \, dp. \tag{19}$$

Based on the above, we may now construct an enthalpic diagram of state such as that in Fig. 1. In this figure, the isotherm a–b shows constancy of H with varying pressure of the ideal gas and c–d the corresponding situation for the solid phase. The isobars f–g and b–e indicate the effect of temperature on the enthalpy at constant pressure and the points T_t and T_c show the nature of the triple point S–L–V and the critical point, respectively.

C. ENTROPIC RELATIONSHIPS AND ENTROPY DIAGRAMS OF STATE

Since it is not our primary function to reiterate general techniques in the manipulation of thermodynamic equations, the treatment of entropy relationships need not be described in detail. Using very similar procedures to those employed in the preceding section, the following approximate relationships may be arrived at to enable a schematic construction of an entropy diagram of state.

For a homogeneous phase we find that

$$(\partial S / \partial T)_p = C_p / T, \tag{20}$$

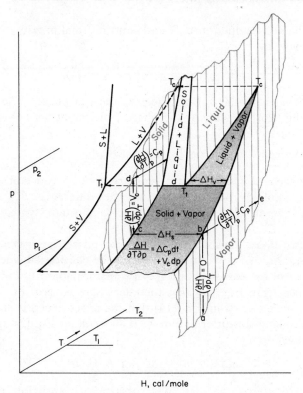

FIG. 1. The enthalpy diagram of state for a one-component system.

while for a two-phase coexistence

$$(\partial \Delta S/\partial T)_p = \Delta C_p/T, \tag{21}$$

where ΔS is the entropy of transition. For an ideal gas phase the effect of pressure at constant temperature is given by

$$(\partial S/\partial p)_T = -R/p \tag{22}$$

while for a condensed phase

$$(\partial S/\partial p)_T \simeq V_c[(T-1)/T] \sim V_c. \tag{23}$$

The pressure effect for the two phase coexistence is given by

$$(\partial \Delta S/\partial p)_T \simeq -(R/p) - V_c. \tag{24}$$

From the above, it is seen that as might have been predicted intuitively, the entropy increases with temperature and decreases with pressure.

C. ENTROPY DIAGRAMS OF STATE

The pressure effects are not large, however, so that in general the entropy diagram of state qualitatively has the appearance of its enthalpy analog. Another difference between the two diagrams is that the change in the entropy of transition with temperature is less pronounced since the ΔC_p term is divided by the temperature. Figure 1, then, may be taken as an adequate representation of the S–T–p diagram of state.

10

Multicomponent Systems, Homogeneous Systems, and the Equilibrium Constant

A. INTRODUCTION

In later chapters we will be concerned with a general description of heterogeneous behavior in multicomponent, multiphase systems. As it is not the primary intent of this book to provide only a descriptive introduction to the phase rule and the implications thereof, an attempt will be made to develop the topic in terms of analytical models where feasible or appropriate. In this chapter and Chapter 11 then, it is our purpose to render a detailed account of the techniques to be employed as well as to provide a set of ground rules within whose frame of reference we shall operate.

It should become evident as we delve more deeply, that the ability to depict heterogeneous behavior, even in an idealized fashion, is at best primitive. Analytical treatments of relatively simple situations are, however, possible and such treatments are of considerable value in providing insight into the possible nature of processes whose terminus is an equilibrium state. We shall see that the problem that confronts us is not primarily one of which approach to employ, but rather of the cumbersome nature of the mathematics involved, even in simple cases.

B. THE HOMOGENEOUS EQUILIBRIUM CONSTANT

We shall see further that attempts at describing heterogeneous reactions in an idealized fashion, by which we mean invoking the concept of "perfect behavior" leads always into quasi-thermodynamic pathways. For example, in order to develop simple models as a basis for analytical arguments, we must introduce essentially nonthermodynamic concepts such as the perfect gas equation, Raoult's empirical law of solution vapor pressures, and the concept of the species.

Since the phase rule interrelates only the parameters temperature, pressure, and component concentration, we will have to develop the bridge between the species and the component in order to enable translation of answers from the working quantity, the species, to the graphic quantity, the component.

Because our focus will be in terms of idealized models, notwithstanding their tenuous applicability to real systems, we will present a cursory discussion of more phenomenological approaches aimed at providing the tools for treating real systems. These approaches use correction terms that provide the necessary answers to satisfy observed phenomena. Since such arguments do not invoke models, the correction terms involved are sometimes referred to as "thermodynamic concentrations," activities, and "thermodynamic pressures," fugacities. Their use enables more rigid adherence to more normal thermodynamic techniques for treating systems. The reader is encouraged to refer to standard texts on thermodynamics for more precise statements on the activity and fugacity arguments than those we will offer, a full treatment lying outside the scope of our presentation.

B. THE SPECIES AND THE HOMOGENEOUS EQUILIBRIUM CONSTANT

If our system of interest is homogeneous, and, for purposes of simplicity, if this homogeneous system is a gas, then not less than two parameters must be specified to define completely the state of the system. In fact, it is only when one component is present that this minimum variance occurs. If the number of components rises to two, the variance rises to three. Thus, in addition to specifying the temperature and pressure, one concentration term must also be designated. Unfortunately, the specification of a component concentration only provides us with a restriction on the overall stoichiometry of the homogeneous system. It does not explicitly provide us with the wherewithal to analytically define the properties of the phase unless we invoke a model defining the nature of the molecular aggregation of both components. In other words, knowing the component stoichiometries and concentrations, we must define the species that are derived from the components.

Consider for example, a unary or pseudounary component whose chemical designation is A_2. By specifying the component stoichiometry via the notation A_2 we are not stating that in the vapor phase all of the molecules are composed of two atoms of A although this may indeed be the case. We may find upon inspection that the component A_2 when present in the vapor is represented by a mixture of species having the molecularities A, A_2, A_3, A_4, A_x. The component designation A_2 is only used for purposes of counting. Thus, an A species only adds $\frac{1}{2}$ of an A_2 unit to the total count, an A_2 species adds a whole unit, etc. Having counted all of the species contributions, the component concentration is known.

Attempts at invoking idealized models to describe the system, implicitly at least, requires specification of the number of moles of the different species present, a species mole containing $6.02 \cdots \times 10^{23}$ atoms or molecules. Thus, while a species mole contains a defined number of particles of a given kind, a component mole only contains this number in the abstract sense, the mass of the component mole being dependent upon the component designation, and the actual number of particles derived from a component mole depending upon what species are formed from it.

When assessing the ideality of behavior of a gas, it is customary to equate the component mole number m with the species mole number n and determine whether or not the behavior is described by

$$pV = nRT, \qquad (1)$$

where p is the pressure, V the volume, n the number of species moles, R the gas constant, and T the temperature. If it is not, and one cannot obtain agreement by simply varying the number n, the gas is termed nonideal. While this definition of nonideality is functionally as valid as any other, it is based on ignorance rather than knowledge. As the ideal gas equation has its basis in the kinetic theory of gases, the number n depends on the number of idealized particles present, i.e., the number of noninteracting point sources partaking in completely elastic collisions with the walls of the chamber. Consequently, the use of Eq. (1) requires the assumption that a mole of a component generates precisely 1 mole of species. If the species mole number is not a simple multiple or fraction of m, and if, furthermore, this species mole number varies with T, then Eq. (1) will obviously appear inapplicable. In much of what follows we will attempt to invoke Eq. (1) or analogous relationships plus models

B. THE HOMOGENEOUS EQUILIBRIUM CONSTANT

that make possible the use of these relationships in a given context. Since the species lies at the heart of the approach, the models we invoke will be based on assumptions as to the nature of participating species. The starting point for the development of idealized treatments involving species relationships is the concept of the equilibrium constant. Thermodynamically, the equilibrium constant represents a number whose absolute value for any particular system is a function only of the temperature and if defined in special ways, of both the temperature and pressure. It implies a constancy of the final state of a process, independent of the means by which the final state is arrived at. The use of this constant, if it is known, enables calculation of specially defined free energy changes. Alternately, provided with the value of such specially defined free energy changes one is able to specify the value of the constant. In fact, the real utility of the equilibrium constant concept is not realized unless one makes use of it with reference to a particular species model, such involvement being typified in treatments involving the so-called "mass action law," which is generally given its fullest extension in courses in analytical chemistry. Since the equilibrium constant will represent our primary tool for describing variations in species concentrations and ultimately in defining component distributions among coexisting phases, we will redevelop the concept to be consistent with our specific intents. Hopefully, the approach to be offered will provide the needed insight into the nature of the equilibrium constant such as to enable development of the formalism necessary to treat our subject. It is to be emphasized that our treatment is a general one and, as such, is applicable to the other areas in which the concept is valuable.

Given a homogeneous system containing X components, the phase rule predicts that $X + 1$ degrees of freedom are possessed by this system. If these X components give rise to aX species, it is obvious that the simultaneous determination of the concentrations of these aX unknowns will require a set of aX independent relationships. Since the phase rule is couched in terms of the variables, components, temperatures, and pressures, it would seem reasonable that the independent relationships we seek be couched in the same terms, including a bridge between the species and the component.

Consider a homogeneous system of volume V, and temperature T. Suppose we insert a moles of the gaseous component A, and b moles of the gaseous component B into this system, allow them to react and when the reaction is completed readjust the pressure in the system, by changing the volume, until it coincides exactly with the initial value of either of the pressures of A or B at the instant prior to their reacting. Also let us assume that the pressures of A and B were equal prior to reaction.

At the conclusion of the experiment, the temperature is permitted to come to its initial value. Thus, the basic points are that for the system

$$p_A{}^0 - p_B{}^0 \quad \text{at start,} \tag{2}$$

$$p_C{}^0 = p_A{}^0 \quad \text{or} \quad p_B{}^0 \quad \text{at conclusion,} \tag{3}$$

$$T_{\text{start}} = T_{\text{final}}. \tag{4}$$

For the sake of simplicity at this stage, let us assume that the component A gives rise in the gas phase only to the species A, and the component B gives rise only to species having the molecularity B. Let us further assume that A and B react completely to form the species C via the reaction

$$a\text{A} + b\text{B} \to c\text{C}. \tag{5}$$

While a reaction to completion is, in the limit, improbable, we need not concern ourselves with this improbability now, the degree of completeness of reaction being one of the consequences of our approach.

In its initial and final states, the system will possess unique Gibbs free energies $F_i{}^0$ and $F_f{}^0$, respectively. The isothermal change in free energy of the system, $\Delta F_T{}^0$, in moving from its initial to final states may be expressed as

$$\Delta F_T{}^0 = F_f{}^0 - F_i{}^0. \tag{6}$$

If the molar free energy of the species C in the final state is $F_C{}^0$ and those of A and B are $F_A{}^0$ and $F_B{}^0$ in the initial state, respectively, we may rewrite Eq. (6) as

$$\Delta F_T{}^0 = cF_C{}^0 - (aF_A{}^0 + bF_B{}^0).\ [1] \tag{7}$$

The process we have just defined was a completely arbitrary one in terms of initial and final pressure states. It is evident that we could have chosen as an example an infinite number of other sets of pressure conditions for the temperature in question. Having therefore arbitrarily picked one set of conditions let us use this set as our reference or standard set insofar as the free energy change is concerned. All other equally arbitrary

---

[1] Let us assume that the values $F_A{}^0$ and $F_B{}^0$ are not necessarily the molar free energies of pure A and B, respectively in the unmixed state, but rather the values they possess instantaneously prior to reacting after they are mixed. For example, visualize a two partition box containing A and B separated by a shutter. When the shutter is removed, A and B mix but do not react instantaneously. At this instant then, we define $F_A{}^0$ and $F_B{}^0$.

B. THE HOMOGENEOUS EQUILIBRIUM CONSTANT

sets may be called nonstandard. Suppose now we wish to compare the free energy change associated with one of the nonstandard processes to the free energy change accompanying our arbitrary standard or reference process.

Consider again the a moles of A at some pressure p_A atm completely reacting with b moles of the species B initially at pressure p_B atm to yield c moles of C at p_C atm where now p_A, p_B, and p_C are not necessarily equal. For the isothermal free energy change at the same temperature as before we may write equations entirely analogous to Eqs. (6) and (7), i.e.,

$$\Delta F_T = F_f - F_i \tag{8}$$

and

$$\Delta F_T = cF_C - (aF_A + bF_B), \tag{9}$$

respectively.

Since ΔF_T^0 and ΔF_T are dimensionally the same, we can relate them via a third quantity having the same dimensions to yield an identity as follows:

$$\Delta F_T^0 = \Delta F_T + X_T, \tag{10}$$

or

$$X_T = \Delta F_T^0 - \Delta F_T, \tag{11}$$

or

$$\Delta F_T^0 = \Delta F_T + (\Delta F_T^0 - \Delta F_T). \tag{12}$$

Substituting Eqs. (7) and (9) into Eq. (12) gives, upon suitable rearrangement of terms,

$$\Delta F_T^0 = \Delta F_T + (cF_C^0 - cF_C) - \{(aF_A^0 - aF_A) + (bF_B^0 - bF_B)\}. \tag{13}$$

Since the total derivative of the Gibbs free energy is given by

$$dF = -S\,dT + V\,dp, \tag{14}$$

and at constant temperature by

$$dF_T = V\,dp, \tag{15}$$

we see that for an ideal gas, Eq. (1) may be substituted for the volume term in (15) to give

$$dF_T = nRT\,dp/p. \tag{16}$$

From (16) we see that the free energy differences in the parentheses in Eq. (13) accrue simply from the presence of a particular species at two different pressures.

Integrating (16) between the limits F^0 at p^0 and F at p, the exact change in free energy due to changing the partial pressure of a given number of moles n_y of a particular species from an arbitrary pressure to an equally arbitrary reference or standard pressure may be stated as

$$n_y(F_y^0 - F_y) = n_y RT \ln p_y^0/p_y. \tag{17}$$

If we substitute Eq. (17) into Eq. (13) we obtain after rearrangement

$$\varDelta F_T^0 = \varDelta F_T + RT \ln \frac{(p_A^a/p_A^{0a})(p_B^b/p_B^{0b})}{p_C^c/p_C^{0c}}. \tag{18}$$

Since the dimensional units (atmospheres) cancel in the logarithmic expression it is seen that the latter is simply a number. This point is mentioned since, when certain standard state values are used as references, the superscript zero terms do not occur explicitly in the final relationship. This might lead one to the implausible conclusion that one is required to take the logarithm of a dimension.

While we have only considered processes occurring at constant temperature, it is evident that an equation analogous to (18) may be obtained by treating a system at constant temperature and constant total pressure and comparing this system to some defined one having only a temperature constraint, the pressure constraint depending on the standard state pressure choices. For example, consider the process in which the same species A and B react to yield C at the temperature T and where the initial total pressure was P_t and where the final total pressure is P_t. Since this system must possess initial and final free energies nothing has really changed excepting that (18) is defined now by

$$\varDelta F_T^0 = \varDelta F_{T,p} + RT \ln \frac{(p_A^a/p_A^{0a})(p_B^b/p_B^{0b})}{p_C^c/p_C^{0c}}. \tag{19}$$

Since $\varDelta F_T^0$ is independent of the pressure restrictions imposed on the process whose free energy change is $\varDelta F_{T,p}$, the former being the result in an unrelated process, it is immaterial what the choice of the defined standard pressure is. Once it is chosen, however, the corresponding $\varDelta F_T^0$ is unique.

Let us finally consider some arbitrary process for which $\varDelta F_{T,p} = 0$ which implies that the initial and final states are the same and that consequently the system is in a state of equilibrium. From (19) we see that

$$\varDelta F_T^0 = -RT \ln(p_C^c/p_A^a p_B^b)(p_A^{0a} p_B^{0b}/p_C^{0c}) \tag{20}$$

B. THE HOMOGENEOUS EQUILIBRIUM CONSTANT

Since ΔF_T^0, for a unique set of conditions, is a constant at constant temperature it is evident that the right-hand side of (20) is a constant. Furthermore, as the values p_A^0, p_B^0, and p_C^0 are the unique values of the pressures of A, B, and C which lead to ΔF_T^0, their exponented ratio is a constant. Consequently, the arbitrary pressures p_A, p_B, and p_C must also give rise to a constant exponented ratio.

Consequently, (20) may simply be denoted by

$$\Delta F_T^0 = -RT \ln(K/K^0), \quad \text{where} \quad K = p_C^c/p_A^a p_B^b \quad \text{and}$$

$$K^0 = p_C^{0c}/p_A^{0a} p_B^{0b}. \tag{21}$$

As ΔF_T^0 has been defined for a constant temperature process with exactly specified standard or reference pressures it is evident that as long as the same reference pressures are stipulated, ΔF_T^0 is a function only of the temperature. Since the absolute magnitude of K is a consequence of introducing pressure values into the ratio, we shall henceforth refer to the ratio of the constants K/K^0 by the symbol K_p and the description: "pressure (derived) equilibrium constant." Remembering that the condition leading to (21) is that $\Delta F_{T,p} = 0$, it is a consequence that the exponented pressure values comprising K are the equilibrium values for the arbitrary nonreference system in a state of equilibrium. It is also a consequence that K_p is pressure independent in an ideal system as can be seen from the following.

Let us rewrite (19) as

$$\Delta F_{T,p} = \Delta F_T^0 - RT \ln Z, \tag{22}$$

where

$$Z = \frac{(p_A^a/p_A^{0a})(p_B^b/p_B^{0b})}{p_C^c/p_C^{0c}}.$$

For another process at the same temperature T, but a different total pressure p', referred to the same reference process, we have

$$\Delta F_{T,p'} = \Delta F_T^0 - RT \ln Z'. \tag{23}$$

At equilibrium $\Delta F_{T,p}$ and $\Delta F_{T,p'}$ are both equal to zero so that

$$\Delta F_T^0 - RT \ln Z = \Delta F_T^0 - RT \ln Z'; \tag{24}$$

whence it is evident that

$$Z = Z'. \tag{25}$$

Since Z and Z' from (21) are defined by

$$Z = K^0/K = 1/K_p, \qquad (26)$$

it is evident that K_p is independent of pressure. Thus we may write

$$(\partial K_p/\partial p)_T = 0 \quad \text{for perfect gas systems.} \qquad (27)$$

While we will not be concerned to any great extent with processes for which $\Delta F_{T,p} \neq 0$, we will have occasion to introduce the concept of metastable equilibria in conjunction with multicomponent systems. Such a state for a system implies that there exists some other state having a lower free energy under identical temperature and pressure restraints. Thermodynamically, therefore, one may expect a spontaneous process in which a decrease in free energy occurs. Let us therefore consider states of a system for which nonzero free energy changes are possible and how one may predict them. Consider the case where $p_C' \neq p_C$, $p_A' \neq p_A$, and $p_B' \neq p_B$, namely, where the partial pressures are not the equilibrium values and consequently where $\Delta F_{T,p} \neq 0$. The analytical expression for the free energy change associated with the process is that given by (19), which we reiterate below

$$\Delta F_T^0 = \Delta F'_{T,p} + RT \ln \frac{(p_A'^a/p_A^{0a})(p_B'^b/p_B^{0b})}{p_C'^c/p_C^{0c}}. \qquad (28)$$

Substituting for ΔF_T^0 its value defined by (21) and again rearranging we obtain

$$\Delta F'_{T,p} = RT \ln(C/K_p), \qquad (29)$$

where

$$C = \frac{p_C'^c/p_A'^a p_B'^b}{p_A^{0a} p_B^{0b}/p_C^{0c}},$$

and where K_p is defined by Eq. (21).

If the ratio C/K_p is less than one, $\Delta F'_{T,p}$ is negative and the reaction proceeds in the direction shown in Eq. (5). On the other hand, if the ratio C/K_p is greater than one, then the reaction is not a spontaneous one. Since the system has been defined as one which is not in a state of equilibrium, it is evident that some spontaneous process must be possible which will result in a free energy decrease. Obviously, such a process is the reverse of that shown in Eq. (5), namely, the dissociation of C to yield A and B species. For the reverse reaction it is seen that its equilibrium constant $K_p{}'$ may be defined by

$$K_p{}' = 1/K_p. \tag{30}$$

Similarly,

$$C' = 1/C. \tag{31}$$

Making these substitutions in (29) we have for the free energy change of the reverse reaction $\Delta F''_{T,p}$,

$$\Delta F''_{T,p} = RT \ln \frac{1/C}{1/K_p}. \tag{32}$$

When C is greater than K_p, $1/C < 1/K_p$, and the logarithmic term is seen to be negative. Consequently, the reverse reaction is the thermodynamically possible one.

As is evident, the standard state reaction choice, $\Delta F_T{}^0$, has no bearing on the free energy changes $\Delta F'_{T,p}$ or $\Delta F''_{T,p}$ which define completely independent processes.

Thus, we have seen in the present section how one may mathematically relate free energy changes associated with arbitrary processes and that, for convenience, one may choose a perfectly arbitrary process as a reference point for comparison with all other arbitrary processes. The next section will demonstrate the utility of such an approach.

C. Standard States and Their Significance

In the general treatment of homogeneous equilibria presented above, each of the p_x terms associated with the several species carried along with it a corresponding divisor $p_x{}^0$, some reference pressure. It has also been demonstrated that the choice of reference pressures in no way affects the value $\Delta F_{T,p}$ for a real process whose attributes we may be studying. In order that reference state treatments are of general value and may be conveniently tabulated, one frequently utilizes standard states of unity since the logarithm of unity is zero.

Consider, for example, the defined reference reaction of a moles of A at 1 atm with b moles of B at 1 atm to yield c moles of C at 1 atm. Equation (20) then acquires the simple form

$$\Delta F_T{}^0 = -RT \ln(p_C{}^c/p_A{}^a p_B{}^b) = -RT \ln K_p \tag{33}$$

from which it is seen that having a knowledge of ΔF^0 for the standard reaction one can calculate K_p for the desired equilibrium and vice versa.

For any other set of nonunity standard-state choices it is evident that the numerical value of $\Delta F_T{}^0$ will be different since its value will, of necessity, then be defined by

$$\Delta F_T{}^0 = -RT \ln(K/K^0) \tag{34}$$

We will, in future discussions, always utilize unity standard states. Thus standard states for gases will be taken as the pure gaseous species at 1 atm or 1 torr; for solids, the pure solid at 1 atm or 1 torr, and for liquids the pure liquid at 1 atm or 1 mm.

Most standard free energy tables are presented in terms of unity standard state choices. A problem arises, however, because in some instances standard states of 1 atm are chosen while in other instances standard states of 1 torr are chosen. In order to make use of standard free energy data from different sources one must make certain that all of the tabulated values are based on the same unity standard states. Transposition of data from one reference state to a second reference state can be realized as follows:

Consider a system containing the species A, B, and C in a state of equilibrium. As such, therefore, it is the case that the partial pressures are fixed. We may express K_p in terms of 1 atm standard states and values of the partial pressures of A, B, and C in atmospheres. Alternately, K_p could be derived by expressing both the partial pressure values and the standard state values in millimeters of mercury. Thus for the two means of specifying K_p we may write

$$K_p(\text{atm}) = \frac{(\text{atm C})^c}{(\text{atm A})^a (\text{atm B})^b} \bigg/ \frac{(1 \text{ atm})^c}{(1 \text{ atm})^a (1 \text{ atm})^b}, \tag{35}$$

$$K_p(\text{mm Hg}) = \frac{(\text{mm C})^c}{(\text{mm A})^a (\text{mm B})^b} \bigg/ \frac{(1 \text{ mm})^c}{(1 \text{ mm})^a (1 \text{ mm})^b}. \tag{36}$$

Let us multiply each value within the parentheses in Eq. (36) by the unity transform $1 \text{ atm}/760 \text{ mm} = 1$

$$K_p(\text{mm Hg}) = \frac{\left(\dfrac{\text{mm C} \cdot 1 \text{ atm}}{760 \text{ mm}}\right)^c \bigg/ \left[\left(\dfrac{\text{mm A} \cdot 1 \text{ atm}}{760 \text{ mm}}\right)^a \left(\dfrac{\text{mm B} \cdot 1 \text{ atm}}{760 \text{ mm}}\right)^b\right]}{\left(\dfrac{1 \text{ mm} \cdot 1 \text{ atm}}{760 \text{ mm}}\right)^c \bigg/ \left[\left(\dfrac{1 \text{ mm} \cdot 1 \text{ atm}}{760 \text{ mm}}\right)^a \left(\dfrac{1 \text{ mm} \cdot 1 \text{ atm}}{760 \text{ mm}}\right)^b\right]}. \tag{37}$$

Upon canceling the mm units in each pair of parentheses, we see that the first ratio on the right-hand side of Eq. (37) is precisely the same as

D. THE MOLE FRACTION EQUILIBRIUM CONSTANT

the first ratio on the right-hand side of Eq. (35). Consistent with our previous treatment let us term this ratio K. We now have

$$K_p(\text{mm Hg}) = K\Big/\Big[\Big(\frac{1\text{ atm}}{760}\Big)^c\Big/\Big(\frac{1\text{ atm}}{760}\Big)^a\Big(\frac{1\text{ atm}}{760}\Big)^b\Big]. \tag{38}$$

If we simplify the second ratio on the right-hand side of Eq. (37) we see that

$$1\text{ atm}^c/1\text{ atm}^a\ 1\text{ atm}^b\ 760^{c-(a+b)} = K^0 \cdot 760^{\Delta n}, \tag{39}$$

where Δn is the change in the molar stoichiometry of the reaction, i.e., sum of the molar coefficients of products less the sum of the molar coefficients of reactants.

Substituting the result in (38) and again referring to (35) we see that

$$K_p(\text{mm Hg}) = (K/K^0) \cdot 760^{\Delta n} = K_p(\text{atm}) \cdot 760^{\Delta n}. \tag{40}$$

D. THE HOMOGENEOUS EQUILIBRIUM CONSTANT IN TERMS OF MOLE FRACTIONS

Instead of defining the equilibrium constant in terms of pressures it is sometimes more convenient to express concentrations in terms of species mole fractions defined by

$$N_X = n_X/n_t, \tag{41}$$

where N_X is the mole fraction of the species X, n_X is the number of moles of the species X and n_t is the total number of moles present in the homogeneous system.

We shall have occasion also to make use of component mole fractions defined by

$$M_X = m_X/m_t, \tag{42}$$

where the analogous terms have corresponding significances to those defined by Eq. (41).

The equilibrium constant based on species mole fraction values is denoted K_N and is related to K_p in the following manner.

In a perfect gas mixture the partial pressure of any species X is given by

$$p_X V_t = n_X RT, \tag{43}$$

where p_X is the partial pressure due to species X in a system of total volume V_t and temperature T. The total pressure of such a system due to the pressures of the several species whose concentrations are dependent on a defined equilibrium is given by

$$P_t V_t = n_t RT, \tag{44}$$

where P_t is the total pressure exerted by the pertinent species and n_t is the total number of moles due to such species.

Dividing (43) by (44) we obtain

$$p_X = N_X P_t. \tag{45}$$

Making this substitution in (20) we obtain

$$\Delta F_T^0 = -RT \ln \{[(N_C P_t)^c/(N_A P_t)^a (N_B P_t)^b](p_A^{0a} p_B^{0b}/p_C^{0c})\}. \tag{46}$$

Choosing unity standard state values for the p^{0y} terms one obtains

$$\Delta F_T^0 = -RT \ln [(N_C^c/N_A^a N_B^b)(P_t^c/P_t^a P_t^b)]$$
$$= RT \ln [(N_C^c/N_A^a N_B^b)(P_t^{c-(a+b)})]$$
$$= -RT \ln K_N P_t^{\Delta n}, \tag{47}$$

where Δn is again the change in the number of species moles attendant with the reaction defined by Eq. (5). From Eq. (47) it is seen that when $\Delta n = 0$, $K_p = K_N$ and that when $P_t = 1$ atm, $K_p = K_N$. It is to be noted that while the logarithmic expressions in all of the above are dimensionless, the ratio K (pressure) itself may or may not be dimensionless depending on the stoichiometry of the reaction. K (mole fraction) on the other hand is always dimensionless since each term N_X is a ratio $(n_X/n_t)^x$. It is to be emphasized however that in either event the constants K_p and K_N are both dimensionless.

A question may arise in the calculation of ΔF_T^0 values from a knowledge of ΔH^0 and ΔS^0 data as to whether this ΔF_T^0 is related to K_p or K_N. For the general case, the calculated constant is related to K_p. Thus, when the constant total pressure constraint on our arbitrary process is unity, the general equation

$$\Delta F_T^0 = -RT \ln K_N. \tag{48}$$

may be written. Alternately, even if P_t is not unity but Δn is unity, Eq. (48) is the applicable one. Finally we may note that

$$K_p = K_N P_t^{\Delta n}. \tag{49}$$

E. THE VARIATION OF K WITH TEMPERATURE

Since K_p is a constant and the value K_N depends on the ratio $K_p/P_t^{\Delta n}$ it is evident that unlike K_p, K_N is pressure dependent.

E. THE VARIATION OF EQUILIBRIUM CONSTANT WITH TEMPERATURE

In the next chapter, which begins our concern with multicomponent heterogeneous equilibria, the variation of equilibrium constants with temperature will be of great interest to us. Where unity standard state choices are made and where the constant pressure constraint on the arbitrary nonstandard process is also unity we have

$$\Delta F_T^0 = -RT \ln K_N = -RT \ln K_p = -RT \ln K_X. \tag{50}$$

Rearranging, we have

$$\ln K_X = -\Delta F_T^0/RT, \tag{51}$$

and differentiating K with respect to T we obtain

$$d(\ln K)/dT = -(1/R) \, d(\Delta F_T^0/T)/dT. \tag{52}$$

The term $d(\Delta F_T^0/T)/dt$ is of the form $d(y/x)/dx$ whose derivative is

$$d(y/x)/dx = (1/x)(dy/dx) - (y/x^2). \tag{53}$$

Thus, Eq. (52) may be written

$$d(\ln K)/dT = -(1/R)[(1/T) \, d(\Delta F_T^0)/dT - (\Delta F_T^0)/T^2]. \tag{54}$$

The free energy for a system in a state 1 is given by

$$F^{(1)} = H^{(1)} - TS^{(1)}. \tag{55}$$

In some other state at the same temperature and total pressure, the free energy for the system is

$$F^{(2)} = H^{(2)} - TS^{(2)}. \tag{56}$$

The difference between the two states is therefore

$$\Delta F_T^0 = \Delta H^0 - T \, \Delta S^0. \tag{57}$$

Now consider the system in state 1 again. Its total derivative with respect to the Gibbs energy is

$$dF^{(1)} = -S^{(1)} \, dT + V^{(1)} \, dp \tag{58}$$

The system in state 2 (subsequent to reaction for example) is defined by

$$dF^{(2)} = -S^{(2)}\,dT + V^{(2)}\,dp. \tag{59}$$

If at constant pressure the system in state 1 has its temperature varied, this variation will be given by

$$(\partial F^{(1)}/\partial T)_p = -S^{(1)}. \tag{60}$$

Similarly for system 2 whose temperature is varied

$$(\delta F^{(2)}/\delta T)_p = -S. \tag{61}$$

The variation of ΔF, which defines the difference in free energy between the systems in state 1 and state 2, all at the pressure p with temperature is seen from Eqs. (60) and (61) to be

$$\frac{\partial F^{(2)}}{\partial T} - \frac{\partial F^{(1)}}{\partial T} = \frac{\partial (F^{(2)} - F^{(1)})}{\partial T} = -\left(\frac{\partial \Delta F}{\partial T}\right)_p = -\Delta S. \tag{62}$$

When the free energy change in question is a standard state choice, we have

$$(\partial \Delta F^0/\partial T)_p = -\Delta S^0. \tag{63}$$

Substituting this result in Eq. (57) we have

$$\Delta F^0 = \Delta H^0 + T(\partial \Delta F^0/\partial T)_p. \tag{64}$$

Dividing through by T^2 and rearranging yields

$$-\frac{\Delta H^0}{T^2} = \frac{1}{T}\left(\frac{\partial \Delta F_T^0}{\partial T}\right)_p - \frac{\Delta F_T^0}{T^2}, \tag{65}$$

from Eq. (54) it is seen that the right-hand side of Eq. (65) is equal to

$$-R\,d\ln K/dT, \tag{66}$$

thus,

$$d\ln K/dT = \Delta H^0/RT^2. \tag{67}$$

The question arises as to the form of Eq. (67) for K_N in the event the total pressure constraint is not unity.

Since K in Eq. (67) is always equal to K_p we may generalize as follows

$$d\ln K_p/dT = d(\ln K_N P_t^{\Delta n})/dT = \Delta H^0/RT^2. \tag{68}$$

E. THE VARIATION OF K WITH TEMPERATURE

Integrating between K_N and K_N' at T and T' assuming P_t is the same at both temperatures and $\partial(\Delta H^0)/\partial T = 0$ we have

$$\ln \frac{K_N' P_t^{\Delta n}}{K_N P_t^{\Delta n}} = \ln \frac{K_N'}{K_N} = -\frac{\Delta H^0}{R}\left(\frac{1}{T'} - \frac{1}{T}\right), \tag{69}$$

which is identical, in form, to Eq. (67). Thus, the conclusions Eqs. (67) and (69) are completely general so long as the constant pressure constraint applied at one temperature is applied at the other temperature also.

11

Multicomponent Systems, the Equilibrium Constant, and Heterogeneous Systems

A. INTRODUCTION

In Chapter 10 the concept of the equilibrium constant was introduced in terms of homogeneous equilibria. Since the phase rule and its experimental application is of greater import in systems consisting of more than one distinguishable agglomeration of matter, we will, in this chapter, attempt to define analogous constants to those already discussed with emphasis on two component polyphase equilibria. From the phase rule, we note that a condition of univariance obtains in a two-component system when three phases coexist simultaneously. Alternately, for a binary system at constant total pressure, a two-phase equilibrium is isobarically univariant. For the sake of simplicity we shall treat binary interactions subject to a constant total pressure constraint. Since the effects of nominal variations in total pressure on the diagrams of state are slight (in a similar way to the effects of pressure on one component systems), the results are applicable to systems under their equilibrium pressure for the most part. Again, as in Chapter 10, the treatment will be attempted assuming that ideal behavior obtains. Thus, all species terms will be fully defined, and all species will be assumed to behave

B. CHEMICAL POTENTIALS

ideally. For example, isotopic effects will be ignored as will minor deviations from stoichiometry. We will have occasion to consider stoichiometric effects in later chapters, again however, within the context of ideal behavior.

One significant departure from the method used in Chapter 10 is the introduction of the concept of the chemical potential when treating polycomponent-multiphase equilibria. In the preceding chapter our discussions were presented in terms of *in situ* molar free energies, which, as we shall see below, is not general enough, since molar free energies are not constants excepting in unary systems. The results previously obtained are, however, precise since all that is required is to substitute chemical potential (partial molar free energies) for the molar free energy terms used.

B. The Chemical Potential for Components and Species

In a homogeneous system consisting of only a single species, the total free energy $F_{T,p}$, at constant temperature and pressure, can be expressed by

$$F_{T,p} = xF_X, \tag{1}$$

where F_X is the molar free energy and x represents the number of moles of the species in question. In a homogeneous system consisting of several species, the quantity F_X is not a constant, its value depending on the composition of the system as a whole. In treating multicomponent systems, it is more appropriate to define a new term related both to the free energy of a component and to the effects of concentration on this free energy. This concentration dependent molar free energy is termed either the chemical potential or the partial molar free energy. Its significance is seen from the following.

The governing relationship for the single phase, single species (or component) system is that for the total derivative of the Gibbs free energy given by

$$dF = -S\,dT + V\,dp.\,^1 \tag{2}$$

The corresponding equation for the single phase multicomponent system is given by

$$dF = -S\,dT + V\,dp + \sum_i \mu_i\,dm_i. \tag{3}$$

---

[1] This equation is also quite appropriate for a homogeneous polycomponent system of constant composition.

The symbol μ_i is termed the partial molar free energy of the ith component while m_i is the number of component moles of this ith component. In other words, Eq. (3) states that in addition to a dependency of the free energy of the system on the temperature and pressure, a further change can be wrought by changing the composition of the system via the addition or substraction of da moles of the component A and/or db moles of the component B and/or di moles of i, etc. The term μ_i is used instead of the term F_i since we wish to include the possibility of composition itself altering the molar contribution to the free energy of the component A, B, i, etc. At constant total pressure and temperature we see from Eq. (3) that

$$(dF)_{T,p} = \sum_i \mu_i \, dm_i = \mu_A (dm_A)_{T,p} + \mu_B (dm_B)_{T,p} + \cdots + \mu_i (dm_i)_{T,p}. \quad (4)$$

Furthermore from Eqs. (3) and (4) we see that

$$(\partial F / \partial m_i)_{T,p,m_i \cdot \text{s}} = \mu_i. \quad (5)$$

Equation (5) states that for a homogeneous, multicomponent system at constant temperature, pressure, and defined composition, i.e., all other components except m_i, the change in the system free energy due to a change of 1 mol of the ith component is referred to as the chemical potential or partial molar free energy. Physically, Eq. (5) does not refer to an actual mole of such addition to a system except in a special context. Thus if we plot the free energy of the system at constant temperature and pressure as a function of concentration of the ith component, μ_i is the slope of the curve at any particular concentration we look at. At any such point on the free energy versus concentration curve, the free energy of the homogeneous system is given by

$$F_{T,p} = \mu_A m_A + \mu_B m_B + \cdots + \mu_i m_i = \sum_i \mu_i m_i. \quad (6)$$

For a polycomponent, polyphase system in a state of equilibrium, the composition requirement that satisfies this equilibrium is

$$\mu_i^1 = \mu_i^2 = \mu_i^z, \quad (7)$$

where the superscripts refer to the different coexisting phases. This criterion has indeed served as the basis for deriving the phase rule. Of immediate interest, however, is the relationship between species and component chemical potential terms.

Consider the reaction depicted in

$$a\text{A} + b\text{B} \rightleftarrows \text{A}_a \text{B}_b, \quad (8)$$

B. CHEMICAL POTENTIALS

which is not necessarily proceeding homogeneously. At equilibrium we have for this entire system

$$F_f - F_i = \Delta F_{T,p} = 0, \tag{9}$$

where F_f is the final free energy state of the system and F_i is the initial free energy state.

If we denote the species partial molar free energies by the designation $\bar{F}_X$ (as opposed to the component designation μ_X), we may expand (9) as

$$\bar{F}_{A_aB_b} - (a\bar{F}_A + b\bar{F}_B) = 0, \tag{10}$$

or

$$\bar{F}_{A_aB_b} = a\bar{F}_A + b\bar{F}_B. \tag{11}$$

Equation (11) states that the chemical potential of a complex species is equal to the chemical potentials of its constituent species multiplied by their stoichiometric mole numbers. Suppose now that the process in question [Eq. (8)] refers to the transfer of a mole of the species A_aB_b from a phase 1 to a phase 2 in which it dissociates to yield a moles of the species A and b moles of the species B. If our component designation is A_aB_b, it is immediately evident that in order for (7) to be valid it must be the case that

$$a\bar{F}_A^{(2)} + b\bar{F}_B^{(2)} = \bar{F}_{A_aB_b}^{(1)}. \tag{12}$$

Generally, therefore, we may conclude that the chemical potential of the component A_aB_b is given

$$\mu_{A_aB_b} = \bar{F}_{A_aB_b}^{(2)} = a\bar{F}_a^{(2)} + b\bar{F}_b^{(2)} = \bar{F}_{A_aB_b}, \tag{13}$$

and

$$\bar{F}_A^{(1)} = \bar{F}_A^{(2)}. \tag{14}$$

The result stated in Eq. (14) is a rather obvious one. Since the equiphase potential criterion for a component is not predicated upon any restriction concerning the state of molecular aggregation of this component in coexisting phases, it follows that the chemical potential of a component must be the sums of any species chemical potentials from which the component stoichiometry is generated. Equations (8)–(14) are a reaffirmation, in a sense, of the conclusion previously drawn: that chemical potentials of the same component are the same in coexisting phases.

Analogously to the definition of the partial molar free energy, one can define a partial molar volume, i.e., the change in the volume of a system at constant temperature, pressure, and composition with a change in the number of species moles of a particular species.

Analytically, the statement of the partial molar volume may be developed as follows.

Since from Eq. (3)

$$(\partial F/\partial p)_{T,n_i\text{'s}} = V, \tag{15}$$

then

$$(\partial V/\partial n_i)_{T,p,n_{i'}\text{'s}} = (\partial F/\partial p, \partial n_i)_{T,p,n_{i'}\text{'s}} = (\partial \bar{F}_i/\partial p)_{T,n_i\text{'s}} = \bar{V}_i. \tag{16}$$

From (16) we see that

$$d\bar{F}_i = \bar{V}_i\, dp. \tag{17}$$

Analogously

$$d\mu_i = \bar{V}_i\, dp. \tag{18}$$

Since $\bar{V}_i = V_i$ for a perfect gas we have for the special case of a perfect gas mixture

$$d\bar{F}_i = V_i\, dp, \tag{19}$$

and

$$d\mu_i = V_i\, dp. \tag{20}$$

Furthermore as

$$V = n_i RT/p_i, \tag{21}$$

in a perfect gas polyspecies system, where V is the total volume and p_i is the partial pressure of the species i we see that

$$(\partial V/\partial n_i)_{T,p,n_i\text{'s}} = RT/p_i. \tag{22}$$

Consequently for a perfect gas phase

$$d\bar{F}_i = (RT/p_i)\, dp. \tag{23}$$

If the pressure change is due solely to a change in partial pressure of the species i, then

$$d\bar{F}_i = (RT/p_i)\, dp_i. \tag{24}$$

C. Equilibrium Constants in Terms of Partial Molar Quantities

Having introduced the concepts of the species and component chemical potentials we are now in a position to apply this information

C. HETEROGENEOUS EQUILIBRIUM CONSTANTS

to the derivation of relationships applicable to polyphase polyspecies equilibria.

Consider an equilibrium between two phases, one a vapor, the second a liquid solution. For a particular vapor phase species A, we may define its chemical potential relative to some arbitrary chemical potential in similar fashion to that for molar free energies as was done in the preceding chapter. Again as before, the arbitrary potential may be chosen so as to be mathematically convenient. Once having done this, we may henceforth refer to this arbitrary potential as a reference or standard state for the vapor species A. The integrated form of Eq. (24) may be used to relate the chemical potential of the species A in the system under scrutiny to the reference state we have chosen. Integrating (24) for the species A between the limits $\bar{F}_A^{(2)}$ at its system partial pressure $p_A^{(2)}$ and $\bar{F}^{0(2)}$ at its reference partial pressure $p_A^{0(2)}$ we obtain

$$\bar{F}_A^{(2)} = \bar{F}_A^{0(2)} + RT \ln(p_A^{(2)}/p_A^{0(2)}). \tag{25}$$

The superscript (2) in this instance refers to the vapor phase.

In mole fraction terms, for the partial pressure of a species comprising a mixture whose total species pressure is P_t, we may write

$$p_A^{(2)} = N_A^{(2)} P_t, \tag{26}$$

where $N_A^{(2)}$ is the mole fraction of the species A in the vapor phase. P_t need not be the same as the total pressure, p, of the system if part of this total pressure is derived from an inert gas which does not participate in the chemical equilibrium.[2] Substituting the value of $p_A^{(2)}$ given in Eq. (26) into Eq. (25) we obtain

$$\bar{F}_A^{(2)} = \bar{F}_A^{0(2)} + RT \ln(N_A^{(2)} P_t/p_A^{0(2)}). \tag{27}$$

For the species A present in the liquid solution in equilibrium with the vapor, assuming that such species are present in the solution, the partial pressure of this species as a function of its concentration in the solution may be expressed by

$$p_A^{(2)} = N_A^{(1)} p_A^\dagger. \tag{28}$$

$N_A^{(1)}$, in this instance, is the mole fraction of A in the solution and $p_A^\dagger$ is

---

[2] While in the limit no inert gas is completely inert, the same arguments are implied as those for the treatment of unary equilibria subject to inert gas over-pressures.

the vapor pressure of a liquid consisting of only the species A at the same temperature and at a total system pressure P_t.[3]

If we now substitute the value of $p_A^{(2)}$ given by Eq. (28) into Eq. (25), we obtain an equivalent expression to Eq. (27), i.e., Eq. (29), this time having as its basis the solution concentration of A rather than its vapor phase concentration. Note, however, that the reference state in both equations is the pure vapor

$$\bar{F}_A^{(2)} = \bar{F}_A^{0(2)} + RT \ln(N_A^{(1)} p_A^\dagger / p_A^{0(2)}). \tag{29}$$

Expanding Eq. (29) gives

$$\bar{F}_A^{(2)} = \bar{F}_A^{0(2)} + RT \ln(p_A^\dagger / p_A^{0(2)}) + RT \ln N_A^{(1)}, \tag{30}$$

$p_A^\dagger$ has been defined as the vapor pressure of a liquid consisting only of the species A which vaporizes to give only species A in the vapor. Since $p_A^{0(2)}$ is the standard state for pure vapor A species, it is evident that we may relate it to $p_A^\dagger$ arbitrarily as follows:

$$\bar{F}_A^{\dagger(1)} = \bar{F}_A^{0(2)} + RT \ln(p_A^\dagger / p_A^{0(2)}). \tag{31}$$

$\bar{F}_A^{\dagger(1)}$ is the chemical potential of pure liquid A referred to the standard state chosen for the vapor, i.e., referred to $\bar{F}_A^{0(2)}$ when the partial pressure of A is $p_A^{0(2)}$. Since the liquid is pure, its molar free energy is equivalent to its chemical potential and we may write for Eq. (31)

$$F_A^{\dagger(1)} = \bar{F}_A^{0(2)} + RT \ln(p_A^\dagger / p_A^{0(2)}). \tag{32}$$

If we define $p_A^\dagger$ further as the vapor pressure of pure liquid species A at 1 atm pressure, then $F_A^{\dagger(1)}$ is indeed the standard state normally chosen for liquids. We may symbolize it by $F_A^{0(1)}$.

---

[3] Equation (28) is the familiar Raoult equation describing the vapor pressure of a perfect solution. Implicit in this empirical relationship is the condition that the vapor pressure of the pure liquid A (where A is the component rather than the species) be derived from a liquid where the particles appear to be of a single kind, namely where the liquid component A comprises solely species A. If the component designated A tends, for example, to dissociate or associate yielding a larger or smaller number of particles than indicated by the component molecular weight, application at Eq. (28) will not yield experimentally viable values. It is for this reason that the Raoult relation has been given in the manner defined by Eq. (28), since it is applicable only if the pure liquid is not only ideal, but comprises a single species. Since real systems are not in the limit ideal nor do they generally consist of a single species, it is evident that Eq. (28) is only valid to a first approximation for real systems and that $p^\dagger$ represents for real liquids an unattainable experimental value.

C. HETEROGENEOUS EQUILIBRIUM CONSTANTS

Utilizing Eq. (32) in this new context we may substitute its value in Eq. (30) to obtain

$$\bar{F}_A^{(2)} = \bar{F}_A^{0(1)} + RT \ln N_A^{(1)}, \tag{33}$$

i.e., the right-hand side of Eq. (32) represents the first two terms in the right-hand side of Eq. (30).

The analogous free energy equation in terms of vapor phase concentration of A [Eq. (27)] may be regrouped by combining the terms $\bar{F}_A^{0(2)} + RT \ln P_t/p_A^{0(2)}$ to give

$$\bar{F}_A^{(2)} = \bar{F}_A^{\dagger(2)} + RT \ln N_A. \tag{34}$$

The reference term $\bar{F}_A^{\dagger(2)}$ is obviously pressure and temperature dependent.

Applying the criterion for species equilibrium developed in Section B we may now write

$$\bar{F}_A^{(1)} = \bar{F}_A^{(2)} = \bar{F}_A^{\dagger(2)} + RT \ln N_A^{(2)} = \bar{F}_A^{0(1)} + RT \ln N_A^{(1)}. \tag{35}$$

The term $\bar{F}_A^{0(1)} + RT \ln N_A^{(1)}$ may be taken as an alternative statement of Raoult's law and the term $\bar{F}_A^{\dagger(2)} + RT \ln N_A^{(2)}$ as an alternative definition of the perfect gas equation.

For the process of transferring the species A from a liquid to a vapor phase, we have at equilibrium

$$\Delta F_{T,p} = 0 = \bar{F}_A^{(2)} - \bar{F}_A^{(1)}. \tag{36}$$

Substituting for $\bar{F}_A^{(2)}$ and $\bar{F}_A^{(1)}$ the values given in Eq. (35) we have

$$\bar{F}_A^{(2)} - \bar{F}_A^{(1)} = \bar{F}_A^{\dagger(2)} - \bar{F}_A^{0(1)} + RT \ln(N_A^{(2)}/N_A^{(1)}) \tag{37}$$

or

$$\Delta \bar{F}^{0\dagger} = -RT \ln(N_A^{(2)}/N_A^{(1)}). \tag{38}$$

At constant T and p the Gibbs free energy is simply

$$F_{T,p} = \sum_i n_i \bar{F}_i. \tag{39}$$

For a pure material in some arbitrary reference state we may write

$$F_{T,p}^0 = n_i \bar{F}_i^0 = n_i F_i^0, \tag{40}$$

where F_i^0 is the molar free energy of the particular pure species, and is for the pure species equivalent to the partial molar free energy of that species.

Since $\bar{F}^0$ and $\bar{F}^\dagger$ designations refer in our previous arguments to pure state situations they may be replaced by the terms F^0 and $F^\dagger$. If we further specify that the total pressure for these pure states is 1 atm for pure vapor species and 1 atm for pure liquid species, we may write for the process described by Eq. (38),

$$\Delta F_A^0 = -RT \ln(N_A^{(2)}/N_A^{(1)}) = -RT \ln K_{N_A}, \tag{41}$$

since the dagger term in Eq. (37) becomes identical to $F_A^{0(2)}$. Equation (41) then serves as the basic relationship for heterogeneous equilibria.

Although no relation analogous to the Raoult equation exists for solid solutions, one may infer the existence of such a relation. We then define a new quantity p_A^0 which is the partial pressure of vapor species A over solid species A for a solid–vapor pure state equilibrium at a specified temperature and total pressure of 1 atm. Consequently, Eq. (41) may be considered applicable to species equilibrium between any two coexisting ideally behaving phases. For an ideal system, an entirely analogous relation to that expressed by Eq. (41) must be valid when we consider component quantities instead of species quantities since the two are directly related.

12

The Thermodynamic Parameters—Fugacity and Activity

A. INTRODUCTION

Although the arguments to be advanced in this chapter will neither be employed nor, for that matter, discussed subsequently, it is of pedagogical value to present them briefly in order to demonstrate an alternative approach to the one we shall use. While this alternative approach, in general form at least, is thermodynamically correct and requires minimal recourse to specification of models, it has certain disadvantages for our purposes. First, rather than attempting to provide an explanation for observed behavior, it fources correct answers which enable fitting of data. Second, in order to arrive at analytical solutions, a starting point must be invoked whose validity is tenuous. No implication is intended, however, that the alternative method is inferior to the one we are employing, since it is the alternative method which, to a limited extent, has permitted the analytical description of real systems whose nature has been elucidated previously. Our approach, on the other hand, while perhaps of less utility in describing a real set of data, does provide the necessary insight to describe general changes in the behavior of heterogeneous systems.

B. Rationale for a Thermodynamic Approach to the Treatment of Real Systems

Given a set of proper data, one is able to predict the behavior of certain simple systems, termed ideal or perfect, with a fair degree of success. Unfortunately, the majority of systems encountered in nature deviate, to a greater or lesser extent, from the predicted states that are based on considerations of perfect behavior. The causes for departure from such predicted states are frequently obscure and one can either attempt to adjust ideal equations to take into account deviations or one can develop phenomenological treatments such as the virial gas equation to fit the deviations. A generalized method which resorts to adjustments of ideal relationships, and which has been useful in describing observed behavior, after the fact, is based on the following rationale:

Consider once again the total derivative of the Gibbs free energy

$$dF = -S\,dT + V\,dp, \tag{1}$$

which is applicable to a homogeneous system of constant composition. At constant temperature, the free energy of this system is given by

$$F_{(T)} = \int V\,dp + C. \tag{2}$$

For a perfect vapor system, as we have seen, the relationship between V, p, and T is given by the equation of state

$$V = nRT/p. \tag{3}$$

If the value of V defined by Eq. (3) is introduced in Eq. (2) and the integration is effected, we obtain the indefinite integral

$$F_{(T)} = nRT \ln p + C, \tag{4}$$

and its counterpart the definite integral

$$(F_2 - F_1)_{(T)} = \Delta F_{(T)} = nRT \ln(p_2/p_1). \tag{5}$$

For 1 mole of an ideal gas we have

$$\Delta F_{(T)}/n = \Delta F_{(T)} = RT \ln(p_2/p_1). \tag{6}$$

Equation (6), by dint of its derivation, is an ideal free energy relationship for a perfect gas system. Thus, if Eq. (6) is employed to determine the free energy change accompanying an isothermal compression or

B. THE TREATMENT OF REAL SYSTEMS

expansion of 1 mole of a real gas, it is evident that the values p_2 and p_1 (more important, their ratio) must be identically those that 1 mole of an ideal gas would exhibit in a system at the same temperature and volume. More often than not, however, real gases do not obey Eq. (3), namely, at a defined volume and temperature the observed pressure does not coincide with that predicted by Eq. (3). Consequently, insertion of measured pressures for two different volumes will not yield correct answers for the free energy change of the process, $\Delta F_{(T)}$. Recognizing then that Eq. (6) is not necessarily descriptive of the bahavior of real systems, insofar as free energy changes are concerned, it is evident that there must exist an infinity of numbers whose ratio when used in Eq. (6) will provide correct answers for the free energy change accompanying a specified process. Any such proper ratio is obviously a thermodynamically suitable one even if neither of the numbers comprising the ratio bear any resemblance to any measured pressure. For our purposes, let us confine our interests to two such numbers which, in addition to providing the proper ratio, also are of similar magnitude to measured pressures for the conditions employed. These special pressures may be thought of as thermodynamic (as opposed to measured) pressures and are termed fugacities f. By definition then, utilizing these thermodynamic pressures in Eq. (6) yields the exact relationship

$$\Delta F_{(T)} = RT \ln(f_2/f_1). \tag{7}$$

It is to be noted that by generalizing Eq. (6) to encompass real and ideal systems our problem has not been resolved since one must be able to assign numerical values to the quantities f before Eq. (7) has any utility. Assignment of values, in simple cases at least, may be done in the following manner:

It has been observed that for gases such as H_2, He, N_2, etc., their bahaviors are more nearly predictable by Eq. (3) as their pressures tend toward lower and lower values. Based on such observations it is postulated that as

$$p \to 0, \tag{8}$$

$$p = f. \tag{9}$$

While Eq. (9) is not unreasonable for uncomplicated gaseous systems, that is for systems in which the component and species designations are equivalent, it is not necessarily valid for the general case. What is implied in Eq. (9) is that *the pressure of a known number of gaseous particles equals the fugacity of these gaseous particles at low pressures.* Note, however, that inadvertently we have been led, in the attempt to define a point of coincidence between pressure and fugacity, to the specification of a model. The model in this instance relates to the constancy of the number n in Eq. (3). Thus, while Eq. (7) is thermodynamic, attempts to utilize Eq. (7) involve the introduction of an assumption based on the number of particles present. As an example of the difficulty that might arise in using the assumption of Eqs. (8) and (9), consider the case where a component formula weight of the compound AB is permitted to vaporize completely according to

$$AB(s) \rightleftarrows AB(g). \tag{10}$$

If the vaporous species generated by the process depicted in Eq. (10) do conform to the composition AB, then Eq. (9) is probably valid at low pressures since particle–particle interactions will tend to vanish. If on the other hand the vaporization process is succeeded by or attended by the processes

$$AB(g) \rightleftarrows A(g) + B(g), \tag{11}$$
$$AB(g) \rightarrow A(g) + \tfrac{1}{2}B_2(g), \tag{12}$$
$$2AB(g) \rightleftarrows A_2B_2(g), \quad \text{etc.}, \tag{13}$$

the assumption, Eq. (9), is invalid unless these processes are taken into account. Furthermore, as the equilibrium constant is a function of temperature we see that the number of particles derived from a component mole of AB is not constant with temperature. At constant temperature, but varying total pressure we see that as p decreases, K_N, the equilibrium constant reflecting the mole fractions of the several coexisting species must decrease if the term Δn is positive, and must increase if the term Δn is negative, where Δn is the change in the number of moles attending the process. Consequently, if we define a number n for a system at one total system pressure, we have no guarantee that this number will remain constant over any usable range of total system pressures. Thus, we see that excepting for very special cases, the postulate of Eq. (9) does not provide a foundation for the numbers f. Keeping this in mind, we may, for these special cases, present an approximate method for computing fugacities from pressures as suggested by Lewis and Randall.[1]

For the perfect gas, the observed species molar volume v_0 is given by

$$v_0 = RT/p_0 = v_i, \tag{14}$$

where p_0 is the observed pressure, R is the gas constant, and T is the

---

[1] See general reference list in the Appendix.

B. THE TREATMENT OF REAL SYSTEMS

absolute temperature. The value, v_0 is the observed volume and is equal to the calculated volume v_i.

For an imperfect gas, Eq. (14) may be generalized by

$$v_0 = (RT/p_0) - A, \qquad (15)$$

where A is a number having the dimensions of volume.

Since RT/p_0 is identically the ideal volume 1 mole of perfect gas would require at p_0 we see that in general

$$v_0 = v_i - A, \qquad (16)$$

and that

$$A = \Delta v \ll v_0 \quad \text{or} \quad v_i. \qquad (17)$$

From our definitions we have the exact relationship

$$RT \ln (f_2/f_1) = \int_{p_1}^{p_2} v_0 \, dp. \qquad (18)$$

Substituting for v_0 its general value defined by Eq. (15), we have

$$RT \ln (f_2/f_1) = \int_{p_1}^{p_2} [(RT/p_0) - A] \, dp. \qquad (19)$$

The right-hand side of Eq. (19) may be integrated readily if it is assumed that

$$(\partial A/\partial p)_T = 0.$$

This assumption is perhaps the most questionable approximation made in the treatment since it necessitates a constancy in apparent nonideal behavior with pressure and implies a constancy in the number n with pressure. Other treatments resort to graphical techniques in order to determine the dependency. It is to be noted that a plot of v_0 versus $1/p_0$ in Eq. (15) will yield A as an intercept value providing either that A as $f(p) = 0$ or A varies as $1/p$. For a nondissociative gas A, which has dimensions of volume, should to a good first approximation, be a constant. If the gas dissociates or associates, however, the corrective term A will not only include the particle volume values (the constant term referred to), but must now also incorporate the volume changes due to formation or disappearance of particles. It is evident, therefore, that before attempting to use the present treatment (or for that matter any other treatment) for purposes of computing fugacity values, it is necessary that a plot be made of v_0 versus $1/p_0$. If a straight-line relationship is obtained, then two things are satisfied: (1) the definition of fugacity,

Eq. (9), is valid for the particular system, and (2) the treatment being offered is a reasonable one.

Assuming A to be constant we obtain upon integration of Eq. (19)

$$RT \ln (f_2/f_1) = RT \ln (p_2/p_1) - A(p_2 - p_1). \tag{20}$$

Invoking Eq. (9) we see that for the case

$$p_2 \gg p_1, \tag{21}$$

$$p_2 - p_1 \sim p_2, \tag{22}$$

we have

$$\ln[(f_2/f_1)(p_1/p_2)] \sim \ln(f_2/p_2) \sim -A(p_2/RT). \tag{23}$$

In exponential form Eq. (23) may be rewritten

$$f_2/p_2 \sim \exp(-Ap_2/RT) \tag{24}$$

For the case where $p_2 \gg p_1$ but p_2 is itself small, the approximation for e^{-x} with x small leads to

$$f_2/p_2 \sim 1 - (Ap_2/RT). \tag{25}$$

Substituting for A its value as given by Eq. (15) and remembering that $p_2 = p_0$ we obtain

$$f_2/p_0 \sim p_0 v_0/RT. \tag{26}$$

Since

$$v_0/RT = 1/p_i, \tag{27}$$

where p_i is the pressure 1 mole of an ideal gas having the molar volume v_0 would exhibit, we may write

$$f_2/p_0 \sim p_0/p_i, \tag{28}$$

from which it is seen that

$$f_2 \sim p_0^2/p_i. \tag{29}$$

If one measures the volume of one component mole at different pressures p_0 and computes the ideal pressure that should be observed consistent with the observed volume and moles added, approximate values of f for the different values p_0 should be obtainable.

C. The Fugacity, Standard States, and Free Energy Relations

The relationships developed in Chapters 10 and 11 for ideal systems may be made general in the same way as Eq. (7) was made general.

D. THE ACTIVITY

Substituting for the partial pressure terms their equivalent fugacity terms we may readily define equilibrium constants K_f in place of the special constants K_p, which, as stated, are applicable only to ideal systems. The variation of the constants K_f with pressure equals zero. Consequently, K_f is the true equilibrium constant. The terms K_p, on the other hand, do, in the general case, vary with pressure since, in the general case, the system is nonideal. As before, a gas reference state of unit fugacity is normally chosen.

D. Thermodynamic Concentrations—The Activity

In much the same manner as that previously employed in discussing ideal systems, we would like to develop relationships applicable to real systems in terms analogous to mole fraction quantities.

For a particular component A in a vapor phase, we may write

$$\mu_A^{(2)} = \mu_A^{0(2)} + RT \ln f_A^{(2)}, \tag{30}$$

where $\mu_A^{(2)}$ is the chemical potential of the component A when its fugacity is $f_A^{(2)}$, and $\mu_A^{0(2)}$ is the chemical potential for this component in a reference state of unit fugacity. In terms analogous to mole fractions, the fugacity of A in a gas mixture may be defined by

$$f_A^{(2)} = a_A P_t, \tag{31}$$

where the term a_A is called the activity and P_t is the total pressure. Substituting for $f_A^{(2)}$ in Eq. (30) its value as defined by Eq. (31), we obtain

$$\mu_A^{(2)} = \mu_A^{0(2)} + RT \ln a_A^{(2)} P_t. \tag{32}$$

If $P_t = 1$, then we obtain

$$\mu_A^{(2)} = \mu_A^{0(2)} + RT \ln a_A^{(2)}, \tag{33}$$

as the general relationship for the chemical potential. In terms of concentrations, Eq. (33) reduces to its mole fraction counterpart for ideal systems. When the activity of the component A is that in equilibrium with a liquid solution whose activity is $a_A^{(1)}$, we may write for this solution

$$f_A^{(2)} = a_A^{(1)} f_A^\dagger, \tag{34}$$

where $f_A^\dagger$ is the fugacity of the pure liquid component A at 1 atm total pressure.[2]

---

[2] This equation is a generalized form of Raoult's equation and is applicable to ideal and nonideal systems.

Substituting Eq. (34) into Eq. (30), we obtain

$$\mu_A^{(2)} = F_A^{0(2)} + RT \ln f_A^\dagger + RT \ln a_A^{(1)}. \tag{35}$$

The term $F_A^{0(2)} + RT \ln f_A^\dagger$ is, by analogy with its pressure counterpart, equivalent to

$$F_A^{0(1)} = F_A^{0(2)} + RT \ln f_A^\dagger, \tag{36}$$

when we choose, as a reference gas state, the pure gaseous component A at unit fugacity.

Thus, we obtain the interrelationship for two-phase systems at 1 atm total pressure and reference states of unity

$$\mu_A^{(1)} = \mu_A^{(2)} = \mu_A^{0(2)} + RT \ln a_A^{(2)} = \mu_A^{0(1)} + RT \ln a_A^{(1)}. \tag{37}$$

If we define all of our conditions in terms of the species A rather than the component A, we obtain the relationship

$$\bar{F}_A^{(1)} = \bar{F}_A^{(2)} = F_A^{0(2)} + RT \ln a_A^{(2)} = F_A^{0(1)} + RT \ln a_A^{(1)}. \tag{38}$$

For the process of transferring a moles of the component A from the liquid to the vapor phase we have as an equilibrium criterion with the unity constraints applied

$$\Delta F_A^0 = -RT \ln(a_A^{(2)}/a_A^{(1)}) = -RT \ln K_{a_A}. \tag{39}$$

It is to be noted that the F_A^0 terms are not barred, since we choose pure substances as reference states.

Thus, it is seen that the form of the relationships for the true thermodynamic pressures and concentrations are equivalent to their ideal counterparts given in Chapters 10 and 11. In general, fugacity and activity approaches may be applied to components or species terms although neither is simple. In particular, species considerations are more difficult to evaluate and such evaluations are quasi-thermodynamic, since models must be specified. Methods for evaluating activities based on determinations of fugacities or via electromotive force measurements are described in standard works and will not be discussed here.

Since we will consider, mathematically, only ideal model systems subsequently, no attempt will be made to further expand on the topic of thermodynamic pressures and concentrations. It is to be reemphasized, however, that such neglect is not intended to minimize the significance of these topics, but rather to keep the scope of this treatise in line with its original intent.

13

Two-Component Systems—Systems in Which One Phase
Is Pure and in Which Species
and Component Mole Terms
Are Equivalent—Simple Eutectic Interactions

A. Introduction

As noted previously, the concept of the component is a mathematical convenience which enables description of the composition parameter of a system. As mentioned also, a detailed analytical description of any particular system requires a knowledge, or at least a working assumption, as to the nature of the participating species as well as their concentrations. These concentrations are not necessarily independent of one another. Since the quantity that is most easily employed for graphic representation (because of its invariance) is the component mole fraction, it is important, in developing a species model, to define a relationship between component and species terms. Having accomplished this, it is then possible to depict pictorially results that may be tested against experimental results.

The simplest solubility model that one may postulate for an ideal system is one in which, to a first approximation, species and component terms are equivalent. Such a system, as we shall see subsequently, does not require specification of homogeneous equilibrium constants for the system.

Consider a constant total pressure process in a two-component, three-phase system where the following obtains: The system consists of a liquid, a solid, and a vapor. The solid phase is composed completely of either the component A or the component B, while the liquid and gas phases are composed of both A and B. Thus, the system is one in which the solid phases are immiscible in one another. In addition, the vapor pressure and evaporation rates of A and B are negligible so that in the hypothetical time necessary to conduct an experiment, the composition of the condensed phase portion of the system is unchanging even when exposed to the environment at a pressure of 1 atm. The specific process we shall examine is one in which either species, A or B, is transferred from a solid to a liquid phase via a change in temperature of the system, everything else remaining constant.

Since the vapor phase is, to a first approximation, nonexistent, compositional changes that occur are confined only to the liquid phase (the composition of the solid is fixed). Thus, we have two components, coexisting in two phases at constant pressure. From the reduced phase rule we see that such a system is univariant viz.

$$V = 2 - 2 + 1 = 1. \qquad (1)$$

In general, the mole fraction of the component A in either of the coexisting phases is given by

$$M_A = m_A/(m_A + m_B), \qquad (2)$$

where M_A is the component mole fraction of A, and $m_A + m_B$ is the total combined number of moles of the components A and B that are present. Analogously, the mole fraction of the component B is given by

$$M_B = m_B/(m_A + m_B). \qquad (3)$$

From Eqs. (2) and (3) it is observed that

$$M_A + M_B = \frac{m_A + m_B}{m_A + m_B} = 1. \qquad (4)$$

The result, Eq. (4), is applicable to the system as a whole or to any of the coexisting phases.

Suppose now that in each of the coexisting phases species and component concentrations are equivalent, i.e., the component A is always

B. THE ANALYTICAL EXPRESSION OF UNIVARIANCE

present as the species A and the component B is always present as the species B. For the liquid phase composition then, we may write

$$n_t^{(2)} = n_A^{(2)} + n_B^{(2)} = m_A^{(2)} + m_B^{(2)} = m_t^{(2)}, \tag{5}$$

where $n_t^{(2)}$ is the total number of species moles and $m_t^{(2)}$ is the total number of component moles in the liquid phase.

The mole fractions of the different species must satisfy a similar condition to that expressed in Eq. (4). Thus, dividing the first equality in Eq. (5) by n_t yields

$$N_A^{(2)} + N_B^{(2)} = 1. \tag{6}$$

Since

$$m_A^{(2)} = n_A^{(2)} \quad \text{in our system}, \tag{7}$$

and

$$m_B^{(2)} = n_B^{(2)}, \tag{8}$$

where

$$n_A^{(2)} = N_A^{(2)} n_t^{(2)}, \tag{9}$$

and

$$n_B^{(2)} = N_B^{(2)} n_t^{(2)}, \tag{10}$$

it is seen that

$$m_A^{(2)} = N_A^{(2)} n_t^{(2)} \tag{11}$$

and

$$m_B^{(2)} = N_B^{(2)} n_t^{(2)}. \tag{12}$$

Substituting for $m_A^{(2)}$ and $m_B^{(2)}$ in Eq. (2) and (3) the values given in Eq. (11) and (12) we have

$$M_A^{(2)} = \frac{N_A^{(2)} n_t^{(2)}}{N_A^{(2)} n_t^{(2)} + N_B^{(2)} n_t^{(2)}} = \frac{N_A^{(2)}}{N_A^{(2)} + (1 - N_A^{(2)})} = N_A^{(2)}, \tag{13}$$

and similarly

$$M_B^{(2)} = N_B^{(2)}. \tag{14}$$

B. THE ANALYTICAL EXPRESSION OF UNIVARIANCE IN A SIMPLE EUTECTIC INTERACTION

As defined in the preceding section, the system that we are examining is one in which, upon cooling a melt containing A and B, a point is reached at which either solid A or solid B (depending upon the starting concentration of the melt) crystallizes in the pure state. It is when this condition prevails that our binary system becomes isobarically univariant. In the composition region in which pure solid A coexists with liquid, the process taking place upon cooling is described by the reverse of the process

$$A^{(1)} \overset{K_{N_A}}{\rightleftarrows} A^{(2)}, \tag{15}$$

where $A^{(1)}$ refers to the species A in the solid phase, and $A^{(2)}$ refers to the species A in the liquid phase.

The equilibrium constant defining this process [shown in Eq. (15)] is given by

$$K_{N_A} = N_A^{(2)}/N_A^{(1)}. \tag{16}$$

Analogously, in the composition region where pure solid B coexists with liquid solution comprising A and B species we have

$$B^{(1)} \overset{K_{N_B}}{\rightleftarrows} B^{(2)}, \tag{17}$$

$$K_{N_B} = N_B^{(2)}/N_B^{(1)}. \tag{18}$$

The univariant system being described in which a pure solid is in equilibrium with a solution is termed a eutectic system, and the curve defining the first temperature at which a pure solid precipitates (alternately at which the last solid melts), for each starting composition in the mole fraction range 0–1 for either component, is termed a simple solubility curve. From a consideration of the phase rule, as already mentioned, such a system is, under a constant pressure, univariant with the compositions of coexisting phases dependent only on the temperature. Thus, if one specifies a given starting composition for the existence of a specified solid–liquid equilibrium, the temperature at which this specified equilibrium obtains is fixed.

The variation of either K_{N_A} or K_{N_B} with temperature is given in Chapter 10 by Eq. (69). Assuming that ΔH^0, for the process of transferring 1 mole of the component A or B from the solid to the liquid phase, is constant with temperature, then Eq. (69) of Chapter 10 may be integrated readily. This assumption of constancy of ΔH^0 for the process in question is, as has been pointed out previously, not strictly valid, depending on whether or not there is a specific heat change with temperature of the products and reactants in the coexisting phases. However, over reasonable temperature intervals, the assumption is generally palatable. The

B. THE ANALYTICAL EXPRESSION OF UNIVARIANCE

integrated form of Eq. (69) of Chapter 10 between the melting point of A and some other temperature is given by

$$\ln K_{N_A} - \ln K_{N_A}^0 = -\frac{\Delta H_A^0}{R}\left(\frac{1}{T} - \frac{1}{T_A^0}\right), \quad (19)$$

where $K_{N_A}^0$ refers to the equilibrium constant for the process at the melting point of A and T_A^0 is the melting point of A at 1 atm total pressure. In an ideal eutectic system, the only species undergoing a change of phase in the composition interval in which solid A is in equilibrium with solution is the species A; similarly, for the species B in the composition interval in which solid B is in equilibrium with solution. Consequently, the ΔH^0 terms are enthalpic heats for a process involving either A or B, but not both simultaneously. For the processes described by Eqs. (15) and (16), the associated ΔH^0 terms are obviously the molar heats of fusion of either the species A or B and are equivalent to the partial molar heats of fusion since ΔH^0 is not in the ideal case a function of composition. Because of the method of deriving Eq. (69) of Chapter 10, the term ΔH^0 is a standard molar enthalpy of fusion. However, as ΔH^0 is not a function of pressure to any great extent, the terms may be thought of as a general enthalpy of fusion.

If we now expand the K terms in Eq. (19), utilizing Eq. (16) for this purpose, we obtain

$$\ln \frac{N_A^{(2)}}{N_A^{(1)}} - \ln \frac{N_A^{0(2)}}{N_A^{0(1)}} = -\frac{\Delta H_A^0}{R}\left(\frac{1}{T} - \frac{1}{T_A^0}\right). \quad (20)$$

For the B species in its range of interest we have

$$\ln \frac{N_B^{(2)}}{N_B^{(1)}} - \ln \frac{N_B^{0(2)}}{N_B^{0(1)}} = -\frac{\Delta H_B^0}{R}\left(\frac{1}{T} - \frac{1}{T_B^0}\right). \quad (21)$$

The quantities $N_A^{0(2)}$ and $N_B^{0(2)}$ refer to the mole fractions of these species present in the liquid phase at the melting point of the pure solid species A or B. Obviously, these terms are equal to one. Similarly, the $N_A^{0(1)}$ and $N_B^{0(1)}$ terms refer to the compositions of these species in the pure solid phases in equilibrium with liquid at the melting points of pure A or B. These values are also equal to one. Finally, as the system defined only treats of the case where a pure solid is in equilibrium with a solution at all compositions, the terms $N_A^{(1)}$ and $N_B^{(1)}$ also equal one. Equations (20) and (21) may, therefore, be simplified to give

$$\ln N_A^{(2)} = -\frac{\Delta H_A^0}{R}\left(\frac{1}{T} - \frac{1}{T_A^0}\right), \quad (22)$$

and

$$\ln N_B^{(2)} = -\frac{\Delta H_B^0}{R}\left(\frac{1}{T} - \frac{1}{T_B^0}\right). \tag{23}$$

Since our component and species mole numbers are equivalent it follows that

$$\ln M_A^{(2)} = -\frac{\Delta H_A^0}{R}\left(\frac{1}{T} - \frac{1}{T_A^0}\right), \tag{24}$$

and

$$\ln M_B^{(2)} = -\frac{\Delta H_B^0}{R}\left(\frac{1}{T} - \frac{1}{T_B^0}\right). \tag{25}$$

For graphical purposes Eqs. (24) and (25) are the equations sought since we wish to relate component composition and temperature.

To reiterate, Eqs. (24) and (25) define the solubility curves of the components A and B when a univariant equilibrium solid–liquid obtains under 1 atm total pressure. Since the solubilities do not vary to a great extent with limited pressure excursions, the relationships are, to a first approximation, valid for ideal systems over a reasonable range of pressures. Equations (24) and (25) are commonly referred to as van't Hoff freezing point relations, and (for eutectic equilibria) have been utilized in modified form to evaluate molecular weights.

In order to make certain inferences we might draw from Eqs. (24) and (25) more obvious, it is of value to set them in exponential form. This form for these equations is

$$M_X^{(2)} = e^{-x}. \tag{26}$$

We note that for $x \geqslant 0$, M_X may exhibit all values between 0 and 1 which values are allowable as per Eq. (4). On the other hand, for $x \leqslant 0$, M_X will have values ranging between infinity and one, which obviously provides us with physically unreal answers. All fusion processes are observed to take place with the absorption of heat (endothermic) and by convention ΔH_X^0 associated with the absorption of heat is assigned a positive value, the sign of the relevant thermodynamic relation reflecting this convention. If an apposite sign convention were chosen, derived answers would be identical since such sign inclusion would carry through all derivations. It is evident then that for the exponent $x \geqslant 0$, the melting temperature T_A^0 or T_B^0 of the pure solid species whose solubility curve is being considered must be greater than any other temperature at which a solid–liquid coexistence may obtain. Thus, for the equilibrium considered, a simple nondissociative eutectic equilibrium,

C. AN ALTERNATE TREATMENT OF EUTECTIC SYSTEMS

the highest melting temperature in each of the solubility arms (one for solid A and one for solid B) must coincide with the melting point of the pure solid. A further point, to be discussed more fully in Section C of this chapter, is that the solubility curve for a stable equilibrium must exhibit higher two-phase coexistence temperatures at a given system composition than that of a metastable equilibrium. Before continuing further, it is of interest to provide an alternate derivation of the solubility relationships in a eutectic system based on vapor pressure considerations. While this dual method of derivation will not be employed subsequently, it is interesting to carry it out as an exercise at the start of our binary system treatments in order to provide a better feeling for the subject.

C. AN ALTERNATIVE DERIVATION OF UNIVARIANT EQUATIONS FOR EUTECTIC SYSTEMS

Along either of the three-phase coexistences in two component simple eutectic interactions under a constant total pressure we have a pure solid in equilibrium with a solution and a vapor phase. Invoking Raoult's law, it is evident that along these solubility arms the partial pressure of a particular component in the vapor phase, e.g., the component A, at a particular temperature and composition, is given by

$$p_A = N_A^{(2)} p_A^\dagger, \tag{27}$$

where $p_A^\dagger$ for the moment, will remain undefined and $N_A^{(2)}$ is the mole fraction of A in the liquid. Since pure solid A is in equilibrium with liquid and vapor along its entire solubility curve, it is a consequence that the partial pressure of A in the vapor must equal the vapor pressure of solid A at each temperature of three-phase coexistence. In order for Eq. (27) to be consistent with this requirement, it is evident that the quantity $p_A^\dagger$ must be equal to or greater, at any temperature, than the quantity p_A^0 which is the vapor pressure of solid A at this temperature, since N_A will always be equal to or less than unity. From our treatment of unary equilibria it becomes evident that $p_A^\dagger$ must, indeed, be the vapor pressure of pure liquid A. Since pure liquid A exists only metastably, as a supercooled phase below the melting point of A, the quantity $p_A^\dagger$ in Eq. (27) is the vapor pressure of pure supercooled liquid A. This pressure is always greater than that of solid A below the triple point of A. Concomitantly, along the A solubility curve the value N_A is always less than unity below the triple point of A.

Along the solubility curve of solid A, therefore (the solubility curve is more generally referred to as the liquidus, since, upon cooling a particular composition, it represents the first temperature at which solid precipitates, or alternately, the last temperature at which solid occurs upon heating) we have the equivalency

$$p_A = p_A{}^0 = N_A^{(2)} p_A{}^\dagger. \tag{28}$$

In other words, at the liquidus temperature for a particular composition, the mole fraction of A in the liquid phase is directly related to the vapor pressure of pure solid A at that temperature and the vapor pressure of supercooled liquid A at that temperature by

$$N_A^{(2)} = p_A{}^0 / p_A{}^\dagger. \tag{29}$$

Since Raoult's equation is linear with composition, if we plot p_A versus composition as shown in Fig. 1 for a particular temperature and

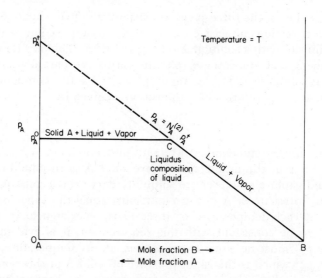

Fig. 1. Determination of the liquidus composition in a eutectic system from vapor pressure data.

then intersect this curve with a line showing the vapor pressure of the pure solid at this temperature, we immediately define the composition of the liquid that must coexist with vapor and solid. With reference to Fig. 1, we note that the composition axis has at its extremes pure A, i.e., $M_A = 1$, and pure B, i.e., $M_B = 1$. This can be done since the sum $M_A + M_B = 1$. The line drawn from the pure A axis at $p_A{}^\dagger$, which is the vapor pressure of supercooled liquid A, terminates at $p_A = 0$ on the pure

C. AN ALTERNATE TREATMENT OF EUTECTIC SYSTEMS

B axis since $M_A = 0$ when $M_B = 1$. Note that in the composition interval c–$p_A{}^\dagger$, the liquid–vapor curve is metastable with respect to a solid–liquid–vapor coexistence. As might be expected, the metastable two-phase equilibrium exhibits a value p_A greater than that for the stable system solid–liquid–vapor at this temperature. Since, in this metastable region, one of the phases that should be present, namely, solid A, has not crystallized, and since, as we move from c to $p_A{}^\dagger$ the composition of A increases, we can conclude that the liquid phase in a metastable system is richer in the component crystallizing than it would be in the stable system. This is, of course, clear if we realize that in order for the vapor pressure p_A to be greater, $N_A^{(2)}$ must be greater. With $N_A^{(2)}$ greater in a metastable system, the solubility curve of A must be depressed. The full significance of this fact will become more apparent when graphical representations of binary eutectic systems are presented. Analogously, we note also that in the composition region, to the right of point c in Fig. 1, the metastable equilibrium solid–liquid–vapor exhibits a higher vapor pressure than the stable system liquid–vapor.

If we plot a series of isotherms for Raoult's equation and intersect each isotherm at the corresponding value of p_A for the solid, it is seen now that we can generate the solubility curve for pure A or pure B. Figure 2

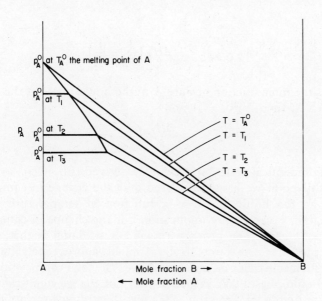

Fig. 2. Generation of a liquidus curve for the component A.

shows such a series for the component A projected upon a common plane. At $T_A{}^0$ which is the melting point of pure A (the triple point of A at 1 atm total pressure), the system is of course invariant and $p_A{}^0 = p_A{}^\dagger$. At all lower temperatures $p_A = N_A^{(2)} p_A{}^\dagger$.

The vapor pressure of solid A and of liquid A may each be expressed as a function of temperature via the expressions

$$\ln \frac{p_A^{(0)T_2}}{p_A^{(0)T_1}} = -\frac{\Delta H_s}{R}\left(\frac{1}{T_2} - \frac{1}{T_1}\right), \tag{30}$$

$$\ln \frac{p_A^{(\dagger)T_2}}{p_A^{(\dagger)T_1}} = -\frac{\Delta H_v}{R}\left(\frac{1}{T_2} - \frac{1}{T_1}\right), \tag{31}$$

previously derived in our treatment of unary systems, i.e., the vapor pressure relationships for the sublimation and vaporization curves. Subtracting (31) from (30) and utilizing Eq. (29) yields

$$\ln p_A^{(0)T_2} - \ln p_A^{(0)T_1} + \ln p_A^{(\dagger)T_1} - \ln p_A^{(\dagger)T_2}$$

$$= \ln\left(\frac{p_A^{(0)T_2}}{p_A^{(\dagger)T_2}} \frac{p_A^{(\dagger)T_1}}{p_A^{(0)T_1}}\right) = \ln \frac{N_A^{(2)T_2}}{N_A^{(2)T_1}} = \frac{\Delta H_v - \Delta H_s}{R}\left(\frac{1}{T_2} - \frac{1}{T_1}\right). \tag{32}$$

Since

$$\Delta H_v - \Delta H_s = -\Delta H_f, \tag{33}$$

we have

$$\ln \frac{N_A^{(2T)_2}}{N_A^{(2)T_1}} = -\Delta H_A\left(\frac{1}{T_2} - \frac{1}{T_1}\right). \tag{34}$$

If $N_A^{(2)T_1}$ is the mole fraction of pure A in the liquid phase at the melting point T_1 of A, then $N_A^{(2)T_1} = 1$ and

$$\ln N_A^{(2)} = -\frac{\Delta H_A}{R}\left(\frac{1}{T_2} - \frac{1}{T_1}\right), \tag{35}$$

which is the identical equation previously derived [Eq. (22)].

Up to this point we have considered only the change in composition occurring in the liquid phase as a function of temperature without much concern, excepting by implication, for the change in composition of the vapor phase. The primary reason for so doing is that, for our constant pressure constraint to be applied appropriately, we have implicity put a constraint on our system: i.e., the total composition of the solid and liquid phases taken together may not change significantly with temperature due to vaporization. This sets the grounds for utilizing the reduced phase rule. In the treatment invoked in the present section,

C. AN ALTERNATE TREATMENT OF EUTECTIC SYSTEMS

therefore, we have implicitly assumed that even though a vapor phase were present, the volume available to this phase would be vanishingly small. Thus when equilibrium vapor pressures obtain, the quantity of the components A and/or B transferred to/or from the condensed phases would be small enough so as not to affect the compositions of the condensed phases to a first approximation. In effect, this condition is also implicit in utilizing Raoult's equation for liquid solutions. If, for example, a solution of mole fraction $N_A^{(2)}$ is permitted to equilibrate with a large vapor available volume, the composition of the solution may change sufficiently such as to disturb the number $N_A^{(2)}$. Partial pressure evaluations in such an instance become dependent upon the vapor available volume, and consequently the use of Raoult's relationship, even for ideal solutions, becomes complicated. In the context, however, that vapor available volume above the condensed phases is vanishingly small, the composition of the three coexisting phases, solid–liquid–vapor may be determined for either a system under a constant total pressure constraint or for one under the equilibrium pressure of the system.

Confusion may arise if an apparent paradox in the predictions of the phase rule seems present; thus, we will amplify on the preceding paragraph.

If we have present a binary three-phase coexistence under a constant total pressure constraint, the phase rule

$$V = C - P + 1 = 2 - 3 + 1 = 0. \tag{36}$$

predicts that the system will be invariant, not univariant. This would not allow for the existence of a univariant region of solubility and indeed this is true. Thus if a solid, a liquid, and a vapor are present at the pressure P_t, there can only be one composition and temperature coincident with this pressure. Any other composition and temperature would be possible for the three-phase coexistence only at some other pressure P_t. The problem resolves itself as follows: In order that a constant total pressure constraint be applicable over a range of temperatures, the vapor phase pressure must be the sum of the partial pressures of the species in the condensed phase plus an inert gas pressure. The sum of these contributions

$$P_{\text{total}} = p_A + p_B + p_{\text{inert}} \tag{37}$$

must be a constant. In addition, we realize that by adding the inert gas to the vapor phase, the total number of components present rises to three even though the inert gas is to a first approximation insoluble in the condensed phases and does not therefore affect the condensed phase behavior. In the limit, of course, the inert gas does exhibit some solubility in the condensed phases and must have some effect on the behavior of the other components even if this effect is trivial.

Thus, if we determine p_A and p_B along either the A liquidus or the B liquidus, the sum of these partial pressures substracted from the total pressure P_t tells us the amount of inert gas pressure necessary to retain the constant total pressure constraint.

On the other hand, if we realize that the effect of limited pressure excursions on both vapor pressures and solubilities is small, we can ignore the constant pressure artificiality, and treat a system under its equilibrium pressure. This requires the assumption that the Raoult equation is reasonably valid in the range of equilibrium pressures encountered. Certainly this is as good an assumption as the Raoult equation itself when applied to real systems. In our preceding treatments, therefore, we have really considered systems under their equilibrium pressures, or alternatively, systems in which an inert gas pressure is added to the component number, but where the added component has played no part in relevant equilibria. Since the mole fraction of A in the liquid phase along the liquidus curve for A is given by Eq. (35) and we know that N_B at any specified temperature is equal to one minus N_A, the mathematics for determining vapor phase compositional variation along the condensed phase univariant curve is essentially at hand. At any temperature along the liquidus curve, Eq. (28) determines the partial pressure of A. Taking the natural logarithm of this equation gives

$$\ln p_A{}^0 = \ln N_A^{(2)} + \ln p_A{}^\dagger. \tag{38}$$

Substituting for $\ln N_A^{(2)}$ its value as given by Eq. (35) we have

$$\ln p_{A_{(T_2)}}^0 = -\frac{\Delta H_A}{R}\left(\frac{1}{T_2} - \frac{1}{T_A}\right) + \ln p_{A_{(T_2)}}^\dagger. \tag{39}$$

The variation in vapor pressure of pure supercooled A with temperature in this same temperature interval is

$$\ln p_{A_{(T_2)}}^\dagger = -\frac{\Delta H_{V_A}}{R}\left(\frac{1}{T_2} - \frac{1}{T_A}\right) + \ln p_{A_{(T_A)}}^\dagger, \tag{40}$$

where $p_{A_{(T_A)}}^\dagger$ is the vapor pressure of A at its melting point T_A, $p_{A_{(T_2)}}^\dagger$ is the vapor pressure of supercooled liquid A at the temperature T_2 and ΔH_{V_A} is the molar heat of vaporization of liquid A. In the same

C. AN ALTERNATE TREATMENT OF EUTECTIC SYSTEMS

region of interest, the component B is always confined to the liquid and gas phases. Its mole fraction is given by

$$N_B^{(2)} = 1 - N_A^{(2)}. \tag{41}$$

The partial pressure of B along the liquidus is at any temperature given by Raoult's law:

$$p_B^{(3)} = N_B^{(2)} p_B^\dagger = (1 - N_A^{(2)}) p_B^\dagger = \left\{1 - \exp\left[-\frac{\Delta H_A}{R}\left(\frac{1}{T_2} - \frac{1}{T_A}\right)\right]\right\} p_B^\dagger, \tag{42}$$

where $N_B^{(2)}$ is the mole fraction of B in the liquid in equilibrium with pure solid A and $p_B^\dagger$ is the vapor pressure of pure liquid B at the temperature in question. $p_B^\dagger$ in the interval from the melting point of A, T_A to T_2 is given by an equation analogous to (40), i.e.,

$$\ln p_{B(T_2)}^\dagger = -\frac{\Delta H_{V_B}}{R}\left(\frac{1}{T_2} - \frac{1}{T_A}\right) + \ln p_{B(T_A)}^\dagger. \tag{43}$$

In exponential form, Eq. (43) has the appearance

$$p_{B(T_2)}^\dagger = \left\{\exp\left[-\frac{\Delta H_{V_B}}{R}\left(\frac{1}{T_2} - \frac{1}{T_A}\right)\right]\right\} p_{B(T_A)}^\dagger. \tag{44}$$

Therefore,

$$p_{B(T_2)} = \left\{1 - \exp\left[-\frac{\Delta H_A}{R}\left(\frac{1}{T_2} - \frac{1}{T_A}\right)\right]\right\}$$
$$\times \left\{\exp\left[-\frac{\Delta H_{V_B}}{R}\left(\frac{1}{T_2} - \frac{1}{T_A}\right)\right]\right\} p_{B(T_A)}^\dagger. \tag{45}$$

Since

$$N_A^{(3)} = p_A^0/(p_A^0 + p_B), \tag{46}$$

$$N_A^{(3)} = \frac{p_{A(T_A)}^\dagger \left\{\exp\left[-\frac{\Delta H_A}{R}\left(\frac{1}{T_2} - \frac{1}{T_A}\right)\right]\right\}\left\{\exp\left[-\frac{\Delta H_{V_A}}{R}\left(\frac{1}{T_2} - \frac{1}{T_A}\right)\right]\right\}}{\left[\begin{array}{c} p_{A(T_A)}^\dagger \left\{\exp\left[-\frac{\Delta H_A}{R}\left[\frac{1}{T_2} - \frac{1}{T_A}\right]\right]\right\}\left\{\exp\left[-\frac{\Delta H_{V_A}}{R}\left(\frac{1}{T_2} - \frac{1}{T_A}\right)\right]\right\} \\ + \left\{1 - \exp\left[-\frac{\Delta H_A}{R}\left(\frac{1}{T_2} - \frac{1}{T_A}\right)\right]\right\} \\ \times \left\{\exp\left[-\frac{\Delta H_{V_B}}{R}\left(\frac{1}{T_2} - \frac{1}{T_A}\right)\right]\right\} p_{B(T_A)}^\dagger \end{array}\right]}. \tag{47}$$

Equation (47) can be simplified somewhat by combining terms

$$N_A^{(3)} = \frac{p_{A(T_A)}^\dagger \exp\left[-\frac{(\Delta H_A + \Delta H_{V_A})}{R}\left(\frac{1}{T_2} - \frac{1}{T_A}\right)\right]}{\left[\begin{array}{c} p_{A(T_A)}^\dagger \exp\left[-\frac{(\Delta H_A + \Delta H_{V_A})}{R}\left(\frac{1}{T_2} - \frac{1}{T_A}\right)\right] \\ + \left\{1 - \exp\left[-\frac{\Delta H_A}{R}\left(\frac{1}{T_2} - \frac{1}{T_A}\right)\right]\right\} \\ \times \left\{\exp\left[-\frac{\Delta H_{V_B}}{R}\left(\frac{1}{T_2} - \frac{1}{T_A}\right)\right]\right\} p_{B(T_A)}^\dagger \end{array}\right]}. \quad (48)$$

Since

$$\Delta H_A + \Delta H_{V_A} = \Delta H_{S_A}, \quad (49)$$

where ΔH_{S_A} is the heat of sublimation of solid A, a further simplification is obvious. From the preceding, the analytic expression for $N_B^{(3)}$ may be written immediately and need not be specified here.

In a subsequent chapter we will examine the behavior of a solid–gas analog to the solid–liquid eutectic interaction, namely, that in which a pure solid is in equilibrium with a vapor of variable composition. For the present, however, we will confine our interests to the solid–liquid binary interactions, that we have treated in this chapter. Before proceeding with representation of our analytical findings in graphical form (the phase diagram of a binary system), one further significant feature of ideal binary eutectic interactions is worth pointing out.

It will be noted that for each of the liquidus curves defined in the binary system A–B, the equations describing the two solubility curves contain only terms related to the solid precipitating out. Thus, the ΔH term is either that for the heat of fusion of A or B, and the melting point specified is that for either A or B. Consequently, it is seen that in an ideal interaction, the nature of the second component is immaterial insofar as the solubility curve of the other is concerned. Thus, in ideal binary eutectic interactions between the component A and B, or A and C, or A and D, etc, the specifics of the A liquidus curve are unaltered. In real systems, of course, this behavior is adhered to, to a lesser or greater extent, depending upon the deviation from ideal behavior evinced by the system. (See for example Chapter 15.)

The next chapter will be devoted to graphical description of our derived solutions, first to introduce the idea of the binary phase diagram and second to demonstrate the effects of quantities like heats of fusion, melting points, and metastability, on the shapes of the liquidus curves.

14

Graphical Representations of Simple Eutectic Interactions

A. Introduction

In Chapter 13, the analytical description of simple condensed-phase eutectic interactions was presented with only perfunctory explanation of the appearance of the system in question. Such explanation is most easily visualized in the context of a discussion of the so-called diagram of state or phase diagram, as it is alternatively termed. If we consider a two-component equilibrium at constant total pressure, we may predict, with the aid of the phase rule, that when two phases coexist, the state of the system is defined once we specify the temperature. Conversely, if we specify the composition of the system, the temperature at which two phases may coexist is defined. While any pair of suitable intensive parameters may be chosen to provide a pictorial representation of the state of the system, i.e., temperature versus density, etc., it is frequently most convenient to depict a condensed phase diagram graphically in terms of the variables temperature and composition.

To reiterate, if we consider a binary system under a constraint of a constant total pressure, then the phase rule acquires the form

$$V = C - P + 1, \qquad (1)$$

this form being termed the reduced phase rule. The implications inherent in this specification have been discussed fully in the preceding chapter. From Eq. (1) it is seen that a two-phase equilibrium is isobarically univariant and a three-phase equilibrium is isobarically invariant. Since no vapor phase is allowed for, the two or three phases referred to must be condensed phases, solid or liquid, or both. The graphical representation of the several equilibria possible in such a system, as pointed out, can be handled two dimensionally. From such representations, quantitative information of practical consequence in materials preparation may be derived.

B. THE PHASE DIAGRAM—A QUALITATIVE VIEW

Suppose we wish to consider the variation in solubility of the component A as a function of temperature in a melt comprising the ideal components A and B. A graph can be constructed with the ordinate representing the temperature and the abscissa representing the composition. If the composition starting point of this graph represents a mole fraction of one for the component A, then on the temperature axis, a point would be situated at the melting point of the pure component A. If, now, the composition of the system is changed such that the mole fraction of A becomes something less than one, it is a consequence of our theoretical arguments that solid A can coexist with liquid only at a temperature somewhat below the melting point of the pure solid A. Starting with a composition of liquid containing still less A, and cooling until solid first crystallizes, results in a temperature of solid–liquid coexistence still lower, etc. Thus, if we begin with a series of melts each decreasingly poorer in A, and cool each of these, the temperature of first appearance of pure solid A will become lower and lower, until as the composition of A in the melt tends toward zero, the temperature of first crystallization tends toward absolute zero. This is shown in Fig. 1.

Since the two-phase equilibrium possible in the system under scrutiny always consists of a pure solid in equilibrium with a melt, the curve o–x in Fig. 1 shows the composition of the liquid that may coexist with pure solid A at any temperature. The composition of the solid is represented by the intersection of a horizontal drawn from the liquidus curve o–x with the temperature axis. It is observed that such horizontals always intersect this axis at the composition $N_A = 1$ in our system. Such lines, horizontal to the composition axis, i.e., the lines a–b, c–d, e–f, drawn until they intersect an axis or some other univariant line on a phase

B. THE PHASE DIAGRAM—A QUALITATIVE VIEW

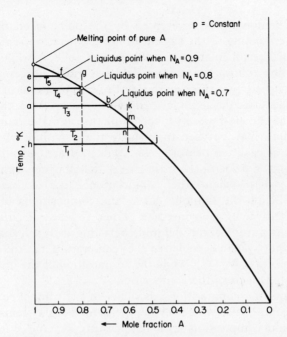

FIG. 1. The change in liquidus temperature as the composition of A in the system is decreased.

diagram, are referred to as tie-lines. They do, in fact, tie together the compositions of the two phases which coexist simultaneously at a specified temperature. Their points of intersection with the lines upon which the phase compositions are depicted represent the compositions of the coexisting phases.

If we cool a melt of starting composition g (Fig. 1) just prior to reaching the point d, the composition of the liquid will obviously still have composition g. When the point d is reached, a solid having the composition pure A will just begin to crystallize. Obviously, when this occurs, the melt will become slightly poorer in A. If the system is cooled to a temperature slightly below that for the point d, more solid A will crystallize in order that the requirements for univariance are met. These requirements, expressed in Eq. (22) of the preceding chapter, are that the composition of A in the liquid phase shall decrease. Since, if the temperature is specified, the state of the system is fixed, it is seen that if we cool the melt of starting composition g to the point i, the composition of the liquid phase must move down the curve d–j since the system is univariant and can exhibit only one equilibrium liquidus curve. If we cool a melt of starting composition k, first appearance of pure A will occur at the temperature represented by the point m on the liquidus curve. Further cooling of the system to the temperature represented by the point l will result in further deposition of solid A while the liquid composition now follows the curve m–j. If we connect horizontal tie-lines through the dashed lines showing the starting compositions of the system as a whole (the latter dashed lines being termed isopleths), the points of intersection of the tie-lines with the isopleths, and the lengths of the intersected lines may be used to evaluate the quantities of liquid and solid existing at each temperature. This will be described subsequently. In any event, it is to be noted from Fig. 1 that starting with either the composition g or k, when either composition is cooled to an identical temperature below the liquidus curve, the compositions of solid and liquid are identical in each instance. Thus, if we cool either melt to T_2, the solid is pure A and the liquid coexisting with it has the composition o. If either system is cooled to the temperature T_1, the solid again has the composition $N_A = 1$ while the composition of the liquid has for both systems changed to the composition $N_A = j$.

Since the starting composition g is richer in the component A than is the starting composition k, it is evident that for both systems to exhibit the same liquid composition at the temperature T_1, the g solution will have had to divorce itself of more of the component A than will have the k solution. As a matter of interest, the relative amounts of solid-to-liquid present in each of these systems at any particular temperature may be determined from the lengths of the arms on the tie-lines at such temperature. Thus for the starting composition g at the temperature T_1, the fraction of the starting number of molecules in the solid, N_s, to that remaining in the liquid, N_ℓ, is given by the ratio

$$|j - i|/|h - i| = N_s/N_\ell. \qquad (2)$$

For the starting composition k, on the other hand, the solid liquid molecular ratio at this same temperature is

$$|j - l|/|h - l| = N_s'/N_\ell'. \qquad (3)$$

It will be noted that the ratio in Eq. (3) is a smaller number than that in Eq. (2).

The use of the ratio of arms, on a tie-line that is intersected by an isopleth showing the starting composition of the system, to determine the solid-to-liquid ratio at any temperature is known as the lever arm or fulcrum principle. The fulcrum is represented by the point of intersection of the isopleth with the tie-line. The mathematical justification of the lever arm principle is presented in Section C of this Chapter.

B. THE PHASE DIAGRAM—A QUALITATIVE VIEW

In the same manner as we have defined the solubility curve for the component A, we may define the solubility curve for the component B. Rather than do this on a separate graph, however, it is much more convenient to construct the B liquidus curve on the same graph. This approach may be effected because of the relationships Eqs. (4) and (6) of Chapter 13 which relate the mole fractions of the two components in a binary mixture. If our graph reads from left to right in terms of mole fraction B, and as shown in Fig. 1 from right to left in terms of mole fraction A, it is evident that at any point

$$N_A = 1 - N_B, \tag{4}$$

and

$$N_B = 1 - N_A. \tag{5}$$

This is shown in Fig. 2 in which the solubility curves of both A and B are simultaneously constructed schematically. The curve T_A–B represents the solubility curve of A and the curve T_B–A represents the solubility curve of B. The values T_A and T_B represent the melting points of pure A and pure B, respectively. If we cool a melt having the composition "a," no change of phase occurs until the point x is reached at the temperature

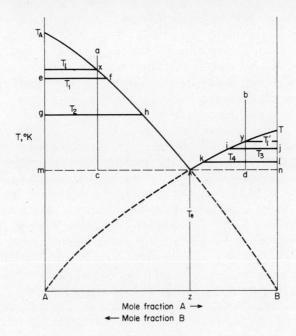

FIG. 2. The solubility curves for A and B in the binary system A–B.

T_ℓ. At this temperature pure solid A first precipitates. Upon cooling to the temperature T_1 the composition of the liquid changes along the line x–f while more solid A crystallizes out. This process of crystallization of pure solid A attends the process of further cooling to the temperature T_e while the liquid composition varies along the line x–T_e. If we start our experiment with the composition "b," nothing occurs upon cooling until the temperature T_ℓ' is reached, at which temperature the first traces of solid B precipitate. As we cool the system, the liquid composition becomes poorer in the component B, the liquid composition moving along the curve from point y to point T_e. At the point T_e a new phenomenon occurs. At this temperature, independent of whether we start at the composition "a" or the composition "b," conditions are correct for both the components A and B to coexist simultaneously with liquid. Thus, in cooling the melt of starting composition "b" to the temperature T_e only pure solid B crystallizes out until the temperature T_e is reached. When this temperature is reached, pure A also starts to settle out. A corresponding phenomenon attends cooling of a melt of composition "a," excepting that pure B starts to crystallize at T_e.

Note now that at the temperature T_e the condition prevails that two pure solids, A and B, coexist with a liquid of composition Z. From the reduced phase rule we see that a three phase coexistence in a constant pressure binary equilibrium is invariant. In other words, no degrees of freedom are available to the system. Thus, if we attempt to cool the system below the temperature T_e, the conglomeration of solid A, solid B, and liquid will not cool further until one of the phases vanishes. With cooling, of course, the phase that vanishes is the liquid phase. As this phase crystallizes isothermally at the temperature T_e, unlike the behavior at higher temperatures, the freezing of the melt occurs with the simultaneous precipitation of solid A and solid B. The temperature at which such simultaneous freezing occurs is called the eutectic temperature, and the composition of the liquid at this temperature is called the eutectic composition. This final temperature at which liquid may exist is for the general case termed the solidus temperature. The tie-line m, T_e, n, while not an actual dividing line in a phase diagram, is frequently drawn as a solid line; this because independent of the starting composition, it represents for any starting composition the last temperature of liquid existence. It is evident, however, from the preceding that it is always the same liquid composition, namely, Z that prevails (independent of starting composition) just prior to final crystallization.

In Fig. 2 the dashed lines A–T_e and B–T_e represent metastable extensions of freezing point curves in much the same way as metastable extensions of solid–vapor or liquid–vapor curves exist in unary

C. THE LEVER ARM PRINCIPLE

systems. In the temperature range from absolute zero to the eutectic temperature it is a consequence of our conclusions about the pressures of metastable phases that these pressures must be greater than those of stable phases in the same temperature interval. That such is the case for the metastable extension $A-T_e$, for example, may be discerned from the following: In this metastable region the equilibrium that obtains is pure solid A in equilibrium with a liquid. If equilibrium were to prevail, the total pressure would simply be given by Eq. (6) since in this range only the solids A and B should exist.

$$P_{\text{total}}^{\text{stable}} = p_A^{(s)} + p_B^{(s)}. \tag{6}$$

The actual pressure, however, is given by

$$P_{\text{total}}^{\text{metastable}} = p_A^{(s)} + p_B^{(\ell)}, \tag{7}$$

where the value of $p_B^{(\ell)}$ is defined by Eq. (44) of Chapter 13. This value of p_B is, at all compositions along the metastable A liquidus, greater than the value of pure solid B as can be seen from Eq. (41) of Chapter 13. Note that as N_B approaches unity, the value of p_B approaches that of pure liquid B. Since the vapor pressure of pure liquid B, below the melting point of B, is always greater than the vapor pressure of solid B it is seen that

$$P_{\text{total}}^{\text{metastable}} > P_{\text{total}}^{\text{stable}}.$$

Figure 3 presents a schematic representation of the binary system A–B similar to that given in Fig. 2, excepting that all pertinent regions are identified. Not shown in the higher temperature reaches are the liquid–vapor curves whose N_A values do not exceed those for which the total pressure exceeds 1 atm in a 1 atm total pressure system.

C. The Lever Arm Principle

One of the important attributes of a diagram of state is the potential afforded to estimate quantitatively the amount of material that has precipitated out during a cooling process. Offered without proof in the preceding section, has been the argument that a knowledge of tie-line arm lengths may be used to good advantage to determine the molecular ratio of solid-to-liquid that exists at any temperature between the liquidus and solidus temperatures. In addition, as will be shown below, a knowledge of tie-line arm lengths and starting mass of a melt may be used to calculate the fraction of that melt that has precipitated as well as the mass

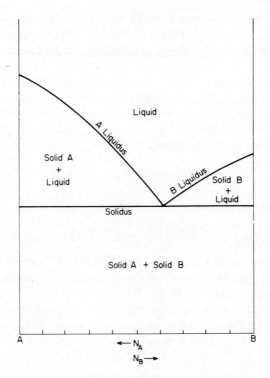

Fig. 3. Labeled diagram of the binary eutectic system A–B.

of one of the components that has precipitated. Such knowledge is valuable in processes involving single crystal growth from the melt as well as in processes for separation of mixtures of components. In purification processes under the heading of "zone refining," such knowledge places a practical upper limit on the degree of refinement one may expect in a single-cycle purification process and offers guide lines for establishing the overall purification process. In many instances it determines whether one attempts purification by eliminating impurities from a solid or, alternately, by eliminating existing impurities from a melt by segregating these impurities in the precipitating solid.

A distinct advantage of the lever arm principle over many of the other theoretical conclusions that may be arrived at is that its validity is independent of the ideality of the system in question. All that is required is an experimentally determined phase diagram as a starting point. The application of the lever arm principle, however, requires an understanding of what conditions must be satisfied in depicting the phase diagram for proper interpretation of results to be obtained. If, for

C. THE LEVER ARM PRINCIPLE

example, the phase diagram is depicted in terms of mass fraction of the components present, then independent of the complexity of the diagram, measured tie-arm lengths are immediately usable for calculating mass solid-to-liquid or other ratios. However, if compositions are represented in terms of molecular fractions, it is frequently advisable, before invoking the lever arm principle, to depict the phase diagram in primitive terms. By this we mean that over the entire range of concentration of the components A and B, the only solids that are ever present are the materials A or B, or a solid solution of the two. If any more complex situation arises, the diagram is termed complex and use of the lever arm principle is best effected after rescaling the complex diagram into its subordinate parts or subdiagrams. When this is accomplished, the situation prevails that over the mole fraction range 0–1 only two pure phases at most exist. More will be said about this subsequently in discussions on coordinate transformations.

Consider now the primitive eutectic system A–B depicted in Fig. 4. If a sample having the composition x is cooled to the temperature T_m, the first traces of solid A appear. Obviously, at this point the ratio fraction liquid-to-fraction solid is very large, the solution having just

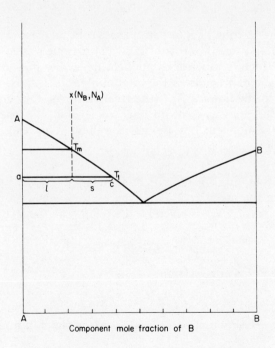

FIG. 4. Derivation of the lever arm principle.

become saturated. As the sample is cooled to T_1 more solid A crystallizes and of course the liquid–solid ratio decreases.

If the original number of component molecules present in both phases at the temperature T_1 is denoted by the symbol M, the fraction of this total number which has crystallized by N_s, and the fraction remaining in the liquid by $(1 - N_s)$ we may write

$$M = N_s M + (1 - N_s)M, \tag{8}$$

for the distribution of material between the two phases at T_1.

The number of B component molecules in the liquid phase is denoted M_B^ℓ. If the mole fraction of B in the liquid phase is N_B^ℓ then

$$M_B^\ell = N_B^\ell(1 - N_s)M. \tag{9}$$

The total number of B molecules in the solid phase is denoted by M_B^s and the mole fraction of B in the solid by N_B^s. This translates into

$$M_B^s = N_B^s N_s M. \tag{10}$$

The total number of B molecules in liquid and solid M_B, is given then by

$$M_B = N_B M = M_B^\ell + M_B^s = N_B^\ell(1 - N_s)M + N_B^s N_s M, \tag{11}$$

where N_B is the fraction of B in both phases collectively.

The second and fourth terms of Eq. (11) may be solved for N_s (the fraction of the total number of component molecules present as a solid) in terms of N_B^ℓ (the fraction of the total number of B component molecules present in the liquid), N_B^s (the fraction of the total number of B component molecules present in the solid), and N_B the molecular fraction of the component B in the whole system. This leads to

$$N_s = (N_B - N_B^\ell)/(N_B^s - N_B^\ell). \tag{12}$$

Note now that, in Fig. 4, the mole fraction N_B is given by the isopleth x, that the quantity N_B^ℓ is given by the point c on the liquidus and that the quantity N_B^s is given by the point a on the isotherm a–c. Thus

$$N_s = |c - x|/|a - c| = s/t, \tag{13}$$

where $l + s$ of Fig. 4 equal t, since

$$N_\ell = 1 - N_s, \tag{14}$$

where $N_\ell = $ the fraction of both components in the liquid phase, i.e.,

$$N_\ell = |x - a|/|a - c| = l/t, \tag{15}$$

D. APPLICATION OF THE LEVER ARM PRINCIPLE

and

$$N_s/N_\ell = |c - x|/|x - a| = s/l. \tag{16}$$

Equations (13), (15), and (16) are explicit expressions for the several statements of the lever arm principle. An analogous set of relationships can be derived starting with the phase diagram depicted in terms of mass fraction of B, f_B.

Thus if g_{total} is the total mass of the system, f_s the mass fraction of this mass in the solid, f_ℓ the mass fraction in the liquid, f_B^s the mass fraction of B in the solid, and f_B^ℓ the mass fraction of B in the liquid, the precise steps outlined in deriving (13), (15), and (16) may be reiterated to provide answers now pertaining to f_s, f_ℓ, and f_s/f_ℓ.

D. Specifics of Application of the Lever Arm Principle

Although we have specifically developed the lever arm principle (LAP) for systems defined in mole fraction terms, it is evident that with an entirely analogous sequence of steps, the principle could have been derived for a system defined in mass fractions. In this section, at the possible expense of belaboring the point, the results of the preceding section will be examined, unencumbered by the mathematical arguments presented there, to point up the physical significance of the results and arguments leading to these results.

First, it is to be noted that a term such as N_s used in the derivation refers to a fraction of a specified total number of molecules M that has precipitated in the solid form. This total number of molecules is based on the specification of a component stoichiometry. In other words, if the specified components are designated by the symbols A and B, then all molecular counting is done on this basis. When the terms M_B^ℓ or M_B^s are employed, these refer to the fraction of the molecules of the component B counted as having the stoichiometry B that are present in the liquid or the solid, respectively. The designation N_ℓ is analogous to that described above for N_s, excepting it refers to the liquid. The solid-to-liquid ratio, i.e., N_s/N_ℓ, defines the fraction of the total number of component molecules based on counting all of the A and all of the B molecules in the solid relative to those in the liquid.

If the system is a primitive one, as has been used for the derivation, no problem is encountered, since the only solids present in the system are defined with the same stoichiometries as was specified for the components. If, however, within the composition interval 0–1 mole fraction B, for a system whose concentration is specified in terms of the components A and B, an intermediate compound of the stoichiometry A_xB_y is generated, a degree of confusion is likely to arise. For example, let us assume that we are analytically examining molecular solid-to-liquid or molecular solid/(solid + liquid) ratios in the liquidus field in which the solid having the stoichiometry A_xB_y is in equilibrium with liquid. When a particular value N_s is determined for the system specified in terms of mole fractions of the components A or B, the value N_s determined by applying the LAP will tell us what fraction of the total number of component molecules has precipitated. Most often, however, we are interested (in a given liquidus field) in knowing what fraction of a specifically designated stoichiometry has settled out. In other words, in the B liquidus field the solid comprises only species derived from the component B. Consequently, when the molecular fraction N_s is determined, this value is due solely to settling out of the B component. If the total number of molecules is M and the fraction of B in this total is N_B, then the number of B molecules in the total mixture is $N_B M = M_B$. If the fraction of solid is N_s, then knowing that the solid is derived solely from the B component in the B liquidus field, we can write that of the M_B original B component molecules with which we started $N_s \times M_B$ of them have precipitated out. Knowing the molecular weight of B, we can calculate the mass of B that has precipitated, etc.

If, however, we are in the A_xB_y field, where the solid is derived from both A and B components, in order to determine the mass of solid that has settled out, we must note that for a given value N_s settled out, x of the A type have dropped out of solution for each y of the B type. Thus, if again our total starting number of molecules is M, then $N_s \times M$ represents the total number deposited M_D, i.e.,

$$M_D = N_s \times M. \tag{17}$$

The number of B molecules referred to the component B in this total number is

$$N_B{}^s = \frac{y}{x+y} \times M_D, \tag{18}$$

and the number of A molecules referred to the component A in this total number is

$$N_A{}^s = \frac{x}{x+y} \times M_D. \tag{19}$$

From this, knowing the starting mole fractions of A and B, the total system mass, and the molecular weights of A and B, we can compute the

D. APPLICATION OF THE LEVER ARM PRINCIPLE

mass of A, the mass of B, and, therefore, the mass of A_xB_y that has settled out.

It would have been much simpler, however, to do the latter if the phase diagram had been defined in terms of the assumed component stoichiometries A and A_xB_y or B and A_xB_y, namely, if the composition interval A to A_xB_y was scaled from 0 to 1 and if the composition A_xB_y to B was scaled from 0 to 1. There would also exist less chance for confusion since the coprecipitation of A and B would immediately be taken into account in the A_xB_y liquidus field.

As we shall see subsequently, this possibility of confusion is not present in systems defined in mass-fraction terms and, in fact, when the use of the LAP is intended, mass-fraction phase diagrams are more easily employed. On the other hand, if one wants to speculate on heats of fusion or the nature of liquid phase species, mole-fraction diagrams are more convenient. With the latter, however, it is of considerable value to perform coordinate transformations so that the concentration range encompassing a single eutectic is from 0 to 1.

15

Eutectic Systems Continued—The Parameter $\Delta H_{\text{fusion}}/T_{\text{melting}}$ and Its Influence on the Contour of Eutectic Liquidus Curves

A. INTRODUCTION

It is taken for granted, in schematizing the appearance of the liquidus curves in a simple eutectic interaction, that they are concave with respect to the composition axis as shown in Fig. 1. This predisposition instills in the experimentalist either a tendency to question the data when curves unlike those of Fig. 1 are obtained (i.e., when the experimental curves are inflected), or to attribute this anomalous behavior to a marked departure from ideality. Such a predisposition is perhaps explainable on the basis that the general statement describing the univariant behavior in an ideal eutectic system is inappropriately thought of as exhibiting a simple exponential form.

Both expected and unexpected results of the types depicted in Figs. 2–4 are obtained if, as an exercise, one chooses nominal values for the latent heat of fusion ΔH_A, e.g., 0.1–23.0 kcal/mole, and nominal values for the temperature of melting T_A, e.g., 500–1300°K, and proceeds to generate hypothetical liquidus curves for the species A using the familiar equation

$$\ln N_A = -\frac{\Delta H_A}{R}\left(\frac{1}{T} - \frac{1}{T_A{}^0}\right), \tag{1}$$

A. INTRODUCTION

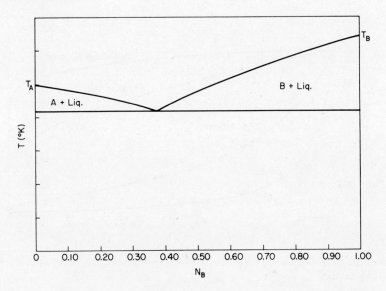

FIG. 1. An ideal eutectic interaction exhibiting "normal" liquidus curves.

where N_A is the liquid phase mole fraction of the species A whose concentration value is assumed equivalent to the concentration of the component A. R is the gas constant in units of cal/mole deg. The frequency of appearance of curves exhibiting points of inflection in the random choices made, and the resemblance of these curves to experi-

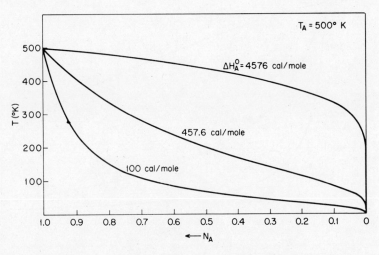

FIG. 2. The effect of varying heats of fusion on the liquidus contour of a component melting at 500°K.

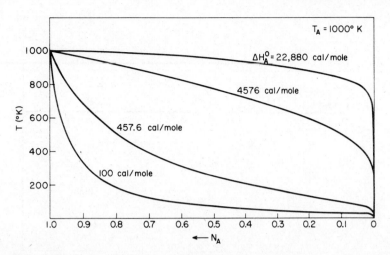

FIG. 3. The effect of varying heats of fusion on the liquidus contour of a component melting at 1000°K.

mental cases attributed to a tendency toward unmixing, leads one to suspect that there is something more fundamental involved than has been assumed previously. It is to be realized, however, that if $\ln N_A$ is plotted as a function of $1/T$, straight lines are always obtained. The examples of Figs. 2–4 are replotted in Figs. 5–7 to demonstrate this.

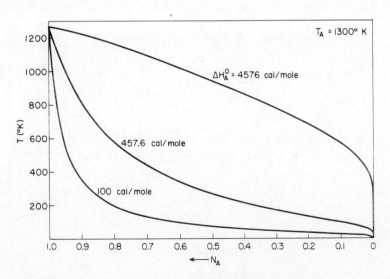

FIG. 4. The effect of varying heats of fusion on the liquidus contour of a component melting at 1300°K.

A. INTRODUCTION

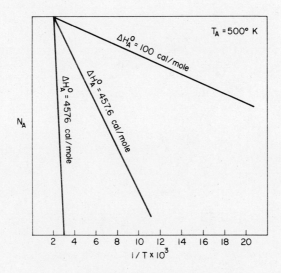

FIG. 5. Liquidus curves for the A–B system.

In the following, an attempt is made to show the underlying cause responsible for the occurrence or nonoccurrence of inflected liquidus curves, and to develop a rationale which enables prediction of both types of behavior in real eutectic binary systems.

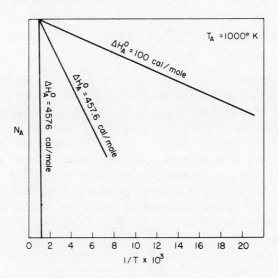

FIG. 6. Liquidus curves for the A–B system.

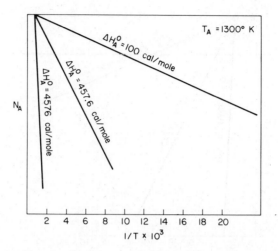

FIG. 7. Liquidus curves for the A–B system.

B. THEORETICAL ANALYSIS

Equation (1), as employed, has certain restrictions built into it. By definition, the mole fraction N_A must vary in the interval $1 \geqslant N_A \geqslant 0$. In addition, the sign of ΔH_A is always positive. These requirements generate the condition that any arbitrary temperature on the liquidus must have a value less than or equal to T_A. In the general form, Eq. (1) has the appearance

$$y = Ae^{-a/x}. \tag{2}$$

If the independent variable chosen is $1/x$, it is found that for any combination of a and x, positive or negative, a plot of y versus $1/x$ will not exhibit a point of inflection. Thus, in the use of Eq. (2) in explicit form for the description of solubility, kinetic or homogeneous equilibria phenomena, graphs of $\ln y$ versus $1/x$ will be linear and graphs of y versus $1/x$ will have a simple exponential appearance. On the other hand, when the independent variable chosen is x rather than $1/x$, then depending upon the signs of a and x, graphs on linear coordinates may or may not exhibit points of inflection. With both a and x positive, a point of inflection occurs at $x = \frac{1}{2}a$. With a negative and x positive, no point of inflection occurs. With both a and x negative, a point of inflection exists at $x = -\frac{1}{2}a$, and with a positive and x negative no point of inflection is present.

For systems of physical interest, therefore, we need not concern

B. THEORETICAL ANALYSIS

ourselves with cases where either x or a is negative. For the remaining cases, where an equation having the form of Eq. (2) is an appropriate description of the observed behavior, a point of inflection is present. Whether it is physically observable depends only on whether any boundary conditions are imposed on the applicable explicit statement of the equation. In the use of Eq. (2) to describe the temperature dependence of either specific rate or equilibrium constants no restrictions are applied, and these physical occurrences, if described unconventionally on linear coordinates, will always exhibit inflection points.

The use of Eq. (2) to describe solubility behavior is subject to the restrictions specified previously. Consequently, whether an inflection point is observable depends solely on whether the inflection point temperature T_i occurs below or above the melting point temperature T_A, namely, whether it occurs in an allowable or hypothetical temperature interval.

The values determined for T_i and N_i, the coordinates of the temperature and mole fraction, respectively, at the inflection point are given by

$$T_i = \Delta H_A/2R, \qquad (3)$$

and

$$\ln N_i = (\Delta H_A/RT) - 2. \qquad (4)$$

In order for N_i to occur in the physically significant interval $1 \geqslant N_A \geqslant 0$, it is evident, from Eq. (4), that

$$\Delta H_A/T_A \leqslant 2R.\,^{[1]} \qquad (5)$$

In other words, if Eq. (5) represents an actual case, the inflection point will occur in the physically meaningful interval and an inflected solubility curve is predicted. Conversely, if

$$\Delta H_A/T_A \geqslant 2R, \qquad (6)$$

the inflection point occurs in the physically meaningless range $N > 1$. This results in the prediction of a noninflected or "normal" liquidus curve, i.e., one that is concave with respect to the composition axis. For the singular case

$$\Delta H_A/T_A = 2R, \qquad (7)$$

it is clear that the inflection point will occur at $N_A = 1$.

[1] Note that if $\Delta H_A/T_A \leqslant 2R$, $(\Delta H_A/RT) - 2 < 0$. Consequently, $\ln N_i < 0$ and N_i lies in the interval 0–1.

Thus, given values of ΔH_A and T_A, it becomes a simple matter to predict the shape of the liquidus curve, at least in ideal systems. The shapes of liquidus curves in real systems are perturbed or obscured by two factors: The first is due to deviations from ideality, the second to the nature of the eutectic interaction itself. If, for example, the components melt at sufficiently different temperatures, the eutectic point will be displaced toward the lower melting component of the two. This will tend to obscure the shape of the liquidus of the lower melting component, to a lesser or greater degree, depending on the disparity in melting points, the deviations from ideality and the values of the heats of fusion of the participating components.

C. Application of the $\Delta H_A/T_A$ Principle to Real Systems

It is interesting, at this point, to test the usefulness of the $\Delta H_A/T_A$ criterion by examining these ratios for common substances, and where possible, comparing predicted liquidus contours based on this criterion with those exhibited in experimentally determined curves. Although some comments will be made on the behavior of pseudobinary systems, a critical analysis will be presented only for elemental systems, data for which are, in general, more available. Frequently, also, elemental systems have been examined more than once. This lends credibility to the proposed construction of their liquidus curves.

In describing the behavior of binary pairs of components, the descriptive technique employed will be as follows: In each elemental category, only the behavior of the element in question is considered, the behavior of the second element being considered in its own category. If the behavior of the element of interest is not discussed, it is understood that either its solubility curve is obscured by eutectic position, in which event the symbol e will be specified, or the number of data points offered are too few to enable categorization of the contour type, in which event the symbol i will be specified.

Table I presents data for the elements based on information in "Metallurgical Thermochemistry" by O. Kubaschewski and E. LL. Evans (1958).[2] Information on the solid–liquid behavior of the elemental systems was obtained from the critical compilation by Hansen and Anderko (1958).[2] In all, some 50 examples of binary elemental interactions generating simple eutectic systems are recorded by the latter.

The most striking feature of Table I is that of the 43 elements for which both heat of fusion and melting point data are available. The

[2] See the general reference list in the Appendix.

C. THE ENTROPY OF FUSION AND REAL SYSTEMS

TABLE I
Selected Values for the Elements

Element	T_A °K	ΔH_A (cal/mole)	$\Delta H_A/T_A$	Element	T_A °K	ΔH_A (cal/mole)	$\Delta H_A/T_A$
Ag	1234	2690	2.2	Mg	923	2100	2.3
Al	932	2500	2.7	Mn	1517	3200	2.1
Au	1336	3050	2.3	Mo	2873	6600	2.3
Ba	983	1830	1.9	Na	371	630	1.7
Be	1557	2800	1.8	Ni	1728	4220	2.5
Bi	544	2600	4.8	Pb	600	1150	1.9
Ca	1123	2100	1.9	Rb	312	525	1.7
Cd	594	1530	2.6	Re	3453	8000	2.5
Ce	1048	2120	2.0	S	392	300	0.8
Co	1768	3750	2.1	Sb	904	9500	10.5
Cr	2123	4600	2.2	Se	493	9000	18.3
Cs	303	500	1.7	Si	1693	12100	7.2
Cu	1356	3100	2.3	Sn	505	1690	3.4
Fe	1812	3700	2.0	Sr	1043	2100	2.0
Ga	303	1336	4.4	Ta	3253	5900	1.8
Ge	1213	7700	6.4	Te	723	8360	11.6
Hg	234	550	2.4	Ti	1933	4500	2.32
I_2	387	3770	9.7	Tl	507	1030	2.03
In	430	780	1.8	U	1403	3000	2.1
K	337	571	1.7	Zn	693	1740	2.5
La	1153	2500	2.2	Zr	2133	4600	2.2
Li	453	700	1.6				

majority of these are expected to exhibit inflected liquidus curve behavior, only the elements Bi, Ga, Ge, I_2, Sb, Se, Si, and Te having $\Delta H_A/T_A$ values in excess of $2R$, i.e., greater than 3.97 cal/mole deg. Of these, the reported accuracy of the heats of fusion is such that no borderline cases are expected. For the remainder of the elements, the $\Delta H_A/T_A$ ratios range around two so that again no borderline cases are anticipated.

1. Silver Systems

The systems Ag–Bi, Ag–Cu, Ag–Ge, Ag–Na, Ag–Pb, Ag$^{(e)}$–Si, and Ag–Tl are all of the simple eutectic type. All of the silver curves, as predicted, are inflected.

2. Aluminum Systems

The systems Al$^{(e)}$–Be, Al–Ga, Al$^{(e)}$–Ge, Al–Hg, Al$^{(e)}$–Si, and Al–Sn are all of the eutectic type. As predicted, the three systems with sufficiently exposed Al solubility curves are inflected.

3. Gold Systems

Only three representative systems are recorded; these are, Au–Ge, Au–Si, and Au–Tl. The Au–Ge system is clearly inflected. The Au–Si system has only three data points on the Au curve, which, if assumed correct, indicate inflection despite the preference on the part of the experimentalist to average out the points in a noninflected curve. A similar paucity of points is exhibited in the Au–Tl system but here the data plot a noninflected curve.

4. Bismuth Systems

Bismuth unlike the preceding three elements, is expected to exhibit noninflected solubility behavior. It forms eutectics with $As^{(e)}$, $Cd^{(e)}$, $Cu^{(e)}$, Ge, Hg, and Sn. The three usable examples all exhibit noninflected Bi curves.

5. Beryllium Systems

Only the systems Be–Al and Be–Si are recorded. Both, as predicted, are inflected.

6. Cadmium Systems

Cadmium forms simple eutectics with Bi, $Ge^{(e)}$, Pb, Sn, Tl, and Zn. The four unambiguous cases, those with Pb, Sn, Tl, and Zn, are inflected as is expected. The first system exhibits almost a straight Cd solubility curve from which it cannot be determined whether inflected or noninflected behavior occurs.

7. Copper Systems

Only the Cu–Li system is available for examination and, as expected, it exhibits an inflected curve.

8. Gallium Systems

Although the systems Ga–Al, Ga–Ge, Ga–Si, Ga–Sn, and Ga–Zn are of the simple eutectic variety, the eutectics all lie very close to the pure Ga axis.

9. Germanium Systems

Germanium, with a $\Delta H_A/T_A$ ratio of 6.4 is the second element in our series that should exhibit noninflected solubility curve contour. It

C. THE ENTROPY OF FUSION AND REAL SYSTEMS

generates simple eutectic systems with Ag, Al, Au, Bi$^{(i)}$, Cd, Ga, In, Pb, Sb, Sn, Tl, and Zn. The constructed curve in the Ge–Ag system is inflected, but the Ge melting point is incorrect by some 10°C. If the melting point is placed at the proper temperature, the data then plot a noninflected curve. The remaining systems, with the exception of the one with Bi, are noninflected as predicted. The Ge contour is ambiguous in the Ge–Bi system because two sets of data presented are in disagreement. One set indicates an inflected curve. The second set, depending on how one constructs the curve, may be used to either validate or contradict the predicted behavior.

10. *Indium Systems*

In the potential examples, i.e., In$^{(e)}$–Ge, In$^{(e)}$–Si, and In$^{(e)}$–Zn, none are usable for evaluation.

11. *Sodium, Rubidium, and Lithium Systems*

Only two potential cases are recorded for the alkali metals, the systems Na–Rb and Li$^{(e)}$–Cu. Both Na and Rb are expected to exhibit inflected curves. The system was examined by two investigators whose data are in good agreement along the Na solubility curve, which is inflected. Neither study defined the Rb curve adequately. In addition, the data provided for Rb are in marked disagreement making evaluation unreliable.

12. *Lead Systems*

Eutectics are formed with Ag$^{(e)}$, As$^{(e)}$, Cd, Ge$^{(e)}$, and Sb$^{(e)}$. The only usable example shows inflected lead solubility behavior as expected.

13. *Sulfur, Selenium, and Tellurium Systems*

Only the system S$^{(e)}$–Te is recorded and has an obscured "S" curve due to eutectic location. The Te curve as expected is the third example of a noninflected behavior.

14. *Antimony Systems*

Three eutectic cases exist: Sb$^{(e)}$–Ge, Sb–Pb, and Sb$^{(e)}$–Si. The fourth test element in the noninflected group, as predicted, gives rise to a noninflected solubility curve.

TABLE II

Selected Values for Compounds

Compound	T_A °K	ΔH_A cal/mole	$\Delta H_A/T_A$	Compound	T_A °K	ΔH_A cal/mole	$\Delta H_A/T_A$
AgCl	728	3100	4.3	CuCl	693	2450	3.5
AgBr	703	2200	3.1	CuBr	761	2300	3.0
AgI	831	2250	2.7	CuI	861	2600	3.0
Ag_2SO_4	928	4300	4.6	Cu_2O	1503	13,400	8.9
Al_2Cl_6	466	17,000	36.5	Fe_2Cl_6	580	20,500	35.4
Al_2Br_6	370	5400	14.6	Fe_3O_4	1870	33,000	17.6
Al_2I_6	464	8000	17.2	FeS	1468	7730	5.2
Al_2O_3	2303	26,000	11.3				
				$GaCl_3$	351	5200	14.8
$AsCl_3$	257	2400	19.3	$GaBr_3$	395	2800	7.1
$AsBr_3$	304	2800	9.2	GaI_3	485	3900	8.0
AsI_3	415	2200	5.3				
As_4O_6	582	8800	15.1	GeO_2	1389	10,500	7.6
BF_3	144	1014	7.0	$HgCl_2$	550	4200	7.6
B_2O_3	723	5300	7.3	$HgBr_2$	501	5000	10.0
				HgI_2	523	4500	8.6
BaF_2	1563	6800	4.4				
$BaCl_2$	1233	5400	4.4	KF	1130	6750	6.0
BaO	2198	13800	6.3	KCl	1045	6100	5.8
$BaSO_4$	1623	9700	6.0	K_2SO_4	1342	8800	6.6
$BiCl_3$	503	2600	5.2	LiF	1121	6400	5.7
Bi_2O_3	1090	6800	6.2	LiCl	887	3200	3.6
				LiBr	823	2900	3.5
CaF_2	1691	7100	4.2	Li_2SO_4	1132	3050	2.7
$CaCl_2$	1055	6800	6.5				
CaO	2873	19,000	6.6	MgF_2	1536	13,900	9.1
$Ca_2P_2O_7$	1626	24,100	14.8	$MgCl_2$	987	10,300	10.4
				$MgBr_2$	983	8300	8.4
CdF_2	1383	5400	3.9	MgO	3073	18,500	6.0
$CdCl_2$	841	5300	6.3	$MgSO_4$	1403	3500	2.5
$CdBr_2$	840	5000	6.0				
CdI_2	663	3700	5.6	$MnCl_2$	923	9000	9.8
				MnO	2058	13,000	6.3
$CrCl_2$	1088	7700	7.1				
				NaF	1265	8000	6.3
CsF	955	3000	3.1	NaCl	1074	6900	6.4
CsCl	918	3600	3.9	NaBr	1023	6100	6.0

C. THE ENTROPY OF FUSION AND REAL SYSTEMS

TABLE II (continued)

Compound	T_A °K	ΔH_A cal/mole	$\Delta H_A/T_A$	Compound	T_A °K	ΔH_A cal/mole	$\Delta H_A/T_A$
NaI	933	5300	5.7	SbI_3	443	4200	9.5
NaOH	593	1520	2.6	Sb_4O_6	929	26,000	28.0
Na_2SO_4	1163	5750	4.9	$SnCl_2$	520	3050	5.9
				$SnBr_2$	505	1720	3.4
PbF_2	1097	1900	1.7	SnI_2	593	3000	5.1
$PbCl_2$	771	5800	7.5				
$PbBr_2$	643	4500	7.0	$SrCl_2$	1146	4100	3.6
PbI_2	685	6000	8.8	$SrBr_2$	916	4800	5.2
PbO	1159	6200	5.3	$TiCl_4$	248	2240	9.0
$PbSO_4$	1363	9600	7.0	TiO_2	2113	15,500	7.3
RbF	1048	4200	4.0	V_2O_5	943	15,600	16.5
RbCl	988	4400	4.5				
RbBr	953	3700	3.9	$ZnCl_2$	591	5500	9.3
RbI	913	3000	3.3	$ZnBr_2$	667	4000	6.0

15. Silicon Systems

Several examples are available in the fifth test case exhibiting expected noninflected curves. These simple eutectic interactions are generated with the elements Ag, Al, Au, Be$^{(i)}$, Ga, In, and Sb. Unfortunately, the Si behavior in a number of the above systems is ambiguous due to the wide spread in Si melting point temperatures that was found by the investigators (1396–1430°C), with Kubaschewski and Evans[3] listing their preferred value as 1420°C. The interactions with Ag and Sb indicate inflected behavior. The remaining examples are noninflected. Whether the nonadherence of the two cited examples to predicted behavior is attributable to extreme deviations from ideality, to the choice of incorrect melting points, or to some other cause cannot be assessed.

16. Tin Systems

Tin, with a $\Delta H_A/T_A$ ratio of 3.4 should exhibit inflected behavior. Eutectic examples are those with Al$^{(e)}$, Bi, Cd, Ga, Ge$^{(e)}$, and Zn$^{(e)}$. The Sn–Bi system exhibits an almost straight line Sn solubility curve whose contour category cannot be established unambiguously. The Sn–Cd system exhibits an inflected curve while the Sn–Ga system exhibits a

---

[3] See Kubaschewski and Evans (1958) in the general reference list in the Appendix.

noninflected curve. Tin represents one of the few cases where the $\Delta H_A/T_A$ ratio is of the order of ± 0.5 cal/mole deg removed from the critical value $2R$. Its system behavior may be indicative of the extent to which nonideality may perturb a solubility curve contour.

17. *Zinc Systems*

Appropriate interactions occur with Cd, Ga, Ge[e], In, and Sn. Zinc, with a $\Delta H_A/T_A$ ratio of 2.5, should exhibit inflected behavior, and does so in the usable cases.

18. *Miscellaneous Systems*

The system Ti–Th is of the eutectic variety and exhibits an inflected Ti curve as expected. Heat of fusion data are not available for Th so the occurrence in this system of a noninflected solubility curve cannot be compared with theory.

From the preceding subsections it is evident that, in the vast majority of instances, the theory enables prediction of solubility curve contour. What is most encouraging is that the predictability appears excellent on the whole, even when marked deviations from ideality are obviously present, i.e., curves of the same element in different systems, while exhibiting the proper contour, exhibit solubility curves whose temperatures at given mole fractions are very different.

D. Pseudobinary Systems

As indicated earlier, information on pseudobinary systems is neither as comprehensive nor as reliable as that for the elements. In general, excepting for a few classes of pseudobinary interactions, simple eutectic behavior is not the rule, intermediate compound or extensive solid solution formation being more normally observed. These intermediate compounds have, in general, not been characterized sufficiently to enable a critical evaluation. The halides often enter into simple eutectic interactions, but the majority of such systems were studied at a time when techniques of thermal analysis were incompletely developed and the implications of the phase rule imperfectly understood. In addition, whole series of halide interactions were examined by single investigators whose studies have not been repeated. Since, within these series, constructions are offered which are thermodynamically invalid, the credibility of the reasonable entries is questionable. Qualitative com-

D. PSEUDOBINARY SYSTEMS

parisons of pseudobinary with elemental system behavior may be made, however, with some level of confidence. $\Delta H_A/T_A$ ratios of a number of oxides, halides, etc., were computed and are presented in Table II. From this table, it is seen that in distinction to the values for the elements, the ratios are in general greater than $2R$. This would lead to the prediction that pseudobinary systems should exhibit, for the most part, noninflected solubility curves. If one peruses a compilation of such systems (e.g., "Phase Diagrams for Ceramists" by E. Levin et al., The American Ceramic Society, Inc., 1964[4]), this prediction is borne out qualitatively. The vast majority of eutectic interactions depicted therein, including those between end members and any intermediate compounds, exhibit noninflected curves. This evidence is not as conclusive as might be desired since many of these systems exhibit melting temperatures above the useful range of thermoelectric element measuring techniques. Liquidus curves, therefore, were frequently deduced by visual observation of melting behavior.

---

[4] The complete reference will be found in the general reference list in the Appendix.

16

Eutectic Interactions Continued—The Effects of Homogeneous Equilibria on Liquidus Contours

A. INTRODUCTION

In the three chapters preceding this one, the quantitative description of eutectic liquidus curves has been developed for the case of a pure solid in equilibrium with a liquid. The derivation of the liquidus equation was based on the stipulation that species and component mole numbers were equivalent. Thus, the component whose stoichiometry was represented by the designation "A" gave rise in the liquid phase only to species having the stoichiometry "A." In depicting liquidus curves graphically for such systems, this equivalency obviated any necessity for differentiating between component and species mole fractions.

For the general case previously considered, it is to be noted that no homogeneous equilibrium constants were involved, the arguments for the liquidus equations being based solely on heterogeneous phenomena. If A or B are either unary or pseudounary components, it is evident, however, that other possibilities exist that may involve homogeneous equilibria. When A or B, or both are unary, for example, the species

A. INTRODUCTION

derived from either or both of them may tend to polymerize in the liquid phase according

$$xA^{(2)} \rightleftarrows A_x^{(2)}. \tag{1}$$

The equilibrium constant for the polymerization is given then by

$$K_A = N_{A_x}/[N_A]^x. \tag{2}$$

On the other hand, A and B species may react to form a binary species in the liquid phase according to

$$xA + yB \rightleftarrows A_xB_y, \tag{3}$$

$$K_{AB} = \frac{[N_{A_xB_y}]}{[N_A]^x [N_B]^y}. \tag{4}$$

In addition to the above possiblities, if A and/or B are pseudounary, i.e., composed of more than one element it may tend to dissociate according to

$$A \rightleftarrows A' + A'', \tag{5}$$

$$K_A = \frac{[N_{A'}][N_{A''}]}{[N_A]}. \tag{6}$$

The products of the dissociation in turn may tend to associate or react, etc., leading to a large number of phenomena which make the simple liquidus equations inappropriate for descriptive purposes.

Another case of considerable interest involves an extremum in that the melting of one or both of the components is attended simultaneously by complete association or dissociation. In this instance, no homogeneous equilibrium is involved. Effectively, however, such a case coincides with the boundary condition for association or dissociation where either process goes to completion. This extreme case is of interest, since it is a starting point for considering so-called incongruent phenomena. The latter refer to a situation where, upon undergoing a change in phase, e.g., melting or vaporization, the compositions of the coexisting phases are not equivalent. For example, if $CuSO_4 \cdot 5H_2O$ dissociates, it does so by evolving vaporous H_2O and simultaneously forming $CuSO_4 \cdot 3H_2O$. The two phases in the resulting equilibrium are obviously of different stoichiometry. The compound $CuSO_4 \cdot 5H_2O$ is said to have vaporized incongruently. On the other hand when a material such as H_2O vaporizes, the stoichiometries of the resulting equilibrium phases are the same. Such a vaporization process is termed congruent. Similar phenomena attend the melting of some pseudounary compounds leading to the terms *congruent* and *incongruent melting*.

While it would be beyond the scope of our treatment to consider many of the involvements which lead to apparent nonideality of behavior, it is instructive to consider some relatively simple cases. Thus, we will present arguments for systems where one of the species dimerizes, one of the species dissociates, the species from each of the end members react to form new species with the stoichiometry AB, and where melting is accompanied by complete dissociation.

This chapter will consider the case of dissociation where the solid AB dissociates in the liquid phase to a degree α while the solid C is present as the species C in the liquid phase. The following chapter will consider a similar occurrence except that the dissociative species AB in this case will be in equilibrium in the liquid with the nondissociative species B giving rise to a common species effect. Chapter 18 will treat association in the liquid phase and Chapter 19 will consider end-member interaction, i.e., A and B form the new species AB. Finally, Chapter 20 will present arguments for the case where the compound AB melts to yield only A and B species, namely, where no homogeneous constant is involved and where no AB species are present in the liquid.

B. Dissociation of One of the Liquid-Phase Species

Consider the quasi-binary system AB–C which is of the simple eutectic type, i.e., no solid solubility of the components and complete liquid miscibility. In the liquid phase, the species AB dissociates to some degree so that the liquid consists of a mixture of the species A, B, AB, and C. Since it is postulated that AB dissociates to yield only A and B species, the mole fractions of the latter two are always equal. Because of the dissociation, the mole fractions of the species AB and C will always be less than for the nondissociative case if equivalent component mole fractions are employed.

Conceptually, the case in question is that in which the solid AB upon melting dissociates to a degree α. For convenience, and in order to establish a reference, the value α will be specified arbitrarily at the melting temperture of solid AB, T_{AB}^0. Any other specification would, of course, be equally acceptable. Furthermore, in order that the treatment have physical significance it is mandatory that the value α at the melting point of AB must lie in the interval $0 < \alpha < 1$. Were α permitted to attain either boundary value, it is evident that no homogeneous equilibrium constant would be involved. Thus, if $\alpha = 0$, then we are not treating a dissociative case, such systems having been considered in the preceding three chapters. If $\alpha = 1$, this coincides with the case yet to be treated

B. DISSOCIATION OF A LIQUID-PHASE SPECIE

in Chapter 20 in which a homogeneous equilibrium constant is again not involved.

The stoichiometric relations describing the melting of solids AB and C and the dissociation of AB in the liquid phase are

$$AB^{(1)} \underset{}{\overset{K_1}{\rightleftarrows}} AB^{(2)} \quad \text{with} \quad K_1 = N_{AB}^{(2)}/N_{AB}^{(1)}, \tag{7}$$

$$AB^{(2)} \underset{}{\overset{K_2}{\rightleftarrows}} A^{(2)} + B^{(2)} \quad \text{with} \quad K_2 = N_A^{(2)} N_B^{(2)}/N_{AB}^{(2)}, \tag{8}$$

$$C^{(1)} \underset{}{\overset{K_3}{\rightleftarrows}} C^{(2)} \quad \text{with} \quad K_3 = N_C^{(2)}/N_C^{(1)}. \tag{9}$$

Examination of Eqs. (7)–(9) reveals that each is representative of a process previously considered. Thus, Eqs. (7) and (9) depict heterogeneous processes in which species and mole numbers are equivalent. Eq. (8), on the other hand, depicts a homogeneous process.

Designation of component mole fractions by the symbol M_X, component mole quantities by m_X, species mole fractions by N_X, species mole quantities by n_X, and the total number of species moles by n_t enables us to generate the following: The component mole fraction of AB in the liquid phase is given by

$$M_{AB}^{(2)} = m_{AB}^{(2)}/(m_{AB}^{(2)} + m_C^{(2)}). \tag{10}$$

The analogous fraction for the component C is given by

$$M_C^{(2)} = m_C^{(2)}/(m_{AB}^{(2)} + m_C^{(2)}). \tag{11}$$

From Eqs. (10) and (11) it is evident, therefore, that

$$M_C^{(2)} + M_{AB}^{(2)} = 1. \tag{12}$$

The total number of species moles in the liquid $n_t^{(2)}$ is given by

$$n_t^{(2)} = n_{AB}^{(2)} + n_A^{(2)} + n_B^{(2)} + n_C^{(2)}. \tag{13}$$

In conjunction with Eq. (13) it is to be noted that the value n_t is greater than it would be for a system in which species derived from the components did not dissociate. Since the value n_t is greater in the case under consideration, it is evident that the mole fraction of the species C, for example, is of necessity less than in a system not involving a dissociative equilibrium. For the very same reason, the species mole fraction $N_{AB}^{(2)}$ is also less.

Dividing Eq. (13) by $n_t^{(2)}$ yields

$$1 = N_{AB}^{(2)} + N_A^{(2)} + N_B^{(2)} + N_C^{(2)} \tag{14}$$

which is the analogous constraint to that expressed for the component concentrations by Eq. (12).

The conservation of component quantities expressed in terms of species quantities may be given by

$$m_{AB}^{(2)} = n_{AB}^{(2)} + n_A^{(2)} = N_{AB}^{(2)} n_t^{(2)} + N_A^{(2)} n_t^{(2)}, \qquad (15)$$

$$m_C^{(2)} = n_C^{(2)} = N_C^{(2)} n_t^{(2)}. \qquad (16)$$

Substituting for $m_{AB}^{(2)}$ and $m_C^{(2)}$ in Eqs. (10) and (11), the values defined by Eqs. (15) and (16) give

$$M_{AB}^{(2)} = \frac{N_{AB}^{(2)} n_t^{(2)} + N_A^{(2)} n_t^{(2)}}{N_{AB}^{(2)} n_t^{(2)} + N_A^{(2)} n_t^{(2)} + N_C^{(2)} n_t^{(2)}} = \frac{N_{AB}^{(2)} + N_A^{(2)}}{N_{AB}^{(2)} + N_A^{(2)} + N_C^{(2)}}. \qquad (17)$$

From Eq. (14) the value of $N_C^{(2)}$ in terms of the other species concentrations may be derived. This value may be substituted for the term $N_C^{(2)}$ in Eq. (17) to give

$$M_{AB}^{(2)} = \frac{N_{AB}^{(2)} + N_A^{(2)}}{N_{AB}^{(2)} + N_A^{(2)} + (1 - N_{AB}^{(2)} - N_A^{(2)} - N_B^{(2)})} = \frac{N_{AB}^{(2)} + N_A^{(2)}}{1 - N_B^{(2)}}. \qquad (18)$$

Finally, as $N_A^{(2)} = N_B^{(2)}$, we may rewrite Eq. (18) in the form

$$M_{AB}^{(2)} = (N_{AB}^{(2)} + N_A^{(2)})/(1 - N_A^{(2)}). \qquad (19)$$

In a similar manner, the conservation equation for the component C may be generated,

$$M_C^{(2)} = N_C^{(2)}/(1 - N_A^{(2)}). \qquad (20)$$

Equations (19) and (20) represent the starting points for the analytical description of the AB and C liquidus curves. They do, indeed, represent the bridges between component and species terms.

C. Solution for the AB Liquidus Curve

From Eq. (19) it is seen that the analytical description of the AB liquidus requires the specification of two independent relationships, one for $N_{AB}^{(2)}$ and one for $N_A^{(2)}$. Since the two-component, two-phase system under a constant pressure constraint is univariant, it is convenient to choose the temperature as the degree of freedom. What we seek then

C. SOLUTION FOR THE AB LIQUIDUS CURVE

are the temperature dependencies of the concentrations of the species AB and A. The independent relationships possessing this attribute are, of course, the equilibrium constants for the melting of solid AB and the dissociation of the AB species.

For the melting of solid AB we may, in explicit form, write

$$\ln \frac{N_{AB}^{(2)}}{N_{AB}^{(1)}} - \ln \frac{N_{AB}^{0(2)}}{N_{AB}^{0(1)}} = -\frac{\Delta H_1^0}{R}\left(\frac{1}{T} - \frac{1}{T_{AB}^0}\right). \quad (21)$$

T_{AB}^0 is the melting point of pure solid AB, ΔH_1^0 is the molar heat of fusion of AB, the superscript 0 terms refer to the mole fractions of AB species in the liquid and solid phases at the melting point of pure AB, and the undesignated N_{AB} values to these mole fractions at some other temperature. As before the superscript 1 refers to the phase of lower heat content and the superscript 2 to the phase of higher heat content.

The superscript 1 terms are each unity since the solid is assumed to be pure. The term $N_{AB}^{0(2)}$, however, is not unity since at all temperatures AB dissociates to some extent in the liquid phase. Equation (21), therefore, has the form

$$N_{AB}^{(2)} = N_{AB}^{0(2)} \exp\left[-\frac{\Delta H_1^0}{R}\left(\frac{1}{T} - \frac{1}{T_{AB}^0}\right)\right]. \quad (22)$$

For the homogeneous reaction described by Eq. (8), the value of the equilibrium constant K_2 at the temperatures T and T_{AB}^0 is given by

$$\ln K_2 - \ln K_2^0 = -\frac{\Delta H_2^0}{R}\left(\frac{1}{T} - \frac{1}{T_{AB}^0}\right). \quad (23)$$

The enthalpy term ΔH_2^0 is the molar heat of dissociation of the species AB at the melting point of the solid AB. Here again we have assumed that this enthalpic term is to a first approximation independent of temperature. Substituting the values of N_X for the values of K_2 in Eq. (23) we have

$$\ln\left[\frac{[N_A^{(2)}]^2}{N_{AB}^{(2)}}\right] - \ln\left[\frac{[N_A^{0(2)}]^2}{N_{AB}^{0(2)}}\right] = -\frac{\Delta H_2^0}{R}\left(\frac{1}{T} - \frac{1}{T_{AB}^0}\right). \quad (24)$$

Equation (24), upon rearrangement, yields

$$[N_A^{(2)}]^2 = \frac{N_{AB}^{(2)}}{N_{AB}^{0(2)}} [N_A^{0(2)}]^2 \exp\left[-\frac{\Delta H_2^0}{R}\left(\frac{1}{T} - \frac{1}{T_{AB}^0}\right)\right]. \quad (25)$$

Substituting for $N_{AB}^{(2)}/N_{AB}^{0(2)}$ its value as defined by Eq. (22), we obtain for the variation of $N_A^{(2)}$ with temperature

$$N_A^{(2)} = N_A^{0(2)} \exp\left[-\left(\frac{\Delta H_1^0 + \Delta H_2^0}{2R}\right)\left(\frac{1}{T} - \frac{1}{T_{AB}^0}\right)\right]. \qquad (26)$$

Equations (22) and (26) are the sought after relationships needed to expand Eq. (19).

When AB dissociates at its melting point to a degree α, we have, starting with m_{AB} moles of the component AB, the following number of moles of the species AB remaining:

$$n_{AB}^{(2)} = m_{AB}^{(2)} - \alpha m_{AB}^{(2)} = m_{AB}^{(2)}(1 - \alpha). \qquad (27)$$

Since each mole of AB that dissociates yields a mole each of the species A and B we see also that

$$n_A^{(2)} = \alpha m_{AB}^{(2)}, \qquad (28)$$

and

$$n_B^{(2)} = \alpha m_{AB}^{(2)}. \qquad (29)$$

At the melting point of the solid AB, therefore, the mole fraction of AB species in the liquid phase $N_{AB}^{0(2)}$ is given by

$$N_{AB}^{0(2)} = \frac{(1 - \alpha) m_{AB}^{(2)}}{(1 - \alpha) m_{AB}^{(2)} + \alpha m_{AB}^{(2)} + \alpha m_{AB}^{(2)}} = \frac{1 - \alpha}{1 + \alpha}. \qquad (30)$$

Under these same conditions, the mole fraction of A species $N_A^{0(2)}$ is given by

$$N_A^{0(2)} = \alpha/(1 + \alpha). \qquad (31)$$

Substituting the results of Eqs. (30) and (31) into Eqs. (22) and (26), respectively, the explicit expressions for $N_{AB}^{(2)}$ and $N_A^{(2)}$ as a function of temperature are

$$N_{AB}^{(2)} = \frac{1 - \alpha}{1 + \alpha} \exp\left[-\frac{\Delta H_1^0}{R}\left(\frac{1}{T} - \frac{1}{T_{AB}^0}\right)\right], \qquad (32)$$

$$N_A^{(2)} = \frac{\alpha}{1 + \alpha} \exp\left[-\left(\frac{\Delta H_1^0 + \Delta H_2^0}{2R}\right)\left(\frac{1}{T} - \frac{1}{T_{AB}^0}\right)\right]. \qquad (33)$$

Equation (19) represents the bridge between component and species concentrations along the AB liquidus curve. It is specified in terms of $N_{AB}^{(2)}$ and $N_A^{(2)}$. Substituting for these terms in Eq. (19) the results given

D. THE NONDISSOCIATING COMPONENT LIQUIDUS

in Eqs. (32) and (33) we obtain, for the variation of the component AB concentration with temperature, the relationship

$$M_{AB}^{(2)} = \frac{\left[\begin{array}{c} \frac{1-\alpha}{1+\alpha} \exp\left[-\frac{\Delta H_1^0}{R}\left(\frac{1}{T} - \frac{1}{T_{AB}^0}\right)\right] \\ + \frac{\alpha}{1+\alpha} \exp\left[-\left(\frac{\Delta H_1^0 + \Delta H_2^0}{2R}\right)\left(\frac{1}{T} - \frac{1}{T_{AB}^0}\right)\right] \end{array}\right]}{1 - \frac{\alpha}{1+\alpha} \exp\left[-\left(\frac{\Delta H_1^0 + \Delta H_2^0}{2R}\right)\left(\frac{1}{T} - \frac{1}{T_{AB}^0}\right)\right]}$$

$$= \frac{(1-\alpha)e^{-A} + \alpha e^{-B}}{1 + \alpha(1 - e^{-B})}, \qquad (34)$$

where

$$A = \frac{\Delta H_1^0}{R}\left(\frac{1}{T} - \frac{1}{T_{AB}^0}\right),$$

and

$$B = \frac{\Delta H_1^0 + \Delta H_2^0}{2R}\left(\frac{1}{T} - \frac{1}{T_{AB}^0}\right).$$

D. The Liquidus for the Nondissociating Component

The heterogeneous equilibrium of the component C may be described by

$$K_3 = K_3^0 e^{-C}, \qquad (35)$$

where

$$C = \frac{\Delta H_3^0}{R}\left(\frac{1}{T} - \frac{1}{T_C^0}\right).$$

ΔH_3^0 is the molar heat of fusion of C and T_C^0 is the melting point of C. As seen from Eq. (20), the solution for $M_C^{(2)}$ is in terms of $N_C^{(2)}$ and $N_A^{(2)}$, requiring specification of two independent relationships. Again, the system is univariant along the C liquidus so that the equilibrium constants as a function of T are the sought after expressions. The two relevant equilibria are Eqs. (8) and (9). Since Eq. (8) also contains the term $N_{AB}^{(2)}$, we may employ Eq. (14) to substitute for $N_{AB}^{(2)}$. The temperature dependency of the homogeneous equilibrium constant [Eq. (8)] is given by

$$K_2 = K_2^0 e^{-D}, \qquad (36)$$

where

$$D = \frac{\Delta H_2^0}{R}\left(\frac{1}{T} - \frac{1}{T_{AB}^0}\right).$$

Writing Eq. (36) in terms of species concentrations yields

$$\frac{N_A^{(2)}N_B^{(2)}}{N_{AB}^{(2)}} = \frac{[N_A^{(2)}]^2}{N_{AB}^{(2)}} = \frac{[N_A^{(2)}]^2}{1 - 2N_A^{(2)} - N_C^{(2)}} = K_2^0 e^{-D}. \tag{37}$$

Equation (35), written in terms of species concentrations, has the form

$$N_C^{(2)}/N_C^{(1)} = (N_C^{0(2)}/N_C^{0(1)})\, e^{-C}. \tag{38}$$

The terms $N_C^{(1)}$, $N_C^{0(2)}$, and $N_C^{0(1)}$ are each unity so that Eq. (38) may be written as

$$N_C^{(2)} = e^{-C}. \tag{39}$$

Substituting in Eq. (37) the value of $N_C^{(2)}$ specified by Eq. (39), we obtain

$$\frac{[N_A^{(2)}]^2}{1 - 2N_A^{(2)} - e^{-C}} = K_2^0 e^{-D}. \tag{40}$$

Equation (40) is a quadratic as shown in Eq. (41):

$$[N_A^{(2)}]^2 + 2K_2^0 e^{-D} N_A^{(2)} - K_2^0 e^{-D}[1 - e^{-C}] = 0. \tag{41}$$

Solving for $N_A^{(2)}$ gives

$$N_A^{(2)} = -K_2^0 e^{-D} \pm ([K_2^0]^2 e^{-2D} + K_2^0 e^{-D}[1 - e^{-C}])^{1/2}. \tag{42}$$

In order for the solution to Eq. (42) to be physically meaningful, the discriminant d must fulfill the condition that

$$d = b^2 - 4ac \geqslant 0. \tag{43}$$

Let us simplify Eq. (42) by denoting $K_2^0 e^{-D} = Y$ so that we may determine the conditions necessary to satisfy Eq. (43). Thus,

$$N_A = -Y \pm (Y^2 + Y[1 - e^{-C}])^{1/2}. \tag{44}$$

If C is negative, then e^{-C} varies between 1 and ∞ and $[1 - e^{-C}]$ is therefore negative. The term $Y[1 - e^{-C}]$ is, therefore, negative and $\sqrt{d} < Y$ even if $\sqrt{d} \geqslant 0$. Thus for C negative, N_A is always negative, which is physically meaningless. On the other hand, with C positive,

D. THE NONDISSOCIATING COMPONENT LIQUIDUS

e^{-C} varies between 0 and 1 and $\sqrt{d} > A$. Since the ΔH term in C is always positive, T_C^0 of Eq. (35), i.e., the melting point of C, must always be greater than T, any other temperature along the liquidus for C. Let us now consider what happens to N_A as K_2^0 varies between 0 and ∞, namely as α, the degree of dissociation varies between 0 and 1, respectively.

When $K_2^0 \to 0$ in Eq. (44), i.e., $\alpha \to 0$, we see that $N_A \to 0$ as expected. Equation (20), then, is of the form

$$M_C^{(2)} \to N_C^{(2)} \to e^{-C}. \tag{45}$$

When $K_2^0 \to \infty$, i.e., $\alpha \to 1$, the following obtains: Rewriting Eq. (41) as

$$1/K_2^0 = e^{-D}(1/[N_A^{(2)}]^2) - (2/N_A^{(2)}) - (e^{-C}/[N_A^{(2)}]^2), \tag{46}$$

we see that as $K_2^0 \to \infty$

$$e^{-D}(1 - 2N_A - e^{-C}) \to 0. \tag{47}$$

Either $e^{-D} \to 0$ and $(1 - 2N_A - e^{-C})$ is finite, or e^{-D} is finite and $(1 - 2N_A - e^{-C}) \to 0$. Realizing that $1 - 2N_A - e^{-C} = N_{AB}$, we see that the first of these possibilities is physically meaningless since as $\alpha \to 1$, $N_{AB}^{(2)} \to 0$ and

$$N_A + N_{AB} + N_C = 2N_A + N_C = 2N_A + e^{-C} = 1. \tag{48}$$

From Eq. (48) we see, therefore, that $1 - 2N_A - e^{-C} \to 0$ making the second possibility the correct one. Solving Eq. (48) for N_A it is seen that N_A may vary between $0 < N_A < 0.5$.

At the melting point of the solid AB, K_2^0 has the form

$$K_2^0 = [N_A^{0(2)}]^2/N_{AB}^{0(2)}. \tag{49}$$

To solve explicitly for $M_C^{(2)}$ in Eq. (20) we need values for $N_A^{0(2)}$ and $N_{AB}^{0(2)}$ in addition to the values for $N_C^{(2)}$ [Eq. (39)] and $N_A^{(2)}$ [Eq. (42)]. Starting with m_{AB} moles of the component AB which dissociates to a degree α at the temperature T_{AB}^0 we have present

$$n_{AB} = m_{AB}(1 - \alpha) \tag{50}$$

species moles of AB and

$$n_A = m_{AB}\alpha \tag{51}$$

species moles of A.

Substituting Eqs. (50) and (51) into Eq. (49) we obtain

$$K_2^0 = \alpha^2/(1 - \alpha^2). \tag{52}$$

The solution for Eq. (20), then, using Eqs. (39), (42), and (52) is

$$M_C^{(2)} = \frac{\exp\left[-\dfrac{\Delta H_3^0}{R}\left(\dfrac{1}{T} - \dfrac{1}{T_C^0}\right)\right]}{\left[1 - \dfrac{\alpha^2}{1-\alpha^2}\exp\left[-\dfrac{\Delta H_2^0}{R}\left(\dfrac{1}{T} - \dfrac{1}{T_{AB}^0}\right)\right]\right.} \\ \left. \pm \left\{\left(\dfrac{\alpha^2}{1-\alpha^2}\right)^2 \exp\left[-\dfrac{2\Delta H_2^0}{R}\left(\dfrac{1}{T} - \dfrac{1}{T_{AB}^0}\right)\right]\right.\right. \\ \left.\left. + \dfrac{\alpha^2}{1-\alpha^2}\exp\left[-\dfrac{\Delta H_2^0}{R}\left(\dfrac{1}{T} - \dfrac{1}{T_{AB}^0}\right)\right]\right.\right. \\ \left.\left. \times \left\{1 - \exp\left[-\dfrac{\Delta H_3^0}{R}\left(\dfrac{1}{T} - \dfrac{1}{T_C^0}\right)\right]\right\}\right\}^{1/2}\right]. \tag{53}$$

17

Common Species Effects in Simple Eutectic Systems

A. INTRODUCTION

Many binary systems of the type A–B give rise within their composition range to so-called "intermediate compounds" of the generalized stoichiometry A_xB_y. Assuming, for our purposes, that only a single intermediate compound is formed and that this compound has the stoichiometry AB, it is not uncommon to find this compound partaking in a simple eutectic interaction with one or both of the end members, A and B. If the pure intermediate compound AB melts congruently, i.e., yields a liquid phase of precisely the same composition as the pure solid, then the systems AB–A or AB–B may be termed subsystems of the binary interaction A–B. The details of construction of such systems will be presented more fully subsequently. At present, however, we wish only to note that if AB dissociates to a degree α to yield A and B species in solution, the extent of this dissociation at any temperature will be affected by the presence of A or B derived from the end member participating in the subsystem interaction.

Since the presence of additional A or B will tend to suppress the dissociation of AB except at pure AB, the AB liquidus curve is expected to lie in an intermediate position between that for the undissociated case, and for the dissociated case where no common species effect is present. The effect on either the A or B liquidus may be anticipated from the following: The equilibrium constants for the melting of either A or B are single-valued functions of the temperature. Since the required liquid composition of A or B species at any particular temperature is partially satisfied via the dissociation of AB species, the equilibrium constants will be satisfied at higher temperatures than normally. Thus, a lower component mole fraction of A or B will suffice, at a given temperature, to satisfy the A or B species concentration in the liquid. This is equivalent to stating that the solubility of A or B is decreased, i.e., the A or B liquidus curve lies above that for the non dissociative case.

As in the preceding chapter the several equilibria are defined by the relationships of the type

$$AB^{(1)} \rightleftarrows AB^{(2)}, \tag{1}$$

$$AB^{(2)} \rightleftarrows A^{(2)} + B^{(2)}, \tag{2}$$

$$B^{(1)} \rightleftarrows B^{(2)}. \tag{3}$$

The species composition of the liquid, as before, is constrained by

$$N^{(2)}_{AB} + N^{(2)}_{A} + N^{(2)}_{B} = 1. \tag{4}$$

Solutions for the concentrations of the three species in equilibrium with pure solid along each of the liquidus curves requires the specification of three independent relationships. As each of the liquidus curves is univariant at constant pressure, the single degree of freedom may be taken as the temperature. For the AB liquidus the set of independent equations may be chosen as Eq. (4) and the equilibrium constant relationships

$$K/K^0_{AB} = \exp\left[-\frac{\Delta H_1^0}{R}\left(\frac{1}{T} - \frac{1}{T^0_{AB}}\right)\right] = e^{-x}, \tag{5}$$

$$K^\dagger/K^0_{Diss} = \exp\left[-\frac{\Delta H_2^0}{R}\left(\frac{1}{T} - \frac{1}{T^0_{AB}}\right)\right] = e^{-y}. \tag{6}$$

For the B liquidus (the A liquidus on the opposite side of the diagram will have an exact counterpart), in addition to Eqs. (4) and (6), Eq. (7) describing the melting equilibrium of solid B may be employed.

$$K^*/K_B^0 = \exp\left[-\frac{\Delta H_3^0}{R}\left(\frac{1}{T} - \frac{1}{T_B^0}\right)\right] = e^{-z}. \tag{7}$$

A. INTRODUCTION

The conservation equations for the quantities of the components AB and B in the liquid phase may be stated as

$$m_{AB}^{(2)} = n_{AB}^{(2)} + n_A^{(2)}, \tag{8}$$

$$m_B^{(2)} = n_B^{(2)} - n_B'^{(2)} = n_B^{(2)} - n_A^{(2)} = n_B^{*(2)}. \tag{9}$$

In Eq. (9) the species term $n_B'^{(2)}$ represents the number of moles of the species B derived from the dissociation of AB. Obviously, $n_B'^{(2)} = n_A^{(2)}$. The quantity $n_B^{(2)}$ represents the total number of moles of the species B in the liquid phase.

In terms of mole fractions, Eqs. (8) and (9) may be written as:

$$m_{AB}^{(2)} = N_{AB}^{(2)} n_t^{(2)} + N_A^{(2)} n_t^{(2)}, \tag{10}$$

$$m_B^{(2)} = N_B^{(2)} n_t^{(2)} - N_A^{(2)} n_t^{(2)}. \tag{11}$$

In terms of species mole fractions, the component mole fractions, $M_{AB}^{(2)}$ and $M_B^{(2)}$ then have the form

$$M_{AB}^{(2)} = \frac{m_{AB}^{(2)}}{m_{AB}^{(2)} + m_B^{(2)}} = \frac{N_{AB}^{(2)} + N_A^{(2)}}{N_{AB}^{(2)} + N_B^{(2)}}, \tag{12}$$

$$M_B^{(2)} = \frac{m_B^{(2)}}{m_{AB}^{(2)} + m_B^{(2)}} = \frac{N_B^{(2)} - N_A^{(2)}}{N_{AB}^{(2)} + N_B^{(2)}}. \tag{13}$$

The denominator of Eqs. (12) and (13) by dint of Eq. (4) is seen to be

$$N_{AB}^{(2)} + N_B^{(2)} = 1 - N_A^{(2)}. \tag{14}$$

Thus, for the component mole fraction conservation conditions in terms of species mole fractions we may write

$$M_{AB}^{(2)} = \frac{N_{AB}^{(2)} + N_A^{(2)}}{1 - N_A^{(2)}} \tag{15}$$

$$M_B^{(2)} = \frac{N_B^{(2)} - N_A^{(2)}}{1 - N_A^{(2)}}. \tag{16}$$

Equations (15) and (16) with the $N_X^{(2)}$ terms as a function of temperature represent the desired solutions.

In expanded form, the equilibrium constants Eqs. (5)–(7) may be written as:

$$\frac{N_{AB}^{(2)}}{N_{AB}^{(1)}} \bigg/ \frac{N_{AB}^{0(2)}}{N_{AB}^{0(1)}} = e^{-x}, \tag{17}$$

$$\frac{N_A^{(2)} N_B^{(2)}}{N_{AB}^{(2)}} \bigg/ \frac{N_A^{0(2)} N_B^{0(2)}}{N_{AB}^{0(2)}} = e^{-y}, \tag{18}$$

$$\frac{N_B^{*(2)}}{N_B^{*(1)}} \bigg/ \frac{N_B^{0(2)}}{N_B^{0(1)}} = e^{-z}. \tag{19}$$

In Eqs. (17) and (19), the superscript (1) terms refer to solid phases which are assumed pure. In addition, the term $N_B^{0(2)}$ in Eq. (19) refers to the pure liquid B at the melting point of pure solid B. All of these, therefore, have a value of unity. Consequently,

$$N_{AB}^{(2)} = N_{AB}^{0(2)} e^{-x}, \tag{20}$$

and

$$N_B^{*(2)} = e^{-z}. \tag{21}$$

If the degree of dissociation of pure liquid AB at the melting point of pure solid AB is α, then the number of moles of the species AB is given by

$$n_{AB}^{0(2)} = m_{AB}^{(2)} - \alpha m_{AB}^{(2)} = m_{AB}^{(2)}(1 - \alpha). \tag{22}$$

The number of moles of A and B species at the melting point of pure AB are each given by

$$n_A^{0(2)} \quad \text{or} \quad n_B^{0(2)} = \alpha m_{AB}^{(2)}. \tag{23}$$

The mole fraction, $N_{AB}^{0(2)}$, of the species AB at the melting point of pure AB then is given by

$$N_{AB}^{0(2)} = (1 - \alpha)/(1 + \alpha), \tag{24}$$

and of A or B species at this same point by

$$N_A^{0(2)} \quad \text{or} \quad N_B^{0(2)} = \alpha/(1 + \alpha). \tag{25}$$

The above provides us with the wherewithal to describe analytically the variation of $M_{AB}^{(2)}$ and $M_B^{(2)}$ as a function of temperature.

B. The Liquidus Curve for AB

From Eq. (15) we see that the solution for M_{AB} is in terms of the concentrations $N_{AB}^{(2)}$ and $N_A^{(2)}$. Equation (18) may be rearranged to give

$$N_A^{(2)} N_B^{(2)} = (N_{AB}^{(2)}/N_{AB}^{0(2)}) N_A^{0(2)} N_B^{0(2)} e^{-y}. \tag{26}$$

If the value for $N_{AB}^{(2)}$ from Eq. (20) is substituted in Eq. (26), this yields

$$N_A^{(2)} N_B^{(2)} = N_A^{0(2)} N_B^{0(2)} e^{-(x+y)}. \tag{27}$$

If for $N_B^{(2)}$ its equivalent value from the constraint imposed by Eq. (4) is substituted in Eq. (27), we obtain

$$N_A^{(2)}(1 - N_{AB}^{(2)} - N_A^{(2)}) = N_A^{0(2)} N_B^{0(2)} e^{-(x+y)}. \tag{28}$$

If the value of $N_{AB}^{(2)}$ from Eq. (20) is substituted in the left-hand side of Eq. (28), we have

$$N_A^{(2)}(1 - N_{AB}^{0(2)} e^{-x} - N_A^{(2)}) = N_A^{0(2)} N_B^{0(2)} e^{-(x+y)}. \tag{29}$$

Equation (29) is a quadratic in $N_A^{(2)}$ of the form

$$-[N_A^{(2)}]^2 + N_A^{(2)}(1 - N_{AB}^{0(2)} e^{-x}) - N_A^{0(2)} N_B^{0(2)} e^{-(x+y)} = 0. \tag{30}$$

The constants $N_{AB}^{0(2)}$, $N_A^{0(2)}$, and $N_B^{0(2)}$ may be introduced in terms of α from Eqs. (21) and (22) to give

$$-[N_A^{(2)}]^2 + N_A^{(2)} \left[1 - \left(\frac{1-\alpha}{1+\alpha}\right) e^{-x}\right] - \left(\frac{\alpha}{1+\alpha}\right)^2 e^{-(x+y)} = 0. \tag{31}$$

The solution for $N_A^{(2)}$ then is

$$N_A^{(2)} = \frac{1}{2}\left\{1 - \left(\frac{1-\alpha}{1+\alpha}\right) e^{-x} \pm \left\{\left[1 - \left(\frac{1-\alpha}{1+\alpha}\right) e^{-x}\right]^2 - 4\left(\frac{\alpha}{1+\alpha}\right)^2 e^{-(x+y)}\right\}^{1/2}\right\}. \tag{32}$$

Substituting Eqs. (32) and (20) into Eq. (15) provides the solution for $M_{AB}^{(2)}$ in terms of the temperature dependencies of the species AB and A.

$$M_{AB}^{(2)} = \frac{\left(\frac{1-\alpha}{1+\alpha}\right) e^{-x} + 1 - \left\{\left[1 - \left(\frac{1-\alpha}{1+\alpha}\right) e^{-x}\right]^2 - 4\frac{\alpha^2}{(1+\alpha)^2} e^{-(x+y)}\right\}^{1/2}}{1 + \left(\frac{1-\alpha}{1+\alpha}\right) e^{-x} + \left\{\left[1 - \left(\frac{1-\alpha}{1+\alpha}\right) e^{-x}\right]^2 - 4\frac{\alpha^2}{(1+\alpha)^2} e^{-(x+y)}\right\}^{1/2}}. \tag{33}$$

When $\alpha = 0$, from Eq. (33), it is found that $M_{AB}^{(2)} = e^{-x}$ providing the negative root of the square root term is taken. In the denominator this negative root becomes positive because the denominator begins as the expression $1 - N_A^{(2)}$ [Eq. (16)].

C. THE LIQUIDUS FOR B

Two factors should be considered before attempting the solution for the B liquidus. First, is to be realized that the value of α, i.e., the fraction of dissociated AB species, specified at the melting point of pure B will change as a function of composition due to the common species effect caused by addition of the component B. Second, in Eq. (18) the term $N_B^{(2)}$ comprises two terms, one due to the dissociation of AB species in the liquid, and the second due to B species derived from the component B. It is the value of this second term alone that is governed by the melting equation defined by Eq. (21).

In Eq. (20), the term $N_{AB}^{0(2)}$ represents the equilibrium concentration of AB species at the melting point of pure solid AB given in Eq. (1). When $N_{AB}^{0(2)}$ is defined in terms of the degree of dissociation α, as in Eq. (24), this value must be satisfied simultaneously by Eq. (18) at the particular temperature and concentration in question, i.e., pure AB at T_{AB}^0. For the general case, however, the value α specified at pure AB and T_{AB}^0 will be different at all other system concentrations. It is nonetheless the case that the ratio $N_A^{0(2)} N_B^{0(2)} / N_{AB}^{0(2)}$ of Eq. (18) is fixed at the temperature T_{AB}^0 since it represents the equilibrium constant for the reaction specified in Eq. (2) at the temperature T_{AB}^0. Since Eq. (20), the solubility equation for AB, cannot be utilized to specify the value $N_{AB}^{0(2)}$ along the B liquidus, this equation being inapplicable in this univariant region, we are compelled to use the ratio $N_A^{0(2)} N_B^{0(2)} / N_{AB}^{0(2)}$ in Eq. (18) as a starting point rather than any of its parts. This ratio may be determined from Eqs. (24) and (25) and is given by

$$K_{\text{Diss}}^0 = \frac{[\alpha/(1+\alpha)]^2}{[(1-\alpha)/(1+\alpha)]} = \frac{\alpha^2}{1-\alpha^2}. \tag{34}$$

Once K_{Diss}^0 is specified,[1] it is evident that $K^\dagger$, the value of the homogeneous equilibrium constant at any other temperature, is uniquely determined via Eq. (6), assuming ΔH_2^0 to be temperature independent.

For the three-species problem along the B liquidus, three independent

---

[1] This specification is made arbitrarily at T_{AB}^0 since the value α is assumed known at this temperature for pure AB.

C. THE LIQUIDUS FOR B

relationships are required for the analytical description of the solubility of B. If the temperature is chosen as the degree of freedom along the isobarically univariant B liquidus, Eqs. (18) and (21), and Eq. (4) in modified form, may be employed.

Examination of Eqs. (9) and (16) reveals that the term $N_B^{(2)}$ in Eq. (16) is composed of terms derived from components AB and B. Remembering that

$$n_x^{(2)} = N_x^{(2)} n_t^{(2)}. \tag{35}$$

The numerator of Eq. (16) may be written

$$M_B = N_B^{*(2)}/(1 - N_A^{(2)}), \tag{36}$$

as the bridge between component B and species terms. This substitution requires a restatement of Eq. (4), taking Eq. (9) into account, which may be written as

$$N_{AB}^{(2)} + N_A^{(2)} + [N_B^{*(2)} + N_A^{(2)}] = N_{AB}^{(2)} + 2N_A^{(2)} + N_B^{*(2)} = 1. \tag{37}$$

We may now substitute Eq. (34) into Eq. (18) to obtain

$$N_A^{(2)} N_B^{(2)} = N_{AB}^{(2)} [\alpha^2/(1-\alpha^2)] e^{-y}, \tag{38}$$

and further substitute the third term on the left-hand side of Eq. (37) for $N_B^{(2)}$ to obtain

$$N_A^{(2)} [N_B^{*(2)} + N_A^{(2)}] = N_{AB}^{(2)} [\alpha^2/(1-\alpha^2)] e^{-y}. \tag{39}$$

From Eq. (37), the value of $N_{AB}^{(2)}$ in terms of $N_A^{(2)}$ and $N_B^{*(2)}$ may be substituted in Eq. (39) to give

$$N_A [N_B^{*(2)} + N_A^{(2)}] = [1 - N_B^{*(2)} - 2N_A^{(2)}][\alpha^2/(1-\alpha^2)] e^{-y}, \tag{40}$$

which is a quadratic in $N_A^{(2)}$ of the form

$$(1-\alpha^2)(N_A^{(2)})^2 + N_A^{(2)}[(1-\alpha^2) N_B^{*(2)} + 2\alpha^2 e^{-y}] + \alpha^2 e^{-y}(N_B^{*(2)} - 1) = 0. \tag{41}$$

The solution for $N_A^{(2)}$ then is

$$N_A^{(2)} = \frac{\left[-[(1-\alpha^2) N_B^{*(2)} + 2\alpha^2 e^{-y}] \pm \{[(1-\alpha^2) N_B^{*(2)} + 2\alpha^2 e^{-y}]^2 - 4[(1-\alpha^2) \alpha^2 e^{-y}(N_B^* - 1)]\}^{1/2} \right]}{2(1-\alpha^2)}. \tag{42}$$

We know that as $\alpha \to 0$, $N_A^{(2)} \to 0$. In order for this to be possible, the positive root of the discriminant in Eq. (42) must be employed.

Substituting for $N_A^{(2)}$ its value from Eq. (42) into Eq. (36) we obtain

$$M_B^{(2)} = \frac{2(1-\alpha^2)N_B^*}{\left[(1-\alpha^2)(N_B^*+2) + 2\alpha^2 e^{-y} - \{[(1-\alpha^2)N_B^{*(2)} + 2\alpha^2 e^{-y}]^2 - 4[(1-\alpha^2)\alpha^2 e^{-y}(N_B^*-1)]\}^{1/2}\right]}. \tag{43}$$

When $\alpha \to 0$ it is seen that $M_B^{(2)} \to N_B^{*(2)}$ as would be expected.

Finally, substituting for N_B^* its value from Eq. (21) and for e^{-z} and e^{-y} their expanded forms we obtain

$$M_B^{(2)} = \frac{2(1-\alpha^2)\exp\left[-\frac{\Delta H_3^0}{R}\left(\frac{1}{T}-\frac{1}{T_B^0}\right)\right]}{\left[\begin{array}{l}(1-\alpha^2)\left\{\exp\left[-\frac{\Delta H_3^0}{R}\left(\frac{1}{T}-\frac{1}{T_B^0}\right)\right]+2\right\} \\ + 2\alpha^2 \exp\left[-\frac{\Delta H_2^0}{R}\left(\frac{1}{T}-\frac{1}{T_{AB}^0}\right)\right] \\ - \left\{\left[(1-\alpha^2)\exp\left[-\frac{\Delta H_3^0}{R}\left(\frac{1}{T}-\frac{1}{T_B^0}\right)\right]\right.\right. \\ \left.\left. + 2\alpha^2 \exp\left[-\frac{\Delta H_2^0}{R}\left(\frac{1}{T}-\frac{1}{T_{AB}^0}\right)\right]\right]^2 \right. \\ - 4\left[(1-\alpha^2)\alpha^2 \exp\left[-\frac{\Delta H_2^0}{R}\left(\frac{1}{T}-\frac{1}{T_{AB}^0}\right)\right]\right. \\ \left.\left.\times \exp\left[-\frac{\Delta H_3^0}{R}\left(\frac{1}{T}-\frac{1}{T_B^0}\right)\right] - 1\right]\right\}^{1/2}\end{array}\right]}. \tag{44}$$

18

Eutectic Interactions Continued—The Effects of Association on Liquidus Contours

A. Introduction

The preceding two chapters have presented arguments which have led to analytical expressions describing relatively minor perturbations of the simplest eutectic systems. It is evident that even these minor perturbations lead to relatively involved descriptions of solubility behavior in ideal systems. This is even more evident if we recall that in each of the perturbations considered only one of the components was involved. The topic for the present chapter is somewhat related to that of Chapter 16 in that a homogeneous equilibrium constant is involved in the liquidus equations. However, instead of this constant describing a dissociative process it refers to an associative phenomena which for obvious reasons will be taken as the case where the component A in the binary system A–B melts to yield species having the stoichiometry A. These species then associate according to

$$2A^{(2)} \rightleftarrows A_2^{(2)}, \tag{1}$$

the degree of association β being specified at the melting point of pure solid A. This process is a first step toward the phenomenon of intermediate compound formation (to be treated in the next chapter).

As previously employed, a constraint on the system may be offered in the form of an equation of the following type:

$$N_A^{(2)} + N_{A_2}^{(2)} + N_B^{(2)} = 1. \tag{2}$$

The reactions defining the solubility of the solids A and B are

$$A^{(1)} \rightleftarrows A^{(2)}, \tag{3}$$

$$B^{(1)} \rightleftarrows B^{(2)}. \tag{4}$$

The equilibrium constants defining Eqs. (1), (3), and (4) are, respectively,

$$K/K_{\text{Assoc}} = \exp\left[-\frac{\Delta H_1^0}{R}\left(\frac{1}{T} - \frac{1}{T_A^0}\right)\right], \tag{5}$$

$$K/K_A = \exp\left[-\frac{\Delta H_2^0}{R}\left(\frac{1}{T} - \frac{1}{T_A^0}\right)\right], \tag{6}$$

$$K/K_B = \exp\left[-\frac{\Delta H_3^0}{R}\left(\frac{1}{T} - \frac{1}{T_B^0}\right)\right]. \tag{7}$$

Since the solutions in equilibrium with either solid A or solid B comprise three species, it is seen that Eqs. (2), (5), and (6) provide the independent set necessary to describe the A liquidus and Eqs. (2), (5), and (7) are satisfactory for solution of the B liquidus.

B. The A Liquidus

Equations (5) and (6) may be written in expanded form to yield, respectively,

$$\frac{N_{A_2}^{(2)}}{(N_A^{(2)})^2} \bigg/ \frac{N_{A_2}^{0(2)}}{(N_A^{0(2)})^2} = \exp\left[-\frac{\Delta H_1^0}{R}\left(\frac{1}{T} - \frac{1}{T_A^0}\right)\right] = e^{-x} \tag{8}$$

and

$$\frac{N_A^{(2)}}{N_A^{(1)}} \bigg/ \frac{N_A^{0(2)}}{N_A^{0(1)}} = \exp\left[-\frac{\Delta H_2^0}{R}\left(\frac{1}{T} - \frac{1}{T_A^0}\right)\right] = e^{-y}. \tag{9}$$

The solid phases, represented as usual by the superscript "1" terms, are assumed pure so that Eq. (9) has the form

$$N_A^{(2)}/N_A^{0(2)} = e^{-y}. \tag{10}$$

B. THE A LIQUIDUS

The superscript 0 terms refer to the pure phases at the melting point of solid A and are not equal to unity since A_2 species are also present due to the association described by Eqs. (1) and (5). If the degree of association at the melting point of pure solid A is β, then starting with m_A moles of the component A we have at this point

$$m_A^{(2)} - \beta m_A^{(2)} = m_A^{(2)}(1 - \beta) = n_A^{(2)} \tag{11}$$

moles of the species A remaining.

Obviously, the quantity of the species A_2 formed simultaneously is given by

$$n_{A_2}^{(2)} = \beta m_A^{(2)}/2. \tag{12}$$

The component mole conservation conditions applicable to this system in terms of species moles may be written as

$$m_A^{(2)} = n_A^{(2)} + 2n_{A_2}^{(2)}. \tag{13}$$

The mole fraction of the component A is given by

$$M_A^{(2)} = \frac{m_A^{(2)}}{m_A^{(2)} + m_B^{(2)}} = \frac{n_A^{(2)} + 2n_{A_2}^{(2)}}{n_A^{(2)} + 2n_{A_2}^{(2)} + n_B}$$

$$= \frac{N_A^{(2)} n_t^{(2)} + 2N_{A_2}^{(2)} n_t^{(2)}}{N_A^{(2)} n_t^{(2)} + 2N_{A_2}^{(2)} n_t^{(2)} + N_B^{(2)} n_t^{(2)}}$$

$$= \frac{N_A^{(2)} + 2N_{A_2}^{(2)}}{2N_{A_2}^{(2)} + 1 - N_{A_2}^{(2)}} = \frac{N_A^{(2)} + 2N_{A_2}^{(2)}}{1 + N_{A_2}^{(2)}}. \tag{14}$$

From the last equality in Eq. (14) we see that the solution for $M_A^{(2)}$ is required in terms of the mole fractions of the species A and A_2. At the melting point of pure A, where A associates to a degree β, if we start with m_A moles of the component A, the mole fraction of the species A in equilibrium with solid A is given by

$$N_A^{0(2)} = n_A^{(2)}/(n_A^{(2)} + n_{A_2}^{(2)}). \tag{15}$$

Substituting from Eqs. (11) and (12) for n_{A_2} and n_A, respectively, into Eq. (15) we obtain

$$N_A^{0(2)} = \frac{m_A^{(2)}(1 - \beta)}{m_A^{(2)}(1 - \beta) + (\beta m_{A_2}^{(2)}/2)} = \frac{2(1 - \beta)}{2 - \beta} \tag{16}$$

and

$$N_{A_2}^{0(2)} = \frac{\beta m_A^{(2)}/2}{(\beta m_A/2) + m_A(1-\beta)} = \beta/(2-\beta). \tag{17}$$

Equation (8) may be rewritten as

$$N_{A_2}^{(2)} = \frac{[N_A^{(2)}]^2}{[N_A^{0(2)}]^2} N_{A_2}^{0(2)} e^{-x}. \tag{18}$$

Substituting Eq. (10) for the ratio $[N_A^{(2)}]^2/[N_A^{0(2)}]^2$ in Eq. (18) yields

$$N_{A_2}^{(2)} = e^{-2y} e^{-x} N_{A_2}^{0(2)} = e^{-(2y+x)} N_{A_2}^{0(2)}. \tag{19}$$

Substituting $N_{A_2}^{0(2)}$ from Eq. (17) into Eq. (19) yields

$$N_{A_2}^{(2)} = [\beta/(2-\beta)] e^{-(2y+x)}. \tag{20}$$

Finally, substituting the value of $N_A^{0(2)}$ from Eq. (16) into Eq. (10) leads to

$$N_A^{(2)} = [2(1-\beta)/(2-\beta)] e^{-y}. \tag{21}$$

These last two relationships enable us to expand Eq. (14) in terms of the temperature dependencies of the species A and A_2 as

$$M_A^{(2)} = \frac{2e^{-y}(1-\beta) + 2\beta e^{-(2y+x)}}{2-\beta + \beta e^{-(2y+x)}} = \frac{2e^{-y} - 2e^{-y}\beta + 2\beta e^{-(2y+x)}}{2-\beta(1-e^{-(2y+x)})}$$

$$= \frac{2e^{-y} - 2\beta(e^{-y} - e^{-(2y+x)})}{2-\beta(1-e^{-(2y+x)})}$$

$$= \frac{\left[2\exp\left[-\frac{\Delta H_2^0}{R}\left(\frac{1}{T} - \frac{1}{T_A^0}\right)\right] - 2\beta\left(\exp\left[-\frac{\Delta H_2^0}{R}\left(\frac{1}{T} - \frac{1}{T_A^0}\right)\right]\right.\right.}{2-\beta\left(1 - \exp\left[-\left(\frac{2\Delta H_2^0 + \Delta H_1^0}{R}\right)\left(\frac{1}{T} - \frac{1}{T_A^0}\right)\right]\right)}$$

$$\left.\left.- \exp\left[-\left(\frac{2\Delta H_2^0 + \Delta H_1^0}{R}\right)\left(\frac{1}{T} - \frac{1}{T_A^0}\right)\right]\right)\right]$$

$$\tag{22}$$

In the limit, as $\beta \to 0$, Eq. (22) assumes the simple form

$$M_A^{(2)} \to e^{-y}. \tag{23}$$

Also in the limit, as $\beta \to 1$, Eq. (22) assumes the form

$$M_A^{(2)} \to 2e^{-(2y+x)}/(1 + e^{-(2y+x)}). \tag{24}$$

C. The Liquidus for the Component B

Equation (7) in expanded form, taking into account the assumptions that solid B is pure and pure liquid B consists only of B species, may be written as

$$N_B^{(2)} = \exp\left[-\frac{\Delta H_3^0}{R}\left(\frac{1}{T} - \frac{1}{T_B^0}\right)\right] = e^{-z}. \tag{25}$$

Together with Eqs. (2) and (8), Eq. (25) forms the set of simultaneously applicable equations for the quantitative description of the B species liquidus. The conservation conditions for moles of the component B in terms of moles of the species B is in this instance

$$m_B^{(2)} = n_B^{(2)}, \tag{26}$$

and the conservation relationship for the mole fraction of the component B in terms of the mole fractions of the several species present is given by

$$M_B^{(2)} = N_B^{(2)}/(N_B^{(2)} + N_A^{(2)} + 2N_{A_2}^{(2)}). \tag{27}$$

From Eq. (2) it can be seen that the denominator of Eq. (27) enables restatement of this latter equation as

$$M_B^{(2)} = N_B^{(2)}/(1 + N_{A_2}^{(2)}). \tag{28}$$

The ratio $N_{A_2}^{0(2)}/[N_A^{0(2)}]^2$ in Eq. (8) is a constant at any chosen temperature. Its value may be assessed at the melting point of pure A from Eqs. (16) and (17) and is defined as

$$N_{A_2}^{0(2)}/[N_A^{0(2)}]^2 = C_1 = \beta(2-\beta)/4(1-\beta)^2. \tag{29}$$

Utilizing Eqs. (8) and (29) we may rewrite Eq. (28) as

$$M_B^{(2)} = N_B/(1 + [N_A^{(2)}]^2 C_1 e^{-x}). \tag{30}$$

Since from Eq. (2) we see that

$$N_{A_2}^{(2)} = 1 - N_A^{(2)} - N_B^{(2)}. \tag{31}$$

Then again from Eqs. (8) and (29) we may rewrite Eq. (31) as

$$[N_A^{(2)}]^2 C_1 e^{-x} = 1 - N_A^{(2)} - N_B^{(2)} \tag{32}$$

which is a quadratic in $N_A^{(2)}$ of the form

$$[N_A^{(2)}]^2 C_1 e^{-x} + N_A^{(2)} + (N_B^{(2)} - 1) = 0. \tag{33}$$

Solving for $N_A^{(2)}$ yields

$$N_A^{(2)} = \frac{-1 \pm [1 - 4C_1 e^{-x}(N_B^{(2)} - 1)]^{1/2}}{2C_1 e^{-x}}. \tag{34}$$

Substituting Eq. (34) for N_A in Eq. (30) and the value of N_B from Eq. (25) into the result yields for $M_B^{(2)}$ in terms of the species equivalents.

$$M_B^{(2)} = \frac{e^{-z}}{1 + \left\{ \dfrac{-1 \pm [1 - 4C_1 e^{-x}(e^{-z} - 1)]^{1/2}}{2C_1 e^{-x}} \right\}^2 C_1 e^{-x}}. \tag{35}$$

If β in Eq. (29) is zero, it is seen that the constant $C_1 = 0$ and $M_B^{(2)}$ is indeterminate. However, we can apply the well-known theorem of L'Hospital to evaluate $M_B^{(2)}$ as $\beta \to 0$.

This theorem states that if

$$y = f(x) \quad \text{and} \quad f(x) = g(x)/h(x) \tag{36}$$

then as $x \to 0$

$$\lim_{x \to 0} f(x) = \frac{\lim_{x \to 0} (dg(x)/dx)}{\lim_{x \to 0} (dh(x)/dx)} \tag{37}$$

Equation (35) may be rearranged to give

$$M_B^{(2)} = \frac{e^{-z}}{2 - e^{-z} + \dfrac{2 \pm 2\{1 - 4C_1 e^{-x}(e^{-z} - 1)\}^{1/2}}{4C_1 e^{-x}}}. \tag{38}$$

Thus, taking the derivative of g with respect to C_1, where g is the numerator of the fraction and h is the denominator of the fraction in the denominator of Eq. (38), we obtain

$$dg/dC_1 = \pm [1 - 4C_1 e^{-x}(e^{-z} - 1)^{-1/2}] \cdot [-4e^{-x}(e^{-z} - 1)]. \tag{39}$$

Similarly, taking the derivative of h with respect to C_1 yields

$$dh/dC_1 = 4e^{-x}. \tag{40}$$

The $\lim_{x \to 0} f(x)$ then is

$$[1 - 4C_1 e^{-x}(e^{-z} - 1)^{1/2}] \cdot [-4e^{-x}(e^{-z} - 1)]/4e^{-x}. \tag{41}$$

When $\beta = 0$, Eq. (41) reduces to

$$-4e^{-x}(e^{-z} - 1)/4e^{-x} = 1 - e^{-z}. \tag{42}$$

C. THE LIQUIDUS FOR THE COMPONENT B

Substituting this result in Eq. (38) for the negative root gives for M_B as $\beta \to 0$

$$M_B^{(2)} \to e^{-z}/(2 - e^{-z} - 1 + e^{-z}) = e^{-z}, \tag{43}$$

the normal van't Hoff relationship.

Finally, replacing the symbols in Eq. (38) by their equivalent values yields as the complete answer for $M_B^{(2)}$

$$M_B^{(2)} = \exp\left[-\frac{\Delta H_3^0}{R}\left(\frac{1}{T} - \frac{1}{T_B^0}\right)\right]$$

$$\div \left[2 - \exp\left[-\frac{\Delta H_3^0}{R}\left(\frac{1}{T} - \frac{1}{T_B^0}\right)\right] + \left[\frac{ 2 - 2\left\{1 - \frac{\beta(2-\beta)}{(1-\beta)^2}\exp\left[-\frac{\Delta H_1^0}{R}\left(\frac{1}{T} - \frac{1}{T_A^0}\right)\right] \times \left\{\exp\left[-\frac{\Delta H_3^0}{R}\left(\frac{1}{T} - \frac{1}{T_B^0}\right)\right] - 1\right\}\right\}^{1/2} }{ \frac{\beta(2-\beta)}{(1-\beta)^2}\exp\left[-\frac{\Delta H_1^0}{R}\left(\frac{1}{T} - \frac{1}{T_A^0}\right)\right] } \right] \right]. \tag{44}$$

19

Eutectic Interactions Continued—Effects of End Member Species Interactions on Liquidus Contours

A. INTRODUCTION

Up to this point, the simple cases examined in which homogeneous equilibria were involved were those where end member interaction, as such, did not occur. In each of the instances, the homogeneous equilibrium arose as a consequence of the dissociation of one of the components in the liquid phase. Even where the latter does not occur it is possible to generate similar homogeneous equilibria if, for example, the species derived from both components react with each other form a new species. For the simplest case we may visualize the events

$$A^{(1)} \rightleftarrows A^{(2)}, \tag{1}$$

$$B^{(1)} \rightleftarrows B^{(2)}, \tag{2}$$

$$A^{(2)} + B^{(2)} \rightleftarrows AB^{(2)}. \tag{3}$$

It is to be noted that the homogeneous equilibrium depicted in Eq. (3) is identically that considered in Chapters 16 and 17. Now, however, the effect of this equilibrium on each of the liquidus equilibria, i.e.,

A. INTRODUCTION

Eqs. (1) and (2), is a direct one resulting in the depletion of both A and B species at all concentrations.

The interaction among species derived from the end members is clearly one of the means by which intermediate compound formation may take place. Thus, if the solubility of the generated species AB in the liquid exceeds the solubility limit of solid species AB, the latter precipitate out. The solubility limit at the precise composition AB, of course, defines the melting point of solid AB. If such a situation does indeed obtain, then the resulting subsystem is that treated in Chapter 17. Unfortunately, our state of knowledge is not such as to enable prediction as to whether such solubility limits are exceeded for any particular case in question. Consequently, as with all situations considered in this monograph, it is only "after the fact" occurrences that may be analyzed; e.g., Chapter 17 is a case in point. The treatment in this chapter is useful, however, in synthesizing a sequence of events that can be visualized in artificial systems exhibiting new species formation with a continuous series of stages from complete solubility at all liquid phase temperatures to one with very small solubility at all temperatures. The latter instances will, of course, lead to intermediate compound formation.

The temperature dependent equilibria for Eqs. (1)–(3) are, as expected, given by

$$N_A^{(2)} = \exp\left[-\frac{\Delta H_A^0}{R}\left(\frac{1}{T} - \frac{1}{T_A^0}\right)\right] = e^{-x}, \quad (4)$$

$$N_B^{(2)} = \exp\left[-\frac{\Delta H_B^0}{R}\left(\frac{1}{T} - \frac{1}{T_B^0}\right)\right] = e^{-y}, \quad (5)$$

$$\frac{[N_{AB}^{(2)}]/[N_A^{(2)}][N_B^{(2)}]}{[N_{AB}^{0(2)}]/[N_A^{0(2)}][N_B^{0(2)}]} = \exp\left[-\frac{\Delta H_R^0}{R}\left(\frac{1}{T} - \frac{1}{T_A^0}\right)\right] = e^{-z}, \quad (6)$$

where ΔH_A^0 is the heat of fusion of solid A, ΔH_B^0 is the heat of fusion of solid B, and ΔH_R^0 is the heat of reaction of A and B species. T_A^0 represents the melting point of A and T_B^0 the melting point of B. For convenience, the value of the equilibrium constant, Eq. (6), is assumed known at the melting temperature of A, T_A^0. It is to be realized, however, that an assumption as to its value at any temperature is equally useful.

Note that as neither of the end members undergoes a dissociation or some other phenomenon at its melting point, the liquidus equations assume the simple van't Hoff form. Consequently, the values of the mole fraction of the species A or B at the pure compositions is unity for both the solid and liquid.

The component moles of A and B in the liquid phase in terms of species are given, respectively, by

$$m_A^{(2)} = n_A^{(2)} + n_{AB}^{(2)} = N_A^{(2)} n_t^{(2)} + N_{AB}^{(2)} n_t, \qquad (7)$$

$$m_B = n_B^{(2)} + n_{AB}^{(2)} = N_B^{(2)} n_t^{(2)} + N_{AB}^{(2)} n_t. \qquad (8)$$

In terms of the species mole fractions, the component mole fractions may be written, using the results of Eqs. (7) and (8), as

$$M_A^{(2)} = \frac{m_A^{(2)}}{m_A^{(2)} + m_B^{(2)}} = \frac{(N_A^{(2)} + N_{AB}^{(2)}) n_t}{(N_A^{(2)} + N_{AB}^{(2)} + N_B^{(2)} + N_{AB}^{(2)}) n_t}$$

$$= \frac{N_A^{(2)} + N_{AB}^{(2)}}{N_A^{(2)} + N_B^{(2)} + 2N_{AB}^{(2)}}, \qquad (9)$$

$$M_B^{(2)} = \frac{m_B^{(2)}}{m_A^{(2)} + m_B^{(2)}} = \frac{N_B^{(2)} + N_{AB}^{(2)}}{N_A^{(2)} + N_B^{(2)} + 2N_{AB}^{(2)}}. \qquad (10)$$

Since

$$N_A^{(2)} + N_B^{(2)} + N_{AB}^{(2)} = 1 \qquad (11)$$

the denominator of Eqs. (10) and (11) is given by

$$N_A^{(2)} + N_B^{(2)} + 2N_{AB}^{(2)} = 1 + N_{AB}^{(2)}. \qquad (12)$$

Thus, the component mole-fraction conservation conditions [Eqs. (9) and (10)] may be written

$$M_A^{(2)} = (N_A^{(2)} + N_{AB}^{(2)})/(1 + N_{AB}^{(2)}), \qquad (13)$$

$$M_B^{(2)} = (N_B^{(2)} + N_{AB}^{(2)})/(1 + N_{AB}^{(2)}). \qquad (14)$$

Equations (13) and (14) with the terms $N_X^{(2)}$ expanded as a function of T represent the desired solutions for the A and B liquidi.

As before, the liquidus curves are assumed to be isobarically univariant, e.g., two components coexisting in two phases. Since we must solve for three species unknowns, three independent relationships are required for the analytical description of each liquidus. As before, we will choose the temperature as the independent variable. For the A liquidus, Eqs. (4), (6), and (11) represent an independent set, and for the B liquidus Eqs. (5), (6), and (11) represent an independent set.

Let us assume that at the composition AB and a temperature T_A^0, the equal number of A and B species react to a degree α to form the

B. SOLUTION OF THE A LIQUIDUS

species AB. At this location then, starting with m_A moles of the component A and an equal number of moles of the component B, we have, once equilibrium has been obtained, the following number of moles of the species A and B remaining:

$$n_A^{0(2)} = m_A^{(2)} - \alpha m_A^{(2)} = m_A^{(2)}(1 - \alpha) = m_B^{(2)}(1 - \alpha), \tag{15}$$

$$n_B^{0(2)} = m_B^{(2)} - \alpha m_B^{(2)} = m_B^{(2)}(1 - \alpha) = m_A^{(2)}(1 - \alpha). \tag{16}$$

The number of moles of the species AB formed is given by

$$n_{AB}^{0(2)} = \alpha m_A^{(2)} = \alpha m_B^{(2)}. \tag{17}$$

Consequently, the mole fractions $N_A^{0(2)}$, $N_B^{0(2)}$, and $N_{AB}^{0(2)}$ in Eq. (6) are given by

$$N_A^{0(2)} = \frac{n_A^{0(2)}}{n_A^{0(2)} + n_B^{0(2)} + n_{AB}^{0(2)}} = \frac{1 - \alpha}{(1 - \alpha) + (1 - \alpha) + \alpha}$$

$$= \frac{1 - \alpha}{2 - 2\alpha + \alpha} = \frac{1 - \alpha}{2 - \alpha}, \tag{18}$$

$$N_B^{0(2)} = \frac{1 - \alpha}{2 - \alpha} \tag{19}$$

$$N_{AB}^{0(2)} = \frac{\alpha}{2 - \alpha}. \tag{20}$$

The ratio $N_{AB}^{0(2)}/N_A^{0(2)}N_B^{0(2)}$, of course, represents the value of the equilibrium constant for Eq. (3) at the temperature T_A^0. While the values $N_X^{(2)}$ may change as a function of composition for the entire system, it is again the case that the value of the ratio is constant with composition in the ideal system being treated.

From Eqs. (18)–(20), this ratio is given by Eq. (21)

$$\frac{[N_{AB}^{0(2)}]}{[N_A^{0(2)}][N_B^{0(2)}]} = \frac{\alpha/(2 - \alpha)}{[(1 - \alpha)/(2 - \alpha)][(1 - \alpha)/(2 - \alpha)]} = \frac{\alpha[2 - \alpha]}{[1 - \alpha]^2} = \frac{2\alpha - \alpha^2}{(1 - \alpha)^2}. \tag{21}$$

B. Solution of the A Liquidus

From Eq. (13) it is seen that the solution of the A liquidus requires species dependencies with temperature for both $N_A^{(2)}$ and $N_{AB}^{(2)}$. $N_A^{(2)}$ is already separated in Eq. (4), so what is required is the specification of $N_{AB}^{(2)}$.

Using Eqs. (21), (4), (6), and (11) the above may be accomplished with the solution being

$$N_{AB}^{(2)} = \frac{(2\alpha - \alpha^2)\, e^{-(x+z)}(1 - e^{-x})}{(2\alpha - \alpha^2)e^{-(x+z)} + (1 - \alpha)^2}. \tag{22}$$

$M_A^{(2)}$, Eq. (13), then is given by

$$M_A^{(2)} = \frac{\left[\begin{array}{c}(1-\alpha)^2 \exp\!\left[-\dfrac{\Delta H_A^0}{R}\!\left(\dfrac{1}{T} - \dfrac{1}{T_A^0}\right)\right] \\ + (2\alpha - \alpha^2) \exp\!\left[-\!\left(\dfrac{\Delta H_A^0 + \Delta H_R^0}{R}\right)\!\left(\dfrac{1}{T} - \dfrac{1}{T_A^0}\right)\right]\end{array}\right]}{\left[\begin{array}{c}(1-\alpha)^2 + (2\alpha - \alpha^2) \exp\!\left[-\!\left(\dfrac{\Delta H_A^0 + \Delta H_R^0}{R}\right)\!\left(\dfrac{1}{T} - \dfrac{1}{T_A^0}\right)\right] \\ \times \left\{2 - \exp\!\left[-\dfrac{\Delta H_A^0}{R}\!\left(\dfrac{1}{T} - \dfrac{1}{T_A^0}\right)\right]\right\}\end{array}\right]}. \tag{23}$$

When α, the degree of reaction at the stoichiometry AB, and T_A^0 approach zero, Eq. (23) acquires the form of Eq. (4) as expected.

C. Solution of the B Liquidus

Because of the symmetry of the system, it is obvious that aside from the changes due to the species being different, and the melting points being different, an analogous equation to (23) will be generated for the B liquidus. This is given by

$$M_B^{(2)} = \frac{\left[\begin{array}{c}(1-\alpha)^2 \exp\!\left[-\dfrac{\Delta H_B^0}{R}\!\left(\dfrac{1}{T} - \dfrac{1}{T_B^0}\right)\right] \\ + (2\alpha - \alpha^2) \exp\!\left\{-\left[\dfrac{\Delta H_B^0}{R}\!\left(\dfrac{1}{T} - \dfrac{1}{T_B^0}\right) + \dfrac{\Delta H_R^0}{R}\!\left(\dfrac{1}{T} - \dfrac{1}{T_A^0}\right)\right]\right\}\end{array}\right]}{\left[\begin{array}{c}(1-\alpha)^2 + (2\alpha - \alpha^2) \exp\!\left\{-\left[\dfrac{\Delta H_B^0}{R}\!\left(\dfrac{1}{T} - \dfrac{1}{T_B^0}\right)\right.\right. \\ \left.\left. + \dfrac{\Delta H_R^0}{R}\!\left(\dfrac{1}{T} - \dfrac{1}{T_A^0}\right)\right]\right\} \times \left\{2 - \exp\!\left[-\dfrac{\Delta H_B^0}{R}\!\left(\dfrac{1}{T} - \dfrac{1}{T_B^0}\right)\right]\right\}\end{array}\right]}.$$
(24)

Equation (24) takes into account the fact that ΔH_B^0 and ΔH_R^0 are not related to the same temperature point of reference, i.e., T_A^0.

20

Eutectic Interactions Continued—Effects of Complete Dissociation on Liquidus Contours

A. Introduction

The last general case to be considered in our discussions of elementary eutectic interactions exhibiting slight perturbations, is that case in which one of the participating solids upon melting either dissociates or associates completely. We shall treat the case where in the subsystem, AB–B, the solid AB, upon melting, dissociates completely to form A and B species according to

$$AB^{(1)} \rightleftarrows A^{(2)} + B^{(2)}. \tag{1}$$

As expected, the solid B melting equation is of the simple form previously described a number of times to be

$$B^{(1)} \rightleftarrows B^{(2)}. \tag{2}$$

In order that Eq. (1) be valid, it is a consequence that the equilibrium constant for a hypothetical homogeneous equilibrium among A, B, and AB species be infinite. In other words, the stipulation of Eq. (1) demands that an equation of the type (2) of Chapter 17 not exist. It is, of course, possible that some other homogeneous equilibrium obtains, i.e.,

$$2A^{(2)} + B^{(2)} \rightleftarrows A_2B^{(2)};\qquad(3)$$

but this is more complex than we care to consider. The reader, as an exercise, may care to treat this possibility as well as the case alluded to above where, for example, in the binary system A–B, solid A associates completely upon melting.

Equation (1) represents a second possible pathway by which an intermediate compound may be generated in the binary system A–B. If, for example, the free energy of the system $AB_{(solid)}$–liquid is lower than that of the system, liquid A–B, the solid AB will precipitate if equilibrium obtains. As mentioned in the last chapter, however, the tools necessary for predicting such an occurrence are not available. In a practical sense, many compounds are known which, upon melting or vaporizing, appear to dissociate completely in the phase of higher heat content.

Materials such as CdSe and other II–VI compounds in the vapor phase are present as the free metal atoms and a dimer or higher polymer of the Group VI element. In general, such vaporization processes are of the incongruent type, that is, the solid and vapor phases in equilibrium are of different composition. A drastic example of such incongruency of vaporization is the volatilization of the aqueous portion of a hydrate, i.e.,

$$CuSO_4 \cdot 5H_2O_{(solid)} \rightleftarrows CuSO_4 \cdot 3H_2O_{(solid)} + 2H_2O_{(vapor)}.\qquad(4)$$

The reason the vaporizations of the type being discussed tend to be incongruent can be visualized from the following.

When the hypothetical compound AB vaporizes (or melts), the process may be visualized as proceeding in a sequence of steps, i.e.,

$$AB_{(solid)} \rightleftarrows A_{solid} + B_{solid}\qquad(5)$$

$$A_{solid} \rightleftarrows A_{vapor}\qquad(6)$$

$$B_{solid} \rightleftarrows B_{vapor}.\qquad(7)$$

Normally, the vapor pressures of solid A and solid B will be different so that one or the other will predominate in the vapor. As a consequence, the condensed phase will become richer in the other component. The question of incongruent vaporization will be considered much more fully subsequently, and the implications on the pressure–composition diagram of a binary system will be dealt with then.

A. INTRODUCTION

The equilibrium constant, describing the dissociative melting of AB is given by

$$\frac{[N_A^{(2)}][N_B^{(2)}]}{[N_{AB}^{(1)}]} \Big/ \frac{[N_A^{0(2)}][N_B^{0(2)}]}{[N_{AB}^{0(1)}]} = \exp\left[-\frac{\Delta H_{AB}^0}{R}\left(\frac{1}{T} - \frac{1}{T_{AB}^0}\right)\right]. \quad (8)$$

The superscript 0 terms refer to the pure solid at the melting point. The solid is assumed to be pure so that the terms $N_{AB}^{(1)}$ and $N_{AB}^{0(1)}$ are each unity. Thus, Eq. (8) takes the form

$$\frac{[N_A^{(2)}][N_B^{(2)}]}{[N_A^{0(2)}][N_B^{0(2)}]} = \exp\left[-\frac{\Delta H_{AB}^0}{R}\left(\frac{1}{T} - \frac{1}{T_{AB}^0}\right)\right]. \quad (9)$$

Furthermore, as the dissociation of the pure component at the melting point is complete

$$[N_A^{0(2)}] = [N_B^{0(2)}] = 0.5. \quad (10)$$

Consequently,

$$[N_A^{(2)}][N_B^{(2)}] = (0.5)^2 \exp\left[-\frac{\Delta H_{AB}^0}{R}\left(\frac{1}{T} - \frac{1}{T_{AB}^0}\right)\right],$$

$$= 0.25 \exp\left[-\frac{\Delta H_{AB}^0}{R}\left(\frac{1}{T} - \frac{1}{T_{AB}^0}\right)\right]. \quad (11)$$

In the system under scrutiny we know that the A species are derived only from the component AB. The mole conservation conditions for the component AB then can be expressed in terms of the A species alone since each mole of the component AB gives rise to precisely 1 mole of the species A. Thus, the component A mole conservation condition in terms of species quantities is given simply by

$$m_{AB}^{(2)} = n_A^{(2)} = N_A n_t^{(2)}. \quad (12)$$

The equivalent conservation condition for the component B is not as evident since B species are derived from both components. Since for each mole of A species derived from the component AB, 1 mole of B species is also derived, the total number of moles of the species B, $n_B^{(2)}$ may be defined in terms of the moles of this species due to the component B, $n_B^{*(2)}$, and the moles of this species derived from the component AB, $n_A^{(2)}$, utilizing the equivalency $n_A^{(2)} = n_B^{(2)}$ (from component AB).

Thus,

$$n_B^{(2)} = n_B^{*(2)} + n_A^{(2)}. \quad (13)$$

The mole conservation condition of the component B then may be expressed by the equalities

$$m_B^{(2)} = n_B^{(2)} - n_A^{(2)} = [n_B^{*(2)} + n_A^{(2)}] - n_A^{*(2)} = n_B^{(2)}. \tag{14}$$

The total number of species moles n_t then may be expressed by

$$n_t^{(2)} = n_B^{(2)} + n_A^{(2)} = n_B^{*(2)} + n_A^{(2)} + n_A^{(2)} = n_B^{*(2)} + 2n_A^{(2)}. \tag{15}$$

Dividing n_t and the last equality each by $n_t^{(2)}$, we obtain

$$n_t^{(2)}/n_t^{(2)} = 1 = (n_B^{*(2)} + 2n_A^{(2)})/n_t^{(2)} = N_B^{*(2)} + 2N_A^{(2)} \tag{16}$$

as the constraint on species mole fractions. It is evident also from Eq. (15) that

$$1 = N_B^{(2)} + N_A^{(2)} = N_B^{*(2)} + 2N_A^{(2)}. \tag{17}$$

In terms of bridging the gap between component and species mole fractions, we may generate the following relationships:

$$M_{AB}^{(2)} = m_{AB}^{(2)}/(m_{AB}^{(2)} + m_B^{(2)}). \tag{18}$$

Utilizing Eqs. (12) and (14) together with the relationship

$$n_X = N_X n_t \tag{19}$$

we see that

$$M_{AB}^{(2)} = N_A^{(2)} n_t^{(2)}/(N_A^{(2)} n_t + N_B^{*(2)} n_t) = N_A^{(2)}/(N_A^{(2)} + N_B^{*(2)}). \tag{20}$$

Furthermore, utilizing Eq. (17), we may substitute for $N_B^{*(2)}$ in Eq. (20) its equivalent value to give

$$M_{AB}^{(2)} = N_A^{(2)}/(N_A^{(2)} + 1 - 2N_A^{(2)}) = N_A^{(2)}/(1 - N_A^{(2)}). \tag{21}$$

The sequence of equations leading to the component mole fraction conservation condition for the component B in terms of species mole fractions is

$$M_B^{(2)} = \frac{m_B^{(2)}}{m_{AB}^{(2)} + m_B^{(2)}} = \frac{N_B^{*(2)} n_t^{(2)}}{N_B^{*(2)} n_t^{(2)} + N_A^{(2)} n_t^{(2)}} = \frac{N_B^{*(2)}}{N_B^{*(2)} + N_A^{(2)}}, \tag{22}$$

and again using Eq. (17) we may now substitute for $N_A^{(2)}$ in Eq. (22) to give

$$M_B^{(2)} = \frac{N_B^{*(2)}}{N_B^{*(2)} + [(1 - N_B^{*(2)})/2]} = \frac{2N_B^{*(2)}}{1 + N_B^{*(2)}}. \tag{23}$$

B. THE AB LIQUIDUS

Equations (21) and (23) represent the conservation equations that need be expanded to define the AB and B liquidi.

Finally, before attending to this task it is necessary to provide the analogous equation to Eq. (11) as one of our tools to provide sufficient information to resolve the B liquidus. This is in the form

$$N_B^{*(2)} = \exp\left[-\frac{\Delta H_B^0}{R}\left(\frac{1}{T} - \frac{1}{T_B^0}\right)\right], \tag{24}$$

which is the equilibrium constant for the reaction defined by Eq. (2).

Note that Eq. (24) is defined only in terms of the mole fraction of B species in the liquid obtained from the component B rather than in terms of all the B species. This is so because Eq. (24) relates only B species that are capable of crystallizing along the B liquidus. The quantity of B species derived from AB can only crystallize along the AB liquidus and then only after they have combined with A species. Another way of considering the B liquidus equilibrium is to realize that once T is specified in Eq. (24), the quantity N_B^* is fixed. In other words, Eq. (24) defines the amount of B species that are derived from the component B that are contributed to the liquid phase. The effect of the additional B species, due to the complete dissociation of solid AB upon melting, is taken into account in setting up the mole conservation conditions for the component B using the constraint Eq. (17) to define $M_B^{(2)}$ in Eq. (23).

B. THE AB LIQUIDUS

In defining the equilibrium constant for the component B via Eq. (24), one must realize that the value $N_B^{*(2)}$ is completely determined by ΔH_B^0 and T. Consequently, the difference between $M_B^{(2)}$ and $N_B^{*(2)}$ cannot occur simply as an alteration in the quantity N_B^*. On the other hand, the dissociative melting of the solid AB is represented by the product of two concentrations. Thus, while the product of these two terms is a constant, consistent with the concept of the equilibrium constant, the terms $[N_A^{(2)}]$ and $[N_B^{(2)}]$ are not necessarily equal, and never so in practice when a common species is present. The net effect of additional B species being present is that additional A must be removed from the solution at any particular temperature T. Since A species can only be removed in the form of solid AB, the effect of the additional B species derived from the component B will be to lower the concentration of the component AB in the liquid at each temperature relative to the simple van't Hoff solubility prediction. From the first equality of Eq. (17), the value of $N_B^{(2)}$ in Eq. (11), may be substituted for to give

$$N_A^{(2)}[1 - N_A^{(2)}] = 0.25 \exp\left[-\frac{\Delta H_{AB}^0}{R}\left(\frac{1}{T} - \frac{1}{T_{AB}^0}\right)\right]. \tag{25}$$

Equations (17) and (25), in fact, define the necessary set of independent relationships required to specify the two-species problem along the AB liquidus. As previously, the system AB solid–liquid is isobarically univariant and for convenience the temperature is chosen as the degree of freedom [Eq. (25)]. The latter is a quadratic of the form

$$[N_A^{(2)}]^2 - N_A^{(2)} + 0.25 \exp\left[-\frac{\Delta H_{AB}^0}{R}\left(\frac{1}{T} - \frac{1}{T_{AB}^0}\right)\right] = 0. \tag{26}$$

The solution for $N_A^{(2)}$ then is

$$N_A^{(2)} = \left\{1 \pm \left\{1 - \exp\left[-\frac{\Delta H_{AB}^0}{R}\left(\frac{1}{T} - \frac{1}{T_{AB}^0}\right)\right]\right\}^{1/2}\right\}/2. \tag{27}$$

Substituting this result in Eq. (21) provides the desired temperature dependency of the component AB concentration along the solid AB–liquid univariant equilibrium, i.e.,

$$M_{AB}^{(2)} = \frac{1 \pm \left\{1 - \exp\left[-\frac{\Delta H_{AB}^0}{R}\left(\frac{1}{T} - \frac{1}{T_{AB}^0}\right)\right]\right\}^{1/2}}{1 \mp \left\{1 - \exp\left[-\frac{\Delta H_{AB}^0}{R}\left(\frac{1}{T} - \frac{1}{T_{AB}^0}\right)\right]\right\}^{1/2}}. \tag{28}$$

In order for the discriminant of Eq. (27) to have values greater than or equal to zero, it is necessary that with ΔH_{AB}^0 positive, $T \leqslant T_{AB}^0$. In this instance e^{-x} with x positive will vary between 0 and 1. Furthermore, having satisfied this, and in order that $M_{AB}^{(2)}$ exhibit allowable values, i.e., 0–1, it is necessary that the negative root of the discriminant be chosen. This leads to

$$M_{AB}^{(2)} = \frac{1 - \left\{1 - \exp\left[-\frac{\Delta H_{AB}^0}{R}\left(\frac{1}{T} - \frac{1}{T_{AB}^0}\right)\right]\right\}^{1/2}}{1 + \left\{1 - \exp\left[-\frac{\Delta H_{AB}^0}{R}\left(\frac{1}{T} - \frac{1}{T_{AB}^0}\right)\right]\right\}^{1/2}}. \tag{29}$$

It is interesting now to consider the physical significance of Eq. (29) and how well this agrees with our expectation. We know that when $M_{AB}^{(2)}$

C. THE B LIQUIDUS

achieves a value of 1 at pure AB, the value of $N_A^{(2)}$ should have achieved its maximum value, namely, 0.5. This should occur at $T = T_{AB}^0$. When $T = T_{AB}^0$ it is evident from Eq. (29) that $M_{AB}^{(2)}$ does indeed equal 1. (Incidentally, when $T = 0$, $M_{AB}^{(2)}$ should equal 0 and from Eq. (29) this is also seen to be the case.) From Eq. (27), furthermore, it is seen that when $T = T_{AB}^0$, $N_A^{(2)}$ does acquire its maximum value of 0.5 and that when $T = 0$, $N_A^{(2)}$ does acquire its minimum value of 0.

C. THE B LIQUIDUS

Equation (23) implicitly requires specification of two independent relationships to solve for the B liquidus [note Eq. (22)]. Equations (17) and (24) satisfy this need and the latter may be substituted into Eq. (23) directly to obtain the desired answer, i.e.,

$$M_B^{(2)} = \frac{2\exp\left[-\frac{\Delta H_B^0}{R}\left(\frac{1}{T} - \frac{1}{T_B^0}\right)\right]}{1 + \exp\left[-\frac{\Delta H_B^0}{R}\left(\frac{1}{T} - \frac{1}{T_B^0}\right)\right]}. \quad (30)$$

When $T = T_B^0$, it is seen that $M_B^{(2)} = 1$ and when $T = 0$, $M_B^{(2)} = 0$. We know also that even though $N_B^{*(2)}$ varies between 1 and 0 for these same conditions [from Eq. (24)], the value $N_B^{(2)}$ varies between 1 and 0.5 since at $M_B^{(2)} = 0$ where $N_B^{*(2)} = 0$, $N_B^{(2)} = N_A^{(2)} = 0.5$, and where $N_B^{*(2)} = 1$, $N_A^{(2)} = 0$ and $N_B^{(2)} = N_B^{*(2)}$.

One further point of interest is, that if we compare Eq. (30) with the simple van't Hoff solubility relationship, i.e.,

$$M_B^{(2)} = N_B^{*(2)}, \quad (31)$$

we see that $M_B^{(2)}$, for our case is,

$$M_B^{(2)} \geqslant N_B^*, \quad (32)$$

i.e., in Eq. (30) the denominator is less than or equal to 2; therefore, the numerator is greater than or equal to $\exp(-\Delta H_B^0/R)[(T)^{-1} - (T_B^0)^{-1}]$. This means that in the phase diagram under consideration, the value $M_B^{(2)}$ along the liquidus at any chosen temperature will be greater than that for the simple case. Thus, a given amount of the component B will be more effective due to the additional B species liberated by the AB dissociation.

21

Eutectic Interactions Continued—Graphical Description of the Results of Chapters 16–20

A. INTRODUCTION

In the five preceding chapters, a number of simple assumptions have been made relating to dissociative, associative, and reaction phenomena in which one or both of the primary end-member species were involved. Since even the small perturbations introduced lead to a marked increase in the complexity of the simplest solubility equations, it is at once evident why phenomenological approaches (fugacities and activities) have been used to define systems which cannot be described by the simplest equations. Nonetheless, the development of simple models provides us with a means of examining potential first-order effects on the contours of liquidus curves. These first-order effects will undoubtedly be perturbed in real systems by other phenomena which may occur simultaneously. Interestingly enough, however, as is evident from the results of Chapter 15, a first-order effect need not be swamped by other effects. It was seen in that chapter that when the entropy of fusion was low enough, the occurrence of inflected liquidus behavior could be predicted fairly accurately for real systems, even though other effects served to alter the quantitative aspects of the inflection behavior.

B. TEST CASES

Whether the effects of phenomena such as postulated in Chapters 16–20 are strong enough to override other phenomena is not ascertainable. We will see, however, in dealing with solid solution behavior that the postulating of liquid phase species dimerization provides us with an excellent qualitative description of the phase diagram of a known solid solution, even though the quantitative aspects are in some disagreement. This leads one to believe that, in general, when a system deviates from the very simplest model one can think of, this deviation is attributable largely to a single major phenomenon even if secondary phenomena also occur, and cause additional perturbations.

B. Test Cases

Each of the succeeding sections uses as a reference case for comparison purposes the following simple eutectic interaction.

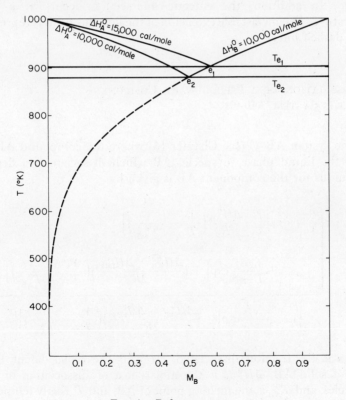

Fig. 1. Reference cases.

For the component AB, B, or C undergoing no perturbation, the liquidus is given simply by

$$M_X^{(2)} = \exp\left[-\frac{\Delta H_X^0}{R}\left(\frac{1}{T} - \frac{1}{T_X^0}\right)\right], \qquad (1)$$

where $M_X^{(2)}$ is the component mole fraction of X in the liquid, ΔH_X^0 is the molar latent heat of fusion and T_X^0 the melting point of the solid having the stochiometry of the component.

ΔH_X^0 has been tested for values of 10,000 and 15,000 cal/mole and T_X^0 has been taken in all cases to be 1000°K.

Figure 1 shows the case where both end members have $\Delta H_X^0 = 10{,}000$ and also where the left hand end member has a ΔH_X^0 value of 15,000 cal/mole. The effect of increased heat of fusion with constant melting point is seen to decrease the solubility of the component in question and shift the eutectic point toward the opposite end of the diagram. In addition, the eutectic intersection occurs at a higher temperature. The metastable extension of the liquidus for B is represented by the dashed line.

C. Dissociation of a Pseudounary Component in the System AB–C

In the system AB–C (see Chapter 16) where the compound AB gives rise in the liquid phase to species AB which dissociate to a degree α, the liquidus for the component AB is given by

$$M_{AB}^{(2)} = \frac{\left[\begin{array}{l}\dfrac{1-\alpha}{1+\alpha}\exp\left[-\dfrac{\Delta H_{AB}^0}{R}\left(\dfrac{1}{T} - \dfrac{1}{T_{AB}^0}\right)\right] \\ \\ + \dfrac{\alpha}{1+\alpha}\exp\left[-\left(\dfrac{\Delta H_{AB}^0 + \Delta H_{Diss}^0}{2R}\right)\left(\dfrac{1}{T} - \dfrac{1}{T_{AB}^0}\right)\right]\end{array}\right]}{\left\{1 - \dfrac{\alpha}{1+\alpha}\exp\left[-\left(\dfrac{\Delta H_{AB}^0 + \Delta H_{Diss}^0}{2R}\right)\left(\dfrac{1}{T} - \dfrac{1}{T_{AB}^0}\right)\right]\right\}} \qquad (2)$$

where α is the fraction dissociated, ΔH_{AB}^0 is the molar latent heat of fusion of solid AB, ΔH_{Diss}^0 is the molar heat of dissociation of liquid AB species, and T_{AB}^0 is the melting point of AB, and T is any temperature on the liquidus of AB.

C. DISSOCIATION WITHOUT A COMMON SPECIES

The liquidus for component C, the undissociated end member, is given by

$$M_C^{(2)} = \frac{\exp\left[-\frac{\Delta H_C^0}{R}\left(\frac{1}{T} - \frac{1}{T_C^0}\right)\right]}{\left[\begin{array}{c}1 - \left\{\left(\frac{\alpha^2}{1-\alpha^2}\right)\exp\left[-\frac{\Delta H_{Diss}^0}{R}\left(\frac{1}{T} - \frac{1}{T_{AB}^0}\right)\right]\right. \\ - \left\{\left(\frac{\alpha^2}{1-\alpha^2}\right)^2 \exp\left[-\frac{2\Delta H_{Diss}^0}{R}\left(\frac{1}{T} - \frac{1}{T_{AB}^0}\right)\right]\right. \\ + \left(\frac{\alpha^2}{1-\alpha^2}\right)\exp\left[-\frac{\Delta H_{Diss}^0}{R}\left(\frac{1}{T} - \frac{1}{T_{AB}^0}\right)\right] \\ \left. \times \left\{1 - \exp\left[-\frac{\Delta H_C^0}{R}\left(\frac{1}{T} - \frac{1}{T_C^0}\right)\right]\right\}\right\}^{1/2}\right\}\end{array}\right]}. \quad (3)$$

where ΔH_C^0 is the heat of fusion of C, T_C^0 is the melting point of C, and ΔH_{Diss}^0 and T_{AB}^0 coincide with the definitions given under Eq. (2).

Solutions were obtained for this system with ΔH_{AB}^0 and ΔH_C^0 each equal to 10,000 cal/mole. The molar heat of dissociation for the process

$$AB^{(2)} \rightleftarrows A^{(2)} + B^{(2)} \quad (4)$$

was chosen as $\Delta H_{Diss}^0 = 5000$ cal/mole and liquidus curves were plotted for $\alpha = 0.4$ and 0.9995. The latter value, of course, represents the case where dissociation of the AB species at the melting point temperature (see Chapter 16) of solid AB is almost complete. The data are shown in Fig. 2. In order to define the upper and lower limits of the molar heats associated with the fusion and dissociation processes, the reference cases plotted in Fig. 1 (Section B) are also shown. Thus, as the fraction dissociated α approaches 1, the total latent heat involved in the double process, melting followed by dissociation, approaches 15,000 cal/mole. As the value of α approaches 0, the evolved heat approaches 10,000 cal/mole, and the present case approaches that specified in Section B.

In Fig. 2, the top liquidi represent the nondissociative reference case with 15,000 cal/mole latent heats of fusion. The bottom liquidi represent the nondissociative case where the molar heats of fusion are each 10,000 cal/mole. It is seen that, with α of the AB species equal to 0.4, both the AB and C liquidi are perturbed. These liquidus curves both exhibit decreased solubility of the respective solids. At $\alpha = 0.9995$, the AB and C liquidi are further perturbed with further decrease in solubility of each of the end member solids.

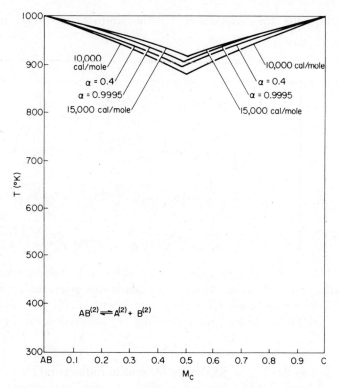

FIG. 2. Dissociation of AB without a common species effect.

What is most evident, however, is that the effect of dissociation in the absence of a common species effect is not very great. It has been a commonly expressed view in the literature, that a liquid phase dissociative phenomenon is signified on a liquidus by the development of a broad melting maxima and conversely, the lack of dissociation is typified by a sharp liquidus ascent to the melting point. As seen from Fig. 2, even drastic dissociation, i.e., $\alpha \rightarrow 1$, does not appear to lead to a flattening of the liquidus in the vicinity of the melting point. In all cases, aside from relative slope, the curves are sharply ascending. Notably, the case for $\alpha = 0.9995$ does not result in as much solubility decrease as a simple melting having a ΔH_{fusion} value of 15,000 cal/mole. A further feature of interest in Fig. 2 is that the effects of dissociation are such as to cause a displacement of the eutectic (about the same amount for very different values of α) toward the dissociating end member. As we shall see in the next section, liquidus curve flattening does occur in a dissociative process if a common species effect is present.

C. DISSOCIATION WITHOUT A COMMON SPECIES

$$M_{AB}^{(2)} = \left[\left(\frac{1-\alpha}{1+\alpha}\right)\exp\left[-\frac{\Delta H_{AB}^0}{R}\left(\frac{1}{T}-\frac{1}{T_{AB}^0}\right)\right] + 1 - \left\{\left\{1 - \left(\frac{1-\alpha}{1+\alpha}\right)\exp\left[-\frac{\Delta H_{AB}^0}{R}\left(\frac{1}{T}-\frac{1}{T_{AB}^0}\right)\right]\right\}^2 - \frac{4\alpha^2}{(1+\alpha)^2}\exp\left[-\frac{(\Delta H_{AB}^0 + \Delta H_{Diss}^0)}{R}\left(\frac{1}{T}-\frac{1}{T_{AB}^0}\right)\right]\right\}^{1/2}\right]$$

$$1 + \left(\frac{1-\alpha}{1+\alpha}\right)\exp\left[-\frac{\Delta H_{AB}^0}{R}\left(\frac{1}{T}-\frac{1}{T_{AB}^0}\right)\right] + \left\{\left\{1 - \left(\frac{1-\alpha}{1+\alpha}\right)\exp\left[-\frac{\Delta H_{AB}^0}{R}\left(\frac{1}{T}-\frac{1}{T_{AB}^0}\right)\right]\right\}^2 - \frac{4\alpha^2}{(1+\alpha)^2}\exp\left[-\frac{(\Delta H_{AB}^0 + \Delta H_{Diss}^0)}{R}\left(\frac{1}{T}-\frac{1}{T_{AB}^0}\right)\right]\right\}^{1/2}\right]$$

(5)

$$M_B^{(2)} = \left[2(1-\alpha^2)\exp\left[-\frac{\Delta H_B^0}{R}\left(\frac{1}{T}-\frac{1}{T_B^0}\right)\right]\right]$$

$$(1-\alpha^2)\left\{\exp\left[-\frac{\Delta H_B^0}{R}\left(\frac{1}{T}-\frac{1}{T_B^0}\right)\right] + 2\right\} + 2\alpha^2\exp\left[-\frac{\Delta H_{Diss}^0}{R}\left(\frac{1}{T}-\frac{1}{T_{AB}^0}\right)\right]$$

$$-\left\{\left\{(1-\alpha^2)\exp\left[-\frac{\Delta H_B^0}{R}\left(\frac{1}{T}-\frac{1}{T_B^0}\right)\right] + 2\alpha^2\exp\left[-\frac{\Delta H_{Diss}^0}{R}\left(\frac{1}{T}-\frac{1}{T_{AB}^0}\right)\right]\right\}^2\right.$$

$$\left. - 4\left\{(1-\alpha^2)\alpha^2\exp\left[-\frac{\Delta H_{Diss}^0}{R}\left(\frac{1}{T}-\frac{1}{T_{AB}^0}\right)\right]\exp\left[-\frac{\Delta H_B^0}{R}\left(\frac{1}{T}-\frac{1}{T_B^0}\right)\right] - 1\right\}\right\}^{1/2}\right]$$

(6)

D. Dissociation with a Common Species Effect

The system to be discussed in this section is represented by AB–B. With AB dissociating, B species are provided by each of the end members. The liquid phase species AB again dissociates according to Eq. (4). The liquidus equations for components AB and B, respectively, are given by Eqs. (5) and (6) [see p. 209].

In these equations ΔH^0_{AB} is the heat of fusion of AB, ΔH_B^0 the heat of fusion of B, ΔH^0_{Diss}, the heat of dissociation of AB, T^0_{AB} is the melting point of AB, T_B^0 the melting point of B, and α is the degree of dissociation of AB species at the melting point of AB. T^0_{AB} and T_B^0 were both taken as 1000°K, ΔH^0_{AB} and ΔH_B^0 were each assumed to have a value of 10,000 cal/mole ΔH^0_{Diss} was taken as 5,000 cal/mole, and α as 0.4 and 0.9995.

As in Section C, the results were compared with test cases of T^0_{AB} and $T_B^0 = 1000°$K and $\Delta H^0_{AB} = 10,000$ and 15,000 cal/mole. The data are plotted in Fig. 3. It is immediately evident, that when dissociation of

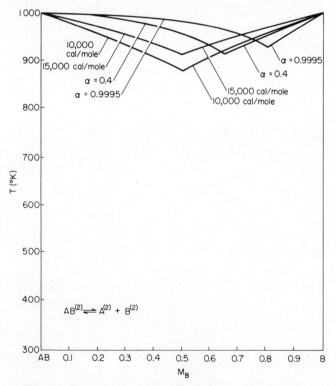

Fig. 3. Dissociation of AB with a common species effect.

E. ASSOCIATION EFFECTS

one of the species occurs with a resultant common species effect, the effect on the liquidi is severe. Now, the flattening of the liquidus curve of the end member whose liquid phase species undergoes dissociation is marked. Increasing dissociation causes increased liquidus flattening and very pronounced decrease in solubility. It is seen that the dissociation liquidus curves for AB lie above both of the test cases rather than between them. Since the compound AB will form in the system A–B, it would appear that the observation of liquidus flattening in intermediate compound forming systems would not be uncommon. On the other hand, if AB interacts with C, its dissociation would not lead to pronounced perturbation of the liquidus.

The liquidi for the nondissociating end member do not tend to flatten, but do exhibit the phenomenon of lying below the test case curves. Thus, the solubility of the nondissociating end member is increased. The location of the eutectics is notably displaced away from the pure dissociative end member. Thus, while a dissociative phenomenon without an attending common species effect creates rather minor perturbations of simple melting curves, the same phenomenon in the presence of a common species leads to drastic changes. A means of detecting such dissociation is via the opposite effects on each of the end members with respect to direction of displacement from simple liquidus behavior.

E. Association of One of the Liquid-Phase Species

The system being considered is A–B where the solid A, upon melting, is present as the species A. The latter associate to a degree β to give dimers according to

$$2A^{(2)} \rightleftarrows A_2^{(2)}. \tag{7}$$

The A and B liquidus equations are given by

$$M_A^{(2)} = \frac{\left[2\exp\left[-\frac{\Delta H_A^0}{R}\left(\frac{1}{T} - \frac{1}{T_A^0}\right)\right] - 2\beta\left\{\exp\left[-\frac{\Delta H_{Assoc}^0}{R}\left(\frac{1}{T} - \frac{1}{T_A^0}\right)\right]\right. \\ \left. - \exp\left[-\left(\frac{2\Delta H_{Assoc}^0 + \Delta H_A^0}{R}\right)\left(\frac{1}{T} - \frac{1}{T_A^0}\right)\right]\right\}\right]}{2 - \beta\left\{1 - \exp\left[-\left(\frac{2\Delta H_{Assoc}^0 + \Delta H_A^0}{R}\right)\left(\frac{1}{T} - \frac{1}{T_A^0}\right)\right]\right\}} \tag{8}$$

$$M_B^{(2)} = \exp\left[-\frac{\Delta H_B^0}{R}\left(\frac{1}{T} - \frac{1}{T_B^0}\right)\right]$$

$$\div \left[\begin{array}{l} 2 - \exp\left[-\dfrac{\Delta H_B}{R}\left(\dfrac{1}{T} - \dfrac{1}{T_B^0}\right)\right] \\ + \dfrac{\left[2 - 2\left\{1 - \dfrac{\beta(2-\beta)}{(1-\beta)^2}\exp\left[-\dfrac{\Delta H_A^0}{R}\left(\dfrac{1}{T} - \dfrac{1}{T_A^0}\right)\right]\right.\right.}{\dfrac{\beta(2-\beta)}{(1-\beta)^2}\exp\left[-\dfrac{\Delta H_A}{R}\left(\dfrac{1}{T} - \dfrac{1}{T_A^0}\right)\right]} \\ \phantom{+\dfrac{\left[2 - 2\left\{1 - \right.\right.}{}\left.\left.\times\left\{\exp\left[-\dfrac{\Delta H_B^0}{R}\left(\dfrac{1}{T} - \dfrac{1}{T_B^0}\right)\right] - 1\right\}\right\}\right]^{1/2} \end{array}\right]$$

(9)

where ΔH_A^0 and T_A^0 are the molar heat of fusion, 10,000 cal/mole, and melting point, 1000°K, of A, respectively; ΔH_B^0 and T_B^0 are the molar heat of fusion, 10,000 cal/mole, and melting point, 1000°K, of B, respectively; $\Delta H_{\text{Assoc}}^0$ is the molar heat of association, 5000 cal/mole, of A species; and β is the degree of association of A species at the melting point of A. Values of 0.4 and 0.9995 were employed.

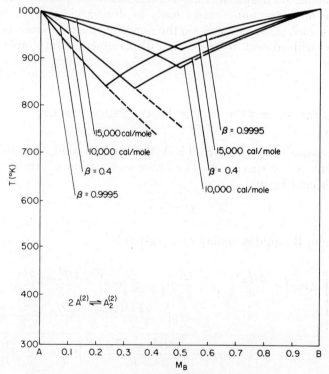

FIG. 4. Association of one of the liquid-phase species.

F. Reaction between Liquid-Phase Species

The results are presented in Fig. 4. Again, as with the dissociative case exhibiting a common species effect, the perturbations of the simple liquidus behavior are striking. Flattening of the liquidi does not, however, occur. In fact, the solubility of the associating end member is greatly enhanced and all associative curves lie below those of the simple liquidi with the eutectics displaced toward the associative end member A. The B liquidus is only slightly perturbed, being only slightly displaced from the test cases. An interesting feature of the A liquidi is the slight tendency toward inflection exhibited with increasing association.

F. Reaction between Liquid-Phase Species

The next to final case to be considered here is where, in the system A–B, the liquid-phase species A and B react to form the new species AB. The fraction of either A or B that reacts is denoted by α. The liquidus equation applicable to A is given by

$$M_A = \frac{\left[\begin{array}{l}(1-\alpha)^2 \exp\left[-\frac{\Delta H_A^0}{R}\left(\frac{1}{T} - \frac{1}{T_A^0}\right)\right] \\ + (2\alpha - \alpha)^2 \exp\left[-\left(\frac{\Delta H_A^0 + \Delta H_{React}^0}{R}\right)\left(\frac{1}{T} - \frac{1}{T_A^0}\right)\right]\end{array}\right]}{\left[\begin{array}{l}(1-\alpha)^2 + (2\alpha - \alpha^2)\exp\left[-\left(\frac{\Delta H_A^0 + \Delta H_{React}^0}{R}\right)\left(\frac{1}{T} - \frac{1}{T_A^0}\right)\right] \\ \times \left\{2 - \exp\left[-\frac{\Delta H_A^0}{R}\left(\frac{1}{T} - \frac{1}{T_A^0}\right)\right]\right\}\end{array}\right]}$$

(10)

and that for B by the symmetrical equation

$$M_B = \frac{\left[\begin{array}{l}(1-\alpha)^2 \exp\left[-\frac{\Delta H_B^0}{R}\left(\frac{1}{T} - \frac{1}{T_B^0}\right)\right] \\ + (2\alpha - \alpha^2) \exp\left\{-\left[\frac{\Delta H_B^0}{R}\left(\frac{1}{T} - \frac{1}{T_B^0}\right) + \frac{\Delta H_{React}^0}{R}\left(\frac{1}{T} - \frac{1}{T_A^0}\right)\right]\right\}\end{array}\right]}{\left[\begin{array}{l}(1-\alpha)^2 + (2\alpha - \alpha^2) \\ \times \exp\left\{-\left[\frac{\Delta H_B^0}{R}\left(\frac{1}{T} - \frac{1}{T_B^0}\right) + \frac{\Delta H_{React}^0}{R}\left(\frac{1}{T} - \frac{1}{T_A^0}\right)\right]\right\} \\ \times \left\{2 - \exp\left[-\frac{\Delta H_B^0}{R}\left(\frac{1}{T} - \frac{1}{T_B^0}\right)\right]\right\}\end{array}\right]}.$$

(11)

The degree of reaction α is specified at the melting point of A which accounts for the difference in appearance between Eqs. (10) and (11). The values ΔH_A^0 and ΔH_B^0 are the heats of fusion of A and B, respectively, at their melting points T_A^0 and T_B^0, and ΔH_{React}^0 is the molar heat of reaction.

Solutions were found for Eqs. (10) and (11) with ΔH_A^0 and ΔH_B^0 = 10,000 cal/mole and ΔH_{React}^0 = 5,000 cal/mole, T_A^0 and T_B^0 = 1000°K.

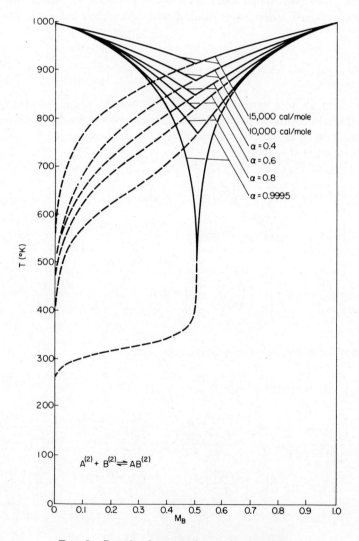

FIG. 5. Reaction between the species A and B.

G. COMPLETE DISSOCIATION EFFECTS

The results are plotted in Fig. 5 for values of $\alpha = 0.4, 0.6, 0.8$, and 0.9995. The B liquidus curves are extended metastably beyond their eutectic intersections to show the increased tendency toward inflection as α tends toward one. All of the liquidi lie below those of the test cases, and, as expected, because of the symmetry in the latent heats and melting points, the eutectics occur at $N_B = 0.5$. This represents the only case thus far considered in which the liquidi for both components lie below the test case values and this may be a guide to detecting species reaction.

G. EFFECTS OF COMPLETE DISSOCIATION ON LIQUIDUS CONTOURS

The final case to be considered is where, in the reaction AB–B, the pseudounary compound AB upon melting dissociates completely. Consequently, no homogeneous equilibrium constant need be involved in solving for the variation of $M_{AB}^{(2)}$ and $M_B^{(2)}$ as a function of T.

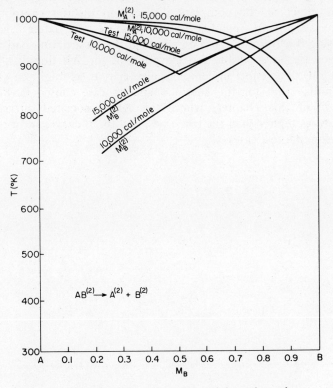

FIG. 6. Complete dissociation of one of the end members.

The analytical equations for the AB and B liquidi are given by

$$M_{AB}^{(2)} = \frac{1 - \left\{1 - \exp\left[-\frac{\Delta H_{AB}^0}{R}\left(\frac{1}{T} - \frac{1}{T_{AB}^0}\right)\right]\right\}^{1/2}}{1 + \left\{1 - \exp\left[-\frac{\Delta H_{AB}^0}{R}\left(\frac{1}{T} - \frac{1}{T_{AB}^0}\right)\right]\right\}^{1/2}}, \quad (12)$$

$$M_B^{(2)} = \frac{2\exp\left[-\frac{\Delta H_B^0}{R}\left(\frac{1}{T} - \frac{1}{T_B^0}\right)\right]}{1 + \exp\left[-\frac{\Delta H_B^0}{R}\left(\frac{1}{T} - \frac{1}{T_B^0}\right)\right]}. \quad (13)$$

As in the previous sections, T–M_X representations are given for the above in addition to those for test cases where ΔH_A^0 and $\Delta H_B^0 = 10$ or 15 kcal/mole. For the complete dissociation case considered, ΔH_{AB}^0 and ΔH_B^0 values of 10 or 15 kcal/mole were also examined, with T_{AB}^0 and T_B^0 both equal to 1000°K. These are shown in Fig. 6. It is seen that the effect of the complete dissociation on the contour of the nondissociating end member is to increase the latter's solubility (depressing the B liquidus). The effect on the dissociating end member is to decrease its solubility (an elevation of the AB liquidus, thereby flattening it). The effect in complete dissociation, as might be expected, is not unlike that for the common species case shown in Fig. 3, in Section D of this chapter, where the AB dissociated curves lie at temperatures higher than the test cases, and the B curves lie at temperatures below the test cases.

22

Eutectic Interactions Continued—Single Compound Formation in Binary Systems and Coordinate Transformations

A. Introduction

Frequently, a binary system A–B gives rise in its composition interval to one or more binary compounds having the generalized stoichiometry A_xB_y. Depending on the nature of these compounds, a subsystem within the primitive system A–B may or may not be treated as binary itself. In this chapter we will consider instances in the temperature composition diagram where either of the subsystems $A-A_xB_y$ or A_xB_y-B are separable and amenable to consideration as binary interactions.

B. A Single Compound Is Generated

In Fig. 1, the temperature–composition diagram is presented for the system A–B in which the single compound AB is formed. It is to be noted that the melting point curve or liquidus of AB reaches a maximum and decreases on either side of this maximum giving rise to eutectics e_1 and e_2 where the AB liquidi intersect those of the end members A

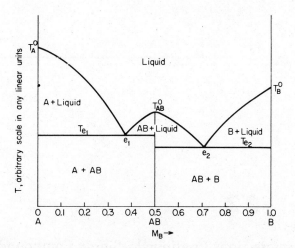

FIG. 1. Formation of the compound AB in the binary system A–B in which pressure is constant.

and B, respectively. The curve $T_A^0-e_1$ represents the liquidus of the solid A, designated in the diagram by the symbol A. Along this curve two phases are in equilibrium, i.e., crystals of pure A and a liquid. This system is isobarically univariant, as can be seen by application of the phase rule. Starting at the composition $M_B = 0.5$, which coincides with the composition AB, it is seen that an analogous curve to $T_A^0-e_1$, namely $T_{AB}^0-e_1$ is present. The latter represents the liquidus for the solid AB. At the dividing line AB it is seen that to the right of AB a similar binary eutectic diagram is generated in which the liquidi $AB-e_2$ and $B-e_2$ are present. It is to be noted that the construction to either side of AB has the appearance of any simple binary eutectic interaction and, indeed, may be treated as such after certain scaling procedures have been attended to. The latter will be considered shortly.

The nature of the A–AB or AB–B constructions are such that each of the participating solids is assumed to be congruently melting over its entire range of existence. This congruency of behavior is evidenced by the fact that from the melting point of each of the solids to the eutectic intersections of their liquidi, the same solid is in equilibrium with liquid. If, for example, solid AB melts incongruently at a particular temperature, two factors would have to be taken into account and these would lead to an anomaly in the AB liquidus. First, above the temperature of incongruency the composition of the solid would no longer be describable by the stoichiometry AB, and this would require suitable representation. Second, as a new solid is in equilibrium with liquid, the liquidus

B. A SINGLE COMPOUND IS GENERATED

would exhibit a change in slope due to the difference in the heat of the process. Such cases will be treated subsequently.

Under equilibrium conditions, as the temperature of the eutectic is reached in cooling, the situation occurs in which two solids and a liquid are simultaneously present. Thus, in the subdiagram A–AB, at the temperature T_{e_1}, solids A and AB coexist with liquid at the temperature T_{e_1}. The system becomes isobarically invariant and a further decrease in temperature cannot occur until one of the coexisting phases vanishes. The phase whose quantity has been decreasing with decreasing temperature is liquid. Obviously, then, the phase that will disappear at the eutectic is the liquid phase. Below this temperature pure solids A and AB may coexist simultaneously.

As stated earlier, if a scaling transformation is performed, then the subsystems A–AB or AB–B may be treated as simple eutectic interactions. As a consequence, the idealized mathematical models presented in earlier chapters may be tested. Graphically, however, we may trace the sequence of cooling or heating experiments on a qualitative level without recourse to changing coordinates.

In Fig. 2 such a tracing is presented for the precipitation of the solid A. Completely analogous descriptions may be made for any of the other solids which exhibit fields of existence.

Let us assume we begin our experiment with a melt at the temperature T_{melt} containing approximately 0.1 mole fraction of the component B and 0.9 mole fraction of the component A relative to the complete system A–B and proceed to cool it. When the point A is reached at

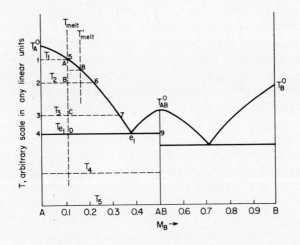

FIG. 2. Cooling effects in a system exhibiting compound formation.

the temperature T_1, the solubility limit of solid A in liquid is reached, namely, the liquidus temperature for that starting composition is reached. At this point, the first traces of pure solid A begin to settle out. The molecular ratio, liquid to solid is very large at this point since pure A has just begun to deposit. If we cool to the temperature T_2, reaching the point B, the tie lines 2–B and B–6 enable us to define the compositions of the solid in equilibrium with liquid simply by inspection. Since the tie line 2–B intersects the pure A axis, it is evident that the solid has the composition A. The liquid, on the other hand, has the composition coinciding with point 6 ($\sim$0.22 mole fraction B) which is much richer in the component B than our starting composition. This is reasonable since if component A is extracted from solution via settling out of a solid having this stoichiometry, the liquid must become richer in the second component. In cooling from temperature T_1 to temperature T_2, it is noted that at any intervening temperature pure solid A is in equilibrium with liquid. The composition of the liquid which follows the curve segment 5–6 becomes richer in component B, its exact composition being fixed by the curve 5–6. If the phase diagram of the system A–B were being deduced, the shape of the diagram could be determined by starting with a series of samples containing different compositions, melting each of these samples in turn, and by some means or other that we will describe in Chapter 36 determining the temperature at which solid first appears upon cooling. For the melt of composition $M_B \simeq 0.1$, this first appearance of solid would occur at temperature T_1 and the coordinate $T = T_1$, $M_B = 0.1$ would be marked as the liquidus for the composition 0.1. If we begin with a sample having the composition $M_B = 0.15$ at temperature T', melt and cool, the first solid will precipitate at point 8 coinciding with the temperature T, and this coordinate would be noted on the graph for a sufficient number of samples to enable plotting of complete curves unambiguously. Although the interpretation of the data points obtained may be complicated by a variety of factors which shall be considered in detail later on, the procedure outlined is applicable, in general, to the resolution of phase diagrams.

Returning to the tracing of the cooling behavior of the starting sample having the composition $M_B \simeq 0.1$, if we cool to the temperature T_3, the tie line at this temperature is seen to intersect the solid composition curve at pure A and the liquid composition curve at point 7 ($M_B \simeq 0.33$). Again, the liquid composition in the temperature interval T_2–T_3 has followed along the curve 6–7, becoming richer in component B as the liquid is depleted of component A via the settling out of pure solid having the composition A.

B. A SINGLE COMPOUND IS GENERATED

Finally, after continued cooling to temperature T_{e_1} during which the solid composition follows the vertical curve A–$T_A{}^0$, coinciding with a constant solid composition, and the liquid composition varies along 7–e_1, the eutectic temperature is attained. From the increase in relative size of the tie lines to the right of the curve T_{melt}–0.1 with decreasing temperature we see qualitatively that the ratio of solid-to-liquid continues to increase as expected since pure A is settling out. By the time T_{e_1} is reached, the equilibrium mixture has a relatively small amount of liquid and this liquid has the composition e_1 ($M_B \cong 0.36$). At this temperature, the solubility limit of the solid AB is reached and in addition to solid A being present solid AB appears. The temperature of the sample will not decrease further, even if the furnace in which the sample is contained continues to cool, until the condition of invariance is changed by disappearance of the liquid phase. Now, however, the liquid vanishes not by the precipitation of solid A alone, but by the simultaneous precipitation of pure crystals of solid A and pure crystals of solid AB. This mixed precipitate is called a eutectic precipitate, and the previously precipitated solid A is embedded in this eutectic matrix.

It is to be noted that in the starting composition interval $M_B = 0$ to $M_B = 0.5$, independent of the initial sample composition and independent of whether pure solid A or pure solid AB was precipitating at temperatures above T_{e_1}, the final solidification will always occur at T_{e_1} and the final liquid composition just prior to complete solidification will have the composition e_1 ($M_B \cong 0.36$). This is necessary since no degrees of freedom are available in a binary system three-phase coexistence under isobaric conditions. In other words, the temperature and composition at which such an occurrence may be generated is unique. On the other hand, in the binary two-phase existence, e.g., for solid + liquid, one degree of freedom is available. Thus, if we specify the composition of the system we implicitly specify that there exists a temperature at which the solid in question may coexist with liquid. This temperature is called the liquidus temperature for that particular system composition. Conversely, if we specify a temperature at which a two-phase coexistence is known to take place, we simultaneously fix the composition of the system as a whole, if this temperature coincides with the first appearance of solid, or of the liquid and solid phase compositions separately if this temperature is lower than the temperature of first precipitation for a particular total system composition. In either event, there exists only a single liquidus curve for a particular solid–liquid coexistence interval and only a single eutectic point for an entire binary system or subsystem. Since the temperature at which the eutectic occurs is that at which, upon cooling, all liquid vanishes or which, upon heating, the first liquid appears, the eutectic temperature is frequently called the solidus temperature. The line $4-e_1-9$ in Fig. 2 is not an actual part of the phase diagram but rather the eutectic tie-line for all compositions in the interval $M_B = 0-0.5$. It indicates that independent of the starting composition in the subsystem A–AB, the bounding liquid compositions along either of the liquidi is $M_B = 0.36$, and that the final freezing for all compositions in this subsystem occurs at temperature T_{e_1}.

Finally, if we trace the cooling from temperature T_{e_1} through T_4 to T_5, we see that at any and all intervening temperatures, the tie-lines we construct intersect the vertical lines $T_A{}^0$–A or T_{AB}^0–AB. This implies that below T_{e_1}, the compositions of the coexisting solids do not vary. Many cases arise, and in the limit it is true for all cases that such a situation does not prevail. Thus, if there exists some slight or major solubility of the component A in the solid AB, or vice versa, the intersection of the tie lines will not be represented as in Fig. 2. Such solid solubility need not be constant with temperature and variations would have to be taken into account. We will not pursue this point now, leaving it until later when it may be discussed in the context of solid solubility.

C. Scaling Factors and Coordinate Transformations

In Chapter 4, we presented the purely geometrical concept of the lever arm or fulcrum principle LAP which enables us, via tie-line construction around a fulcrum centered at a starting system composition (an isopleth), to estimate the molecular or mass solid-to-liquid ratio at a particular temperature, and knowing the starting mass to estimate the actual mass of solid and liquid at any particular temperature. In the present chapter, while alluding to this principle in considerations of cooling behavior within a subsystem, we have purposely avoided its quantitative application.

The reason for avoiding a quantitative use of the lever arm principle is that in a system defined in terms of mole fractions of the end members (as is the case for Figs. 1 and 2), in which subsystems arise, the transformation of compositions from 0–1 mole fraction of the end members A–B into composition terms 0–1 mole fraction for either the subsystem A–AB or AB–B is not a linear one. Consequently, if a molecular solid/liquid fraction is determined directly from tie arm lengths, these refer to counting based on A and B stoichiometries, and do not directly give this ratio in terms of AB plus either A or B stoichiometries.

In general, as detailed in Chapter 14, if we desire to apply the lever arm principle to the practical problem of estimating masses of coexisting

C. COORDINATE TRANSFORMATIONS

solids and liquids, it is more convenient to depict the composition axis in terms of mass fractions. If mole quantities are required, then the composition axis should be depicted in mole-fraction terms. In the latter instance, it is a further convenience, when subsystems are present, to define each subsystem in terms of the mole fraction range 0–1 rather than as part of the unary end-member mole-fraction range 0–1. In the mass fraction case this is not necessary since a simple linear relationship exists. The preceding discussion should become more evident from the following.

1. *Linear Coordinate Transformations*

Consider the subsystem A_3B–B which is part of the larger primitive binary system A–B. Let the mass fraction of B in the subsystem A_3B–B be denoted by the symbol $f_B{}^s$ and the mass fraction of B in the primitive system A–B denoted by the symbol $f_B{}^p$. Correspondingly, in the subsystem A_3B–A let us denote the total mass fraction of A by the symbol $f_A{}^s$ and in the primitive system A–B let us denote the mass fraction of A by $f_A{}^p$. Let us assume the molecular weight of B = mol wt B mass units, the molecular weight of A = mol wt A mass units and the molecular weight of A_3B = mol wt A_3B mass units and the masses of A, B, and A_3B present in any mixture by g_A, g_B, and g_{A_3B}, respectively.

In the subsystem A_3B–B, the mass fraction of B, $f_B{}^s$ is given by

$$f_B{}^s = g_B/(g_B + g_{A_3B}) = g_B/g_{\text{total}}. \tag{1}$$

Starting with some original arbitrary mass of the pseudounary component A_3B, g_{A_3B}, the mass of the component B that must be added to this starting mass g_{A_3B} to provide a desired mass fraction of B, $f_B{}^s$ is obtained by solving Eq. (1) for g_B. Thus,

$$g_B = [f_B{}^s/(1 - f_B{}^s)] g_{A_3B} = f_B{}^s \cdot g_{\text{total}}. \tag{2}$$

Let us now examine the primitive system A–B and attempt to define a relationship between the mass fraction of B in that system, i.e., $f_B{}^p$ and the mass fraction of B in the subsystem A_3B–B, i.e., $f_B{}^s$.

In g_{A_3B} mass units of A_3B we have present (mol wt B/mol wt A_3B) $\cdot g_{A_3B}$ mass units of B. In addition, in any mixture of A_3B and B, we have present g additional mass units of B as specified in Eq. (1). The total mass of B, independent of its mode of aggregation, in terms of masses and mass fractions is given by

$$\begin{aligned}g_B^{\text{total}} &= (\text{mol wt B/mol wt } A_3B) g_{A_3B} + g_B \\ &= (\text{mol wt B/mol wt } A_3B) g_{A_3B} + f_B{}^s \cdot g_{\text{total}}.\end{aligned} \tag{3}$$

The total mass of such a system is given simply by

$$g_{\text{total}} = g_{A_3B} + g_B. \tag{4}$$

The mass fraction of B relative to the system A–B then is the total amount of B present, g_B^{total}, divided by the total mass as defined in Eq. (4). This is

$$f_B^p = \frac{(\text{mol wt B/mol wt A}_3\text{B}) \cdot g_{A_3B} + g_B}{g_{A_3B} + g_B}$$

$$= \frac{(\text{mol wt B/mol wt A}_3\text{B}) \cdot g_{A_3B} + f_B^s \cdot g_{\text{total}}}{g_{\text{total}}}. \tag{5}$$

Since $g_{A_3B}/g_{\text{total}}$ is the mass fraction of A_3B in the system A_3B–B defined by

$$g_{A_3B}/g_{\text{total}} = f^s_{A_3B} = 1 - f_B^s. \tag{6}$$

Equation (5) may be simplified to give

$$f_B^p = (\text{mol wt B/mol wt A}_3\text{B})(1 - f_B^s) + f_B^s. \tag{7}$$

Collecting appropriate terms yields

$$f_B^p = f_B^s[1 - (\text{mol wt B/mol wt A}_3\text{B})] + (\text{mol wt B/mol wt A}_3\text{B}), \tag{8}$$

which is linear in f_B^s.

By a completely analogous procedure, the relationship between f_A^p and f_A^s may be derived and is seen to be

$$f_A^p = f_A^s[1 - 3(\text{mol wt A/mol wt A}_3\text{B})] + 3(\text{mol wt A/mol wt A}_3\text{B}). \tag{9}$$

From Eqs. (8) and (9) we can define the values f_B^s and f_A^s in terms of f_B^p and f_A^p for the cases where data is provided for the primitive systems and the corresponding values in the subsystem are desired. These relationships are given by

$$f_B^s = \frac{f_B^p(\text{mol wt A}_3\text{B}) - \text{mol wt B}}{\text{mol wt A}_3\text{B} - \text{mol wt B}}, \tag{10}$$

$$f_A^s = \frac{f_A^p(\text{mol wt A}_3\text{B}) - 3 \text{ mol wt A}}{\text{mol wt A}_3\text{B} - 3 \text{ mol wt A}}. \tag{11}$$

As a numerical example, let us consider a hypothetical system A_3B–B where the molecular weight of $A_3B = 500$ gm and that of $B = 200$.

C. COORDINATE TRANSFORMATIONS

Applying Eq. (8) for arbitrary 0.1 mass fraction increments defined in terms of f_B^s, we have the values given in Table I.

TABLE I

MASS FRACTION EQUIVALENCIES IN THE SYSTEMS A_3B–B AND A–B

f_B^s	f_B^p	f_B^s	f_B^p
0	0.4	0.6	0.76
0.1	0.46	0.7	0.82
0.2	0.52	0.8	0.88
0.3	0.58	0.9	0.94
0.4	0.64	1.0	1.00
0.5	0.70		

From Table I it is seen that each 0.1 mass fraction change in f_B^s is equivalent to a 0.06 mass fraction change in f_B^p for the hypothetical case in question.

Having done this, let us continue to examine this hypothetical system in terms of the lever arm principle. Let us assume that at a temperature T_1 in the A_3B solid–liquid field starting with a total subsystem composition $f_B^s = 0.2$, the value $f_B^{s(2)} = 0.4$. The value $f_B^{s(1)}$ is zero, since the solid is pure A_3B. The mass solid-to-liquid ratio then, is given by

$$\text{solid/liquid} = |\,0.4 - 0.2\,|/|\,0.2 - 0\,| = 1, \qquad (12)$$

for the subsystem. In the primitive system, the corresponding mole fractions specified are from Table I.

$$\text{solid/liquid} = |\,0.64 - 0.52\,|/|\,0.52 - 0.40\,| = 0.12/0.12 = 1 \qquad (13)$$

for the primitive system.

This result, which indicates that the lever arm principle is applicable directly to a phase diagram represented either in primitive or subsystem mass fraction terms, can be demonstrated for the general case as follows.

In Eq. (8) let us designate the constant ratio mol wt B/mol wt A_3B by the symbol C for simplicity. Eq. (8) then has the form

$$f_B^p = f_B^s(1 - C) + C. \qquad (14)$$

If in the subsystem A_3B–B, the total system composition of the component B is f_B^s, that of B in the liquid in equilibrium with solid A_3B is $f_B^{s(2)}$ and that of B in the solid $f_B^{s(1)}$, the mass solid-to-liquid ratio in the A_3B field is given by

$$\text{solid/liquid} = |\,f_B^{s(2)} - f_B^s\,|/|\,f_B^s - f_B^{s(1)}\,|. \qquad (15)$$

In the primitive system, these mass fractions transform via Eq. (14) into

$$\text{solid/liquid} = \frac{|f_B^{s(2)}(1-C) + C - [f_B^s(1-C) + C]|}{|f_B^s(1-C) + C - [f_B^{s(1)}(1-C) + C]|} = \frac{|f_B^{s(2)} - f_B^s|}{|f_B^s - f_B^{s(1)}|}. \quad (16)$$

From Eqs. (15) and (16) it is evident that the ratios are identical. Having demonstrated this equivalency let us now consider the status in a complex binary system in which the compositions are defined in terms of mole rather than mass fractions.

2. Nonlinear Coordinate Transformations

In the primitive system A–B, consider the subsystem A_xB_y–B. Let the mole fraction of B in the subsystem be denoted by M_B^s and the mole fraction of B in the primitive system be denoted by M_B^p. In the subsystem A_xB_y–B, the mole fraction of B is given by

$$M_B^s = m_B^s / (m_{A_xB_y} + m_B^s). \quad (17)$$

It is to be noted that differentiation is made between B in A_xB_y and that added separately. The mole fraction of the component B in the primitive system A–B counts all of the B present and this fraction is given by

$$M_B^p = \frac{ym_{A_xB_y} + m_B^s}{ym_{A_xB_y} + m_B^s + xm_{A_xB_y}}. \quad (18)$$

Dividing the numerator and denominator of Eq. (18) by $xm_{A_xB_y}$ gives

$$M_B^p = \frac{(y/x) + (m_B^s / xm_{A_xB_y})}{(y/x) + (m_B^s / xm_{A_xB_y}) + 1}. \quad (19)$$

Equation (17) may be solved for the ratio $m_B^s / m_{A_xB_y}$ and this result is given by

$$m_B^s / m_{A_xB_y} = M_B^s / (1 - M_B^s). \quad (20)$$

Substituting this result into Eq. (19) provides us with the transform for converting given values M_B^s into the corresponding values M_B^p, i.e.,

$$M_B^p = \frac{(y/x) + [M_B^s / x(1 - M_B^s)]}{(y/x) + [M_B^s / x(1 - M_B^s)] + 1} = \frac{y(1 - M_B^s) + M_B^s}{(x+y)(1 - M_B^s) + M_B^s}$$

$$= \frac{y + M_B^s(1 - y)}{x + y + M_B^s[1 - (x+y)]} \quad (21)$$

D. SOLUBILITY CURVES FOR COMPOUNDS

Given a phase diagram specified either in terms of M_B^s or corresponding terms M_B^p, let us see whether application of the lever arm principle yields equivalent results. In the A_xB_y field of the subsystem A_xB_y–B with compositions expressed in terms of A_xB_y and B we know, from our earlier derivation of the lever arm principle, that when the mole fraction of B is specified relative to the subsystem, the ratios obtained are correct. Thus, in the subsystem A_xB_y–B where system mole fractions are referred to the components A_xB_y and B we have for the solid-to-liquid ratio in the A_xB_y field the value

$$| M_B^{s(2)} - M_B^s | / | M_B^s - M_B^{s(1)} | = (\text{solid/liquid})^s, \tag{22}$$

where M_B^s is the subsystem mole fraction of B in the starting composition, $M_B^{s(2)}$ is the mole fraction of liquid in equilibrium with solid A_xB_y and M_B^s is the mole fraction B in the solid. Here it is to be noted that $M_B^{s(1)} = 0$.

If we now convert the values $M_B^{s(x)}$ to their equivalent values $M_B^{p(x)}$ with the aid of Eq. (21), it is apparent that application of the lever arm principle yields answers different from Eq. (22), since M_B^p and M_B^s do not differ by only a multiplicative constant, i.e., M_B^p is not linear in M_B^s. Consequently, we have demonstrated that if a phase diagram exhibiting compound formation is expressed in mole fraction terms relative to the primitive system, the lever arm principle will give answers related only to the primitive system and cannot be applied directly to obtain answers relevant to the subsystem in question.

In the next chapter, the transforms for nonlinear and linear representations will be developed for more complex systems, that is, systems where more than a single compound is generated.

D. EXAMINATION OF SOLUBILITY CURVES IN SYSTEMS EXHIBITING COMPOUND FORMATION

In effect, the result obtained in considering nonlinearity of mole fraction transforms is not applicable to lever arm calculations alone. In a subsystem A_xB_y–B or A_xB_y–A, the counting of component quantities may be accomplished in terms of moles of A_xB_y and moles of either A or B since the pseudounary component A_xB_y contains a fixed ratio of A and B in either the liquid or solid phases. In other words, the congruent nature of melting enables us to treat it as a component. Consequently, with reference to Fig. 1, two separate and physically independent phase diagrams are present. Thus, in the subsystem A–AB, the A liquidus will vary in A concentration from 0 to 1 in the interval A–AB and the analogous range will be covered in the AB–B liquidus. In order to utilize the liquidus equations developed in Chapters 16–20, it is necessary that the coordinates for the primitive system A–B be transformed into those for the system AB–A or AB–B depending upon which subsystem is under consideration.

23

Eutectic Interactions Continued—Multiple Compound Formation in Binary Systems and Coordinate Transformations in Multiple Compound Systems

A. Introduction

In the simplest instance of compound formation, a single intermediate solid phase having an ordered structure is generated. Often, in an interaction of the type A–B, several new ordered phases of the generalized stoichiometry A_xB_y are formed. In this chapter we will consider the representation of such compound formation and the generation of a general relationship to enable coordinate transformations from either a subsystem to a primitive one or vice versa. In later chapters, the question of solid solubility will be considered in depth beginning with description of systems in which the species derived from the two components are miscible in all proportions in the solid state, an exact opposite case to that considered up to now. Between these two possible extremes, i.e., complete solid immiscibility and complete solid miscibility, a complete gradation is possible into which all known systems actually fit. This will tend to cause some philosophical consideration of the proper definition for a compound. Also, when is a compound a solid solution and when is it not?

The definitions we propose to employ are, that a compound is a structurally ordered solid phase, and a solid solution is a structurally disordered phase. In addition, a compound is a solid phase known to exhibit a melting maximum in a eutectic interaction. These definitions preclude denoting phases of variable composition as compounds or nonstoichiometric compounds unless either or preferably both of the specified criteria are satisfied.

B. Two Compounds Are Generated

Figure 1 depicts a system in which the interaction A–B gives rise to two intermediate solid phases A_xB_y and A_mB_n. It is to be noted that three independent phase diagrams are generated. These subsystems are all of the eutectic type and each can be considered a separate entity

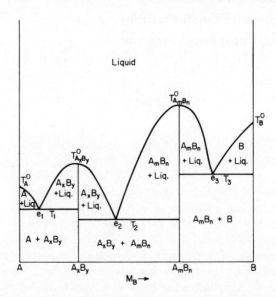

Fig. 1. The binary system A–B in which two intermediate compounds are formed.

in terms of description of the several liquidi. It is pertinent that no simple, known, fixed relationship exists between the stoichiometry of any of the intermediate compounds and their melting points. Similarly, no relationship exists between the melting points of the compounds and those of the end members.

B. TWO COMPOUNDS ARE GENERATED

Figure 2 depicts a similar system to the one we have just described, excepting that the melting points of the compounds and the extent of their two-phase fields are markedly different from those shown in Fig. 1.

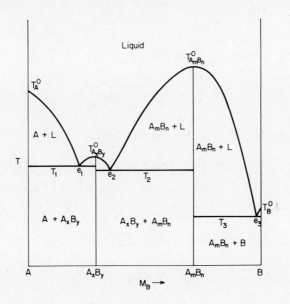

FIG. 2. The binary system A–B in which two intermediate compounds are generated.

The variations in the shapes of such diagrams are many, and the number of intermediate compounds formed may be quite numerous. Pictorial speculations of the types depicted in Figs. 1 and 2 may be constructed *ad infinitum*.

In Chapter 22 we derived equations for coordinate transformations in both mass and mole fraction defined systems. Since it was also demonstrated that application of the lever arm principle to mass fraction defined systems is direct, independent of whether the concentrations refer to the primitive or subsystems, transformation formulas will not be generated for multiple compound systems represented in mass concentration terms.

Let us consider the subsystem A_xB_y–A_mB_n. Let the mole fraction of B in the primitive system be denoted by M_B^p, the mole fraction of A_mB_n in the subsystem by $M_{A_mB_n}^s$. The problem we will define is that of determining M_B^p, the coordinate transform, given $M_{A_mB_n}^s$. The subsystem component mole fraction of A_mB_n is given by

$$M_{A_mB_n}^s = m_{A_mB_n}^s / (m_{A_mB_n}^s + m_{A_xB_y}^s). \tag{1}$$

The mole fraction of the component B in the primitive system in the composition range of this subsystem is

$$M_B^p = \frac{nm_{A_mB_n}^s + ym_{A_xB_y}^s}{nm_{A_mB_n}^s + ym_{A_xB_y}^s + mm_{A_mB_n}^s + xm_{A_xB_y}^s}. \tag{2}$$

Solving Eq. (1) for $m_{A_mB_n}^s$ we obtain

$$m_{A_mB_n}^s = \frac{M_{A_mB_n}^s \cdot m_{A_xB_y}^s}{1 - M_{A_mB_n}^s}. \tag{3}$$

Substituting for $m_{A_mB_n}^s$ in Eq. (2) we obtain

$$M_B^p = \frac{[n(M_{A_mB_n}^s \cdot m_{A_xB_y}^s)/(1 - M_{A_mB_n}^s)] + ym_{A_xB_y}^s}{\left[\begin{array}{c}[n(M_{A_mB_n}^s \cdot m_{A_xB_y}^s)/(1 - M_{A_mB_n}^s)] + ym_{A_xB_y}^s \\ + [m(M_{A_mB_n}^s \cdot m_{A_xB_y}^s)/(1 - M_{A_mB_n}^s)] + xm_{A_xB_y}^s\end{array}\right]}. \tag{4}$$

Dividing each term in Eq. (4) by $m_{A_xB_y}^s$ yields

$$M_B^p = \frac{[nM_{A_mB_n}^s/(1 - M_{A_mB_n}^s)] + y}{[(m+n)M_{A_mB_n}^s/(1 - M_{A_mB_n}^s)] + (x+y)}. \tag{5}$$

Finally, by clearing fractions, we obtain

$$M_B^p = \frac{nM_{A_mB_n}^s + y(1 - M_{A_mB_n}^s)}{(m+n)M_{A_mB_n}^s + (x+y)(1 - M_{A_mB_n}^s)}. \tag{6}$$

Although we can derive an equation for M_A^p in the same manner, use can be made of the relationship

$$M_B^p = 1 - M_A^p; \tag{7}$$

therefore,

$$M_A^p = 1 - \frac{nM_{A_mB_n}^s + y(1 - M_{A_mB_n}^s)}{(m+n)M_{A_mB_n}^s + (x+y)(1 - M_{A_mB_n}^s)}$$

$$= \frac{mM_{A_mB_n}^s + x(1 - M_{A_mB_n}^s)}{(m+n)M_{A_mB_n}^s + (x+y)(1 - M_{A_mB_n}^s)}. \tag{8}$$

Furthermore, since

$$M_{A_xB_y}^s = 1 - M_{A_mB_n}^s, \tag{9}$$

B. TWO COMPOUNDS ARE GENERATED

Eqs. (6) and (8) may be restated as

$$M_B^p = \frac{nM_{A_mB_n}^s + yM_{A_xB_y}^s}{(m+n)M_{A_mB_n}^s + (x+y)(M_{A_xB_y}^s)}, \qquad (10)$$

$$M_A^p = \frac{mM_{A_mB_n}^s + xM_{A_xB_y}^s}{(m+n)M_{A_mB_n}^s + (x+y)M_{A_xB_y}^s}. \qquad (11)$$

Note now the similarity between Eqs. (2) and (10). Remembering that

$$m_x^s = M_x^s m_t^s, \qquad (12)$$

where m_x^s represents the moles of a component and m_t^s represents the total number of component moles (here referred to a subsystem), we see that replacing each m_x term in Eq. (2) by its equivalent from Eq. (12) leads directly to Eq. (10) or Eq. (11).

Thus, for any subsystem, the transformation to primitive system coordinates simply involves the steps of (a) counting the moles of the primitive component so that an equation of the type shown in Eq. (2) is generated and (b) replacing mole terms by mole-fraction terms.

Given the value M_B^p in the primitive system A–B, to transform coordinates into $M_{A_mB_n}^s$ in the subsystem A_xB_y–A_mB_n, solution for $M_{A_mB_n}^s$ in Eq. (6) leads to

$$M_{A_mB_n}^s = \frac{M_B^p(x+y) - y}{M_B^p[(x+y) - (m+n)] + (n-y)}. \qquad (13)$$

Solutions in terms of M_A^p or $M_{A_xB_y}^s$ are readily obtained from one of the preceding, either directly or by first making a substitution of the type given in Eq. (7).

24

Eutectic Interactions Continued—p–T–M_x Representations and Phase Changes

A. Introduction

Up to this point we have concerned ourselves with systems constrained by a constant total pressure imposition. If this constraint is removed, then a two-component system possesses three possible degrees of freedom in the form of the temperature, pressure, and composition. Complete representation of the binary eutectic system, allowing for variation among these variables, requires a three-dimensional or space-model construction. Description of the three-dimensional structure on a sheet of paper creates some problem in visualization, but is made worthwhile by the insight to be gained.

The starting point for the space model is a set of three orthogonal axes as shown in perspective in Fig. 1. These describe a volume open-ended on two of its sides, bounded on its basal plane by a zero pressure T–M_x surface, on two of its sides by unit composition p–T planes, and in front by a zero temperature p–M_x plane. The side bounding unit composition p–T planes are those of the pure components whose interaction(s) in the interior of the enclosed volume give rise to the three-dimensional space model.

A. INTRODUCTION

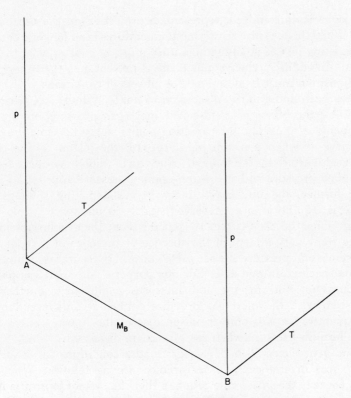

FIG. 1. The three-dimensional space model.

Our treatments of binary systems thus far have been concerned only with condensed phase equilibria. These treatments, in effect, coincided with an experiment in which the condensed phases were contained in a cylinder fitted with a piston that rested on the surface of the condensed phases and exerted a constant pressure on them. The $T-M_x$ diagrams already described then, represented isobaric sections through the $p-T-M_x$ space model, the number of such unique isobaric sections being infinite. As the pressure on the piston is decreased, we finally encounter, at each temperature, a condition under which the hypothetical piston no longer exerts any pressure on the condensed phases and a vapor available space is free to form. Once this happens, the system volatilizes sufficiently to cause the development of the equilibrium pressure for the system under the temperature and composition conditions imposed. This lower, vaporless, pressure limit to the system represents the bounding surface(s) of the three-dimensional space model. A $T-M_x$ diagram, that of the system under its equilibrium pressure, is not isobaric and, indeed, represents a projection on the basal $T-M_x$ plane. While the pressure at varying temperatures is no longer a constant, the coexisting phases at any temperature must, of course, be at the same pressure. In addition, since a new phase is present, i.e., the vapor phase, its composition must be described for all sets of conditions.

As has been noted in the discussion of unary systems (see Chapter 8, Section A), the effect of a so-called inert gas pressure on the pure components is to change their melting points along the same univariant S–L curve as when a piston pressure with the piston resting on the condensed surface(s) is applied. The basic difference between the gas-applied pressure and the piston-applied pressure approaches is that in the former, the univariant unary curves depicting solid–gas and liquid–gas equilibria are preserved, each pseudounary triple point moving along the S–L curve. In general then, the results obtained in treating the binary system in the absence or presence of a vapor phase at constant total pressure are the same excepting that vapor composition must be specified in one case. For our purposes, the inert gas pressure case does not add insight and increases complexity. Furthermore, the system with the inert gas pressure applied is only pseudobinary. Consequently, we will only treat the case of the binary system under its equilibrium pressure when gas phases are discussed.

In the generalized $T-p-M_x$ diagram, the conditions for invariance, univariance, bivariance, and tervariance are as follows: When four phases coexist, the phase rule requires that the system is invariant, i.e., $F = C - P + 2 = 2 - 4 + 2 = 0$. The compositions of the four simultaneously coexisting phases must be represented by four points lying at the same temperature and pressure. When three phases are present, conditions are correct for univariance. Description of three-phase coexistence is via a curve in the interior of the space model. The terminus of each such curve is one of the invariant points referred to previously. Consequently, points of invariance must be associated with four curves of univariance. Two-phase coexistences are bivariant and are represented by a surface. Where two surfaces meet, the lines of univariance are generated. Single phase situations require the delineation of a volume since they represent tervariant cases. The preceding will be appreciated more as our discussion continues.

B. The $T-p-M_x$ Diagram of State

In Fig. 2, the $p-T$ diagrams for the pure components A and B are constructed on the bounding unit composition $p-T$ planes. Since each

B. THE T–p–M_x DIAGRAM OF STATE

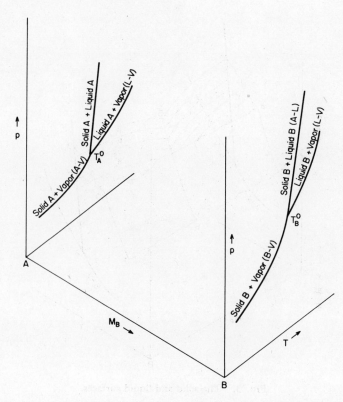

Fig. 2. Location of the unary p–T diagrams on the bounding T–p planes of the space model.

univariant line in the unary diagrams must, because of the introduction of an additional composition variable in the space model, give rise to conditions of bivariance, the extensions of the lines solid–vapor (S–V), solid–liquid (S–L), and liquid–vapor (L–V) into the space model are in the form of surfaces. Each of these one-component two-phase lines must, in addition, give rise to two such surfaces, one representing the composition of one of the phases and the second representing the composition of the coexisting second phase (Fig. 3). Let us first consider the S–L curves of the end members A and B. The compositions of the solid phases in the simple eutectic system are always pure A or pure B so that the solid composition surface must lie in the unit concentration T–p planes of the pure component rather than extending into the space model. The composition of each of the liquid surfaces, on the other hand, varies with system concentration since the components are miscible in the liquid phase. Consequently, the bivariant A and B

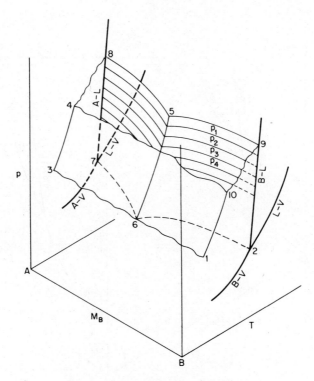

Fig. 3. The solid and liquid surfaces.

liquid composition surfaces emanating from the unary univariant lines A–L and B–L, respectively, intersect in the interior of the diagram to form a univariant eutectic line. Since the solid A–L and solid B–L slopes of the pure components are, in general, negative or positive, the binary eutectic line 5–6 of Fig. 3, will not be perpendicular to the basal T–M_B plane. The lower boundaries of the liquid composition surfaces start at the one component triple points 2 and 7 and intersect at the point 6. This point 6 represents the composition of the liquid phase at the invariant quadruple point of the diagram where solid A, solid B, liquid, and vapor coexist. The three-phase univariant liquidus curves 2–6 and 7–6 intersect at a temperature lower than either of the pure component triple points, in accordance with the liquidus equations previously derived. The value of the total pressure, P_Q, at point 6 is not necessarily below or above that of both triple points, its value depending on the triple point pressures, and the values of the sublimation pressures of the pure components below the triple points, as will become evident. Clearly, however, since solid A, solid B, liquid, and vapor coexist

B. THE T–p–M_X DIAGRAM OF STATE

at the temperature defined by point 6, the total pressure must be given by

$$P_Q = p_A{}^0 + p_B{}^0, \qquad (1)$$

where P_Q is total pressure at the quadruple point, $p_A{}^0$ and $p_B{}^0$ are the vapor pressures of pure solid A and pure solid B at the quadruple point temperature. More will be said about this subsequently.

In Fig. 3, the planes 2–6–5–9 and 5–6–7–8 represent the liquid composition surfaces which intersect in the univariant eutectic curve 5–6, each point along the latter coinciding with a unique pressure, i.e., points of isobaric invariance. A series of isobars on these surfaces define the liquidus curves at different constant total pressure. The plane 1–10–4–3 represents the lower temperature limit of liquid–solid or, in one case, liquid–solid–vapor coexistence. The eutectic line 5–6 lies in this plane and the latter, while perpendicular to the bounding unit composition T–p planes, is not, in general, perpendicular to the basal plane. The invariant composition of the solids A and B is shown by the surfaces 1–2–9–10 and 3–4–8–7 each of which is coplanar with, and lies in a respective bounding unit composition plane. The volumes enclosed by the surfaces described define the p–T–M_X conditions within which solid–liquid equilibria can occur. In order that a constant pressure slice provide us with a complete isobaric T–M_X diagram of normal form, it must coincide at all temperatures and compositions with pressures greater than the maximum vapor pressure in the binary S–L–V equilibria.

If we assume the simplest type of liquid–vapor equilibrium, the appearance of the liquid and vapor composition surfaces may be visualized from the following: Each of the unary L–V curves gives rise to two surfaces, one describing vapor composition and the second describing liquid composition. These surfaces are continuous from one L–V curve to the other and terminate at the critical points 1 and 2 of Fig. 4. The liquid surface is a plane extending from one side of the diagram to the other, each isotherm on this plane being represented by a straight line. The composition A–B of this liquid surface at any liquid composition is

$$M_A^{(3)} = p_A/(p_A + p_B), \qquad (2)$$

$$M_B^{(3)} = p_B/(p_A + p_B). \qquad (3)$$

Where the superscript (3) refers to the vapor phase, and the terms p_A and p_B refer to the partial pressures of A and B in the vapor phase.

For any particular liquid composition $M_A^{(2)}$ the values p_A and p_B are given by Raoult's equation.

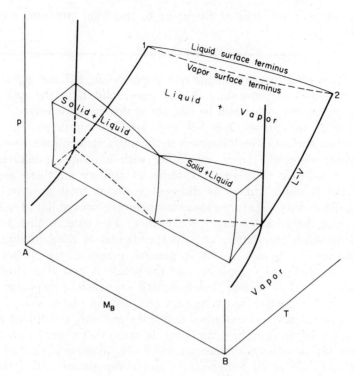

Fig. 4. The S–L and L–V volumes.

Thus,

$$p_A = M_A^{(2)} p_A^\dagger \tag{4}$$

where $M_A^{(2)}$ is the liquid phase composition of A and $p_A^\dagger$ is the vapor pressure of pure liquid A at the temperature in question

$$p_B = M_B^{(2)} p_B^\dagger = (1 - M_A)^{(2)} p_B^\dagger \tag{5}$$

where the terms correspond to those for Eq. (4).

Substituting for p_A and p_B in Eq. (2), the values of Eqs. (4) and (5), we obtain for example,

$$M_A^{(3)} = \frac{M_A^{(2)} p_A^\dagger}{M_A^{(2)} p_A^\dagger + (1 - M_A)^{(2)} p_B^\dagger}. \tag{6}$$

Figure 5 shows a typical isotherm through the liquid vapor surfaces of Fig. 4. Any tie line, such as a–b, shows the compositions of the coexisting phases at any total system pressure. Note that the vapor is richer in

B. THE T–p–M_X DIAGRAM OF STATE

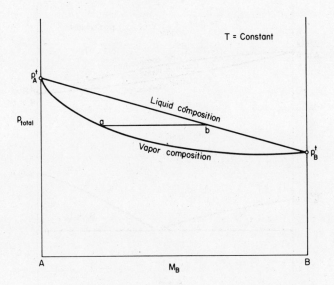

FIG. 5. An isothermal slice through the L–V curves of Fig. 4.

the component having the higher vapor pressure and that the total system pressure varies between $p_A{}^\dagger$ and $p_B{}^\dagger$, the vapor pressures of pure A and pure B. Note also that the vapor composition curve, frequently called the vaporous curve, also specifies the total system pressure for a specified system composition at the dewpoint of the latter.

An isobaric section through the S–L volume will eventually intersect the L–V surfaces, if this section is chosen to lie below either of the unary critical points. Such a section will intersect the liquid surface at lower temperatures than the vapor surface. Consequently, a T–M_X slice will show the liquid composition curve beneath the vapor composition curve. The end points of an isobaric slice, i.e., where $M_A = 0$ or where $M_B = 0$ are the *normal boiling points* of the pure components and the liquid and vapor coexistence curves in between, represent the normal boiling point curves for the solutions. These curves in fact show the temperatures at which the solutions develop the total pressure in question. Such a section is shown in Fig. 6. It is to be noted that the vapor in equilibrium with liquid is richer in that component having the lower boiling point. This is consistent with the expectation that the component with the lower boiling point will exhibit the higher vapor pressure at any temperature.

At all pressures above the maximum value exhibited by the vapor surface of Fig. 4, the system is liquid and in Fig. 7, the liquid and

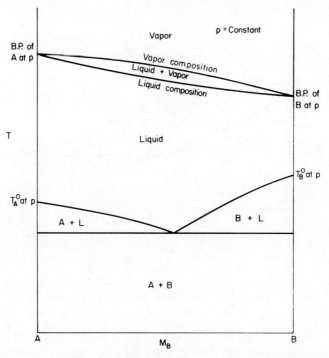

Fig. 6. An isobaric section through the S–L and L–V surfaces of Fig. 4.

solid–liquid volumes are shown on the p–T–M_X diagram. In Fig. 7, the lowest temperature at which liquid may be present is that of the quadruple point T_Q. Below this temperature, the system may contain either solids A and B alone or solid A and vapor, solid B and vapor, or solids A, B, and vapor. Let us begin our discussion of these regions with a consideration of the bivariant systems A–V or B–V.

Clearly, if pure solid A alone is present, at $M_B = 0$, the total pressure in the system is p_A^0, where p_A^0 is the vapor pressure of this pure solid (its sublimation pressure) at a particular temperature. On the other hand, if pure solids A and B coexist with vapor, the total pressure for this univariant condition is

$$P_t = p_A^0 + p_B^0. \tag{7}$$

In a binary region in which solid A alone, for example, coexists with vapor, it is evident that the total pressure will vary between p_A^0 and $p_A^0 + p_B^0$ at a specified constant temperature, and will be given by the sum $p_A^0 + p_B$ (where $p_B < p_B^0$). Furthermore, in any such region, the partial pressure of A at a particular temperature will always be the same since solid A is present.

B. THE T–p–M_X DIAGRAM OF STATE

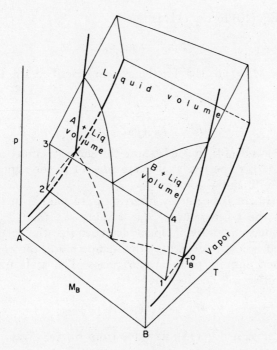

FIG. 7. The volumes A–liquid, B–liquid, and liquid.

It is, therefore, a consequence that

$$p_t - p_A^0 = p_B, \tag{8}$$

where p_B is the partial pressure of B and

$$p_B < p_B^0, \tag{9}$$

at system compositions less than that required for solid A, solid B, and vapor to be present simultaneously.

At constant temperature, a binary system comprising two phases, i.e., solid A–vapor or solid B–vapor is univariant ($F = C - P + 1 = 2 - 2 + 1 = 1$). Thus, if we specify the composition of the system, the total pressure is also fixed. To better understand the relationships involved, let us consider a chamber containing solid A, solid B, and vapor. If the vapor available volume of the chamber is V, then the number of moles of A and B contained in the vapor is

$$n_A^{(3)} = p_A^0 V / RT, \tag{10}$$

$$n_B^{(3)} = p_B^0 V / RT. \tag{11}$$

The ratio of Eq. (10) to Eq. (11) is

$$n_A^{(3)}/n_B^{(3)} = p_A^0/p_B^0, \tag{12}$$

and the values $M_A^{(3)}$ and $M_B^{(3)}$ for the system described are, therefore,

$$M_A^{(3)} = p_A^0/(p_A^0 + p_B^0), \tag{13}$$

$$M_B^{(3)} = p_B^0/(p_B^0 + p_A^0). \tag{14}$$

Equations (13) and (14), indeed, show the mole fractions of the components A and B, respectively, that are present in the vapor when the three phases A, B, and V coexist.

If our experiment is begun by inserting pure solid A in the cylinder maintained at the temperature T (at a value less than T_Q of Fig. 7) and a small enough quantity of B is added such that the partial pressure of B developed is less than p_B^0 for this temperature, the component B will be confined entirely to the vapor phase. The mole fraction of A in the vapor phase will be given by

$$M_A^{(3)} = p_A^0/(p_A^0 + p_B^0). \tag{15}$$

Substituting for p_B in Eq. (15), its value from Eq. (8) gives

$$M_A^{(3)} = p_A^0/(p_A^0 + P_T - p_A^0) = p_A^0/P_T \tag{16}$$

along the A vaporous.

If the experiment is done starting with solid B in the container, the value of M_B along the B vaporous is

$$M_B^{(3)} = p_B^0/P_T. \tag{17}$$

Since all values P_T are allowed between the limits p_A^0, or p_B^0, and $p_A^0 + p_B^0$ in the respective solid–vapor regions, the contours of the vaporous curves are readily defined via Eqs. (16) and (17). The two isothermally univariant vaporous curves intersect at $P_t = p_A^0 + p_B^0$ and the composition of the vapor at this isothermal invariant point is given by either Eq. (13) or Eq. (14). Such an isothermal section is shown in Fig. 8.

Solid–vapor ratios, etc., are deduced from tie lines in precisely the same manner as in a $T-M_X$ condensed-phase diagram and need not be discussed further. The triple point A–B–V is displaced at all temperatures toward that end member exhibiting the higher vapor pressure, and the series of such triple points each at a new temperature generate a univariant three-phase line as a function of the temperature. The

B. THE T–p–M_X DIAGRAM OF STATE

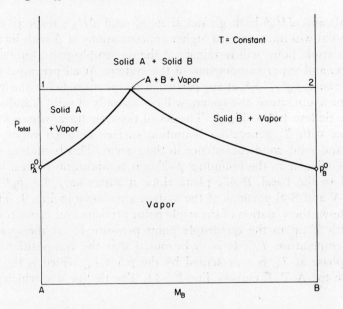

FIG. 8. An isotherm on the solid–vapor regions of the system A–B.

value $M_A^{(3)}$ or $M_B^{(3)}$ along this line which moves to higher total p, with increasing T, is deduced from the following.

We desire the variation

$$dM_A/dT = d[p_A^0/(p_A^0 + p_B^0)] \tag{18}$$

for the triple point A–B–V. At any temperature T, M_A at the triple point is given by Eq. (13). The variation of p_A^0 and p_B^0 with T is given by

$$p_A^0 = C_1 \exp[-(\Delta H_A^s/RT)], \tag{19}$$

$$p_B^0 = C_2 \exp[-(\Delta H_B^s/RT)], \tag{20}$$

where ΔH_A^s and ΔH_B^s are the respective heats of sublimation of A and B, while C_1 and C_2 are constants.

Substituting Eqs. (19) and (20) into Eq. (13) gives

$$\begin{aligned} M_A &= \frac{C_1 \exp[-(\Delta H_A^s/RT)]}{C_1 \exp[-(\Delta H_A^s/RT)] + C_2 \exp[-(\Delta H_B^s/RT)]} \\ &= [1 + (C_2/C_1) \exp\{-((\Delta H_A^s - \Delta H_B^s)/RT)\}]^{-1}. \end{aligned} \tag{21}$$

If the values p_A^0 and p_B^0 are known at a particular temperature T_1, the value for M_A at any other temperature T_2 is given by

$$M_A^{(T_2)} = [1 + (p_B^{T_1}/p_A^{T_1}) \exp\{-((\Delta H_A^s - \Delta H_B^s)/R)(T_2^{-1} - T_1^{-1})\}]^{-1}. \tag{22}$$

With $p_A^{T_1}$ and $\Delta H_A{}^s$ both greater than $p_B^{T_1}$ and $\Delta H_B{}^s$, respectively, the triple point will move toward higher concentrations of A with increasing T. This triple point will terminate at the quadruple-point temperature, pressure, and vapor composition of the system. At all pressures greater than the sum, $p_A{}^0 + p_B{}^0$, at any particular temperature below the quadruple point temperature, the system will consist only of solid A and solid B, i.e., the tie line 1–2 of Fig. 8. The set of these tie lines, whose P_t value increases with T, generate a bounding surface between vapor, solid–vapor, and solid–solid coexistence in the system. This bounding surface is perpendicular to the bounding p–T unit composition planes, but not parallel to the basal T–M_B plane since it varies as $p_A{}^0 + p_B{}^0$ varies. The S–V and S–S portions of the system are shown in Fig. 9. The line 1–V_Q shows the variation of the triple point pressure and vapor composition with T up to the quadruple point pressure P_Q at the quadruple point temperature T_Q. It is to be noted that the composition of the liquid phase at T_Q is represented by the point L_Q which is the lowest point on the A–B–L eutectic line L_Q–11. The tie line 4–6 which occurs

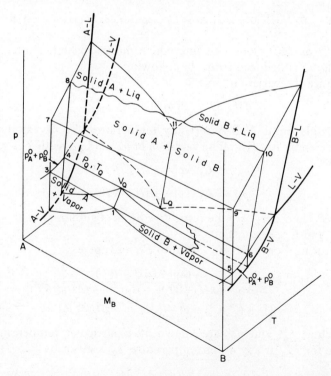

FIG. 9. The regions of S–V, solid A–solid B, solid A–liquid, and solid B–liquid.

B. THE T–p–M_X DIAGRAM OF STATE

at the quadruple point temperature and pressure also shows the composition of solids in equilibrium with vapor and liquid. Since these solids are pure, the termini of the tie line are at the unit concentration p–T planes.

The volume above the bounding plane 3–4–6–5 [the plane that describes the variation $(p_A{}^0 + p_B{}^0)$ with T] contains only solid A and solid B. This volume is bounded by the points 3–10 inclusive. The upper temperature limits of the solid A–solid B volume is the plane 4–8–10–6, which includes the eutectic line L_Q–11, and is perpendicular to the bounding T–p unit concentration planes, but is not necessarily perpendicular to the basal plane, since the pure component A–L and B–L slopes are, in general, positive or negative as mentioned before.

This leaves us now with two areas still to be described, those under the solid A–liquid and the solid B–liquid volumes shown in Figs. 7 and 9.

At temperatures greater than the quadruple point temperature T_Q, two distinct three-phase regions may be defined. In one of these, solid A–L–V may coexist and in the other solid B–L–V may coexist. Since three-phase coexistences are univariant in a binary system $(F = C - P + 2 = 2 - 3 + 2 = 1)$, it is to be anticipated, based on our earlier discussions, that lines showing the composition of vapor in equilibrium with solid A and liquid, the composition of vapor in equilibrium with solid B–liquid, the composition of liquid in equilibrium with solid A and vapor, and the composition of liquid in equilibrium with solid B–vapor all need be defined. The univariant lines showing the compositions of solids A and B, as usual, lie in the plane of the unit concentration T–p planes. The two vapor-composition lines originate at the vapor quadruple point P_Q of Fig. 9, while the two liquid-composition lines (the two liquidi for the system under its equilibrium pressure) originate at the liquid composition quadruple point L_Q. Since p varies with T along each of the four lines, the univariant lines may be seen only in projection onto the basal T–M_X plane. Each point on each of the lines will coincide with a given total pressure due to $p_A{}^0 + p_B$ or $p_B{}^0 + p_A$ depending upon which of the three-phase coexistences is studied. Since solid, liquid, and vapor coexist simultaneously, P_t for a particular temperature is the same for the corresponding points on each of the coexistent univariant curves.

While the terminus at one end of each of the lines is at a binary quadruple point, its terminus at its opposite end is a unary triple point. Thus, as in all instances, one additional degree of freedom arises as we move off of the bounding T–p unit-concentration planes. A unary triple point (invariant) gives rise to binary three-phase lines.

The concentration of the vapor phase in either of the three-phase regions as a function of T can be deduced as follows (as can the variation of total system pressure with T).

In the solid A–L–V equilibrium, the total pressure in the system is given by

$$P_t = p_A{}^0 + p_B, \qquad (23)$$

where $p_A{}^0$ is the sublimation pressure of solid A. The pressure p_B is no longer the sublimation pressure of solid B, and depends on the composition of B in the liquid which is in equilibrium with solid A.

As shown in Chapter 13, Section C, the mole fraction of A in the liquid phase at any total pressure P_t is given by

$$N_A^{(2)} = p_A{}^0/p_A{}^\dagger, \qquad (24)$$

where $p_A{}^\dagger$ is the vapor pressure of pure supercooled A and $p_A{}^0$ that of pure solid A (its sublimation pressure). The variation of $N_A^{(2)}$ with T along the three phase liquidus A–L–V as shown in Chapter 13 is

$$N_A^{(2)} = \exp\left[-\frac{\Delta H_A{}^0}{R}\left(\frac{1}{T} - \frac{1}{T_A{}^0}\right)\right], \qquad (25)$$

where $\Delta H_A{}^0$ is the latent heat of fusion of A and $T_A{}^0$ is the triple point temperature of pure A. Equation (25) is, or course, our familiar liquidus relation. We have invoked the result of Section C, Chapter 13, to demonstrate that the equation is equally applicable to a three-phase system in which total pressure varies with T.

The value p_B at any one temperature along the liquidus for A is given by

$$p_B = N_B^{(2)} p_B{}^\dagger \qquad (26)$$

where p_B is the partial pressure of B, $N_B^{(2)}$ is its mole fraction in solution, and $p_B{}^\dagger$ is the vapor pressure of pure liquid B at the temperature in question. Along the A liquidus it is, of course, the case that

$$N_B^{(2)} = 1 - N_A^{(2)}. \qquad (27)$$

Consequently, at any temperature T along this liquidus

$$p_B = (1 - N_A^{(2)}) p_B{}^\dagger, \qquad (28)$$

the variation of $p_B{}^\dagger$ with T is given by

$$p_B{}^\dagger = C_2 \exp[-(\Delta H_B{}^V/RT)]. \qquad (29)$$

B. THE T-p-M_x DIAGRAM OF STATE

Starting at $T_A{}^0$, the melting point of A where the vapor pressure of pure liquid B is $p_B{}^{\ddagger}$, the value $p_B{}^{\dagger}$ at lower temperatures is

$$p_B{}^{\dagger} = p_B{}^{\ddagger} \exp\left[-\frac{\Delta H_B{}^V}{R}\left(\frac{1}{T} - \frac{1}{T_A{}^0}\right)\right], \qquad (30)$$

where $\Delta H_B{}^V$ is the molar heat of vaporization of pure liquid B.

Similarly, for the partial pressure of A in this interval

$$p_A = p_A{}^0 = p_A^{00} \exp\left[-\frac{\Delta H_A{}^s}{R}\left(\frac{1}{T} - \frac{1}{T_A{}^0}\right)\right] = N_A^{(2)} p_A{}^{\dagger}$$

$$= \exp\left[-\frac{\Delta H_A{}^0}{R}\left(\frac{1}{T} - \frac{1}{T_A{}^0}\right)\right] p_A{}^{\ddagger} \exp\left[-\frac{\Delta H_A{}^V}{R}\left(\frac{1}{T} - \frac{1}{T_A{}^0}\right)\right], \qquad (31)$$

where $\Delta H_A{}^s$ is the molar heat of sublimation of solid A, $\Delta H_A{}^V$ is the molar heat of vaporization of liquid A, p_A is the partial pressure of A and equals the vapor pressure of solid A at that temperature. $p_A{}^{\dagger}$ is the vapor pressure of pure supercooled A at the same temperature, and $p_A{}^{\ddagger}$ is the vapor pressure of liquid A at the triple point of A, and p_A^{00} is the vapor pressure of solid A at the triple point of A. Obviously

$$p_A^{00} = p_A{}^{\ddagger}. \qquad (32)$$

Both of these pressures may, for simplicity of notation, be written as

$$p_A^{00} = p_A{}^{\ddagger} = p_A{}^t. \qquad (33)$$

Equations (28), (30), and (31) may be combined and then substituted into Eq. (23) to show the variation in P_t along the A liquidus with T

$$P_t = p_A{}^t \exp\left[-\frac{\Delta H_A{}^s}{R}\left(\frac{1}{T} - \frac{1}{T_A{}^0}\right)\right]$$
$$+ 1 - \exp\left[-\frac{\Delta H_A{}^0}{R}\left(\frac{1}{T} - \frac{1}{T_A{}^0}\right)\right] p_B{}^{\ddagger} \exp\left[-\frac{\Delta H_B{}^V}{R}\left(\frac{1}{T} - \frac{1}{T_A{}^0}\right)\right]. \qquad (34)$$

The mole fraction of A in the vapor phase along the liquidus is

$$M_A^{(3)} = \frac{p_A{}^0}{p_A{}^0 + p_B}$$

$$= \frac{p_A{}^t \exp\left[-\frac{\Delta H_A{}^s}{R}\left(\frac{1}{T} - \frac{1}{T_A{}^0}\right)\right]}{\left[p_A{}^t \exp\left[-\frac{\Delta H_A{}^s}{R}\left(\frac{1}{T} - \frac{1}{T_A{}^0}\right)\right] + \left\{1 - \exp\left[-\frac{\Delta H_A{}^0}{R}\left(\frac{1}{T} - \frac{1}{T_A{}^0}\right)\right]\right\} \times p_B{}^{\ddagger} \exp\left[-\frac{\Delta H_B{}^V}{R}\left(\frac{1}{T} - \frac{1}{T_A{}^0}\right)\right]\right]} \qquad (35)$$

While the vapor pressure of pure liquid B along the A liquidus is increasing due to an increase in T, the partial pressure of B is decreasing since its mole fraction is decreasing. Thus when $T = T_A^0$, the term $1 - \exp\{-(\Delta H_A^0/R)[T^{-1} - (T_A^0)^{-1}]\}$ in Eq. (35) is equal to zero and $M_A^{(3)} = 1$. Consequently, the total pressure of the system, which equals $p_A^0 + p_B^0$ at the quadruple temperature, becomes equal to p_A^0 at the melting point of A. This is shown in Fig. 10 where the solid A–solid

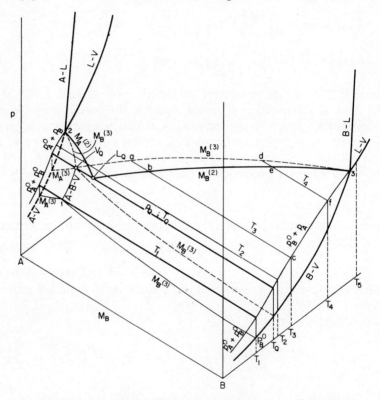

FIG. 10. Triple points for the A–L–V and B–L–V equilibria.

B volumes and solid A–liquid and solid B–liquid volumes of Fig. 9 have been left out for purposes of clarity.

The curves 2–V_Q and 3–V_Q describe the variation of mole fractions of A and B in the vapor phase along the A and B liquidi. The compositions $M_A^{(2)}$ and $M_B^{(2)}$ along the A and B liquidi, respectively, are shown in the curves 2–L_Q and 3–L_Q. The three-phase line, solid A–solid B–vapor extends from 1 to V_Q.

Below the temperature T_Q, condensed-vapor phase equilibria (i.e.,

B. THE T–p–M_X DIAGRAM OF STATE

solid A–vapor or solid B–vapor) are bivariant or isothermally univariant. Consequently, the composition of the vapor phase may be depicted as a function of p on a p–M_X isothermal plane. Above T_Q, the vapor coexists with solid and liquid and the system is isothermally invariant. The specification of a temperature automatically fixes the composition of the vapor phase along 2–V_Q or 3–V_Q and simultaneously fixes the pressure of such a coexistence. An isothermal slice through the A–L–V and B–L–V regions would show six points all at the same total pressure $p_A^0 + p_B$ or $p_B^0 + p_A$ of Fig. 10. Two of these points (those for solid compositions) would be included on the pure component axes. Two of these would show vapor composition (one for each liquidus), two would show liquid composition. A projection of the four univariant curves on the p–M_X, zero-temperature plane would have the appearance shown in Fig. 11. Tie lines connecting solid–liquid and vapor composi-

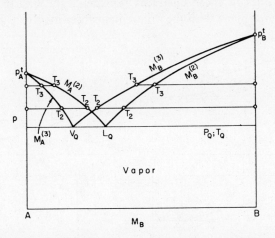

FIG. 11. A projection on the p–M_B plane of the liquidus and vaporous curves of Fig. 10.

tion isothermally invariant triple points are shown for a few temperatures. Each S–L–V equilibrium has associated with it three such triple points at each temperature.

In Fig. 10, the vapor surfaces at temperatures greater than the quadruple point temperature T_Q are not shown. Below T_Q, the isothermal planes T_1 and T_Q enable one to visualize the vapor surfaces for the two-phase equilibria A–V and B–V. In this region the three-phase line 1–V_Q shows the points of intersection of the A–V and B–V vapor composition surfaces. The lines 2–V_Q, 2–L_Q, 3–V_Q, and 3–L_Q are analogous to the line 1–V_Q. The former are generated by intersection of either the A–V or B–V surfaces with the L–V surface above the quadruple-point temperature, each of these regions occupying a given composition interval. Thus, the point a lies on the vapor surfaces at the point of intersection of the B–V surface with the vapor surface for the two-phase equilibrium L–V; consequently, it is an isothermal triple point. The point b represents the corresponding composition of the liquid composition when the liquid composition surface of the L–V equilibrium encounters the liquid composition surface of the B–L equilibrium. The latter surface is more or less perpendicular to the M_B–T basal plane, depending on whether or not the solubility of B in liquid varies with pressure in the two-phase equilibrium B–L. The points d, e, and f have the same significances at the temperature T_3, Fig. 10.

The appearance of isotherms at a series of temperatures through the space model of Fig. 10 are shown schematically in Fig. 12a–f. Each of the lines in these diagrams represents the line of intersection of the isothermal plane with the different surfaces involved. For example,

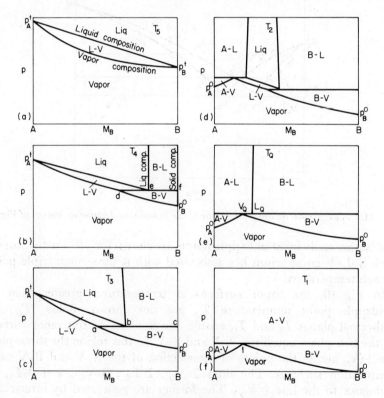

FIG. 12. Isotherms through Fig. 10.

B. THE T–p–M_x DIAGRAM OF STATE

in Fig. 12a the isothermal plane cuts through only two surfaces, that describing the liquid composition and that describing the vapor composition for the two-phase equilibrium L–V. Figure 12b shows the intersection of an isothermal plane with the several surfaces at a temperature below the triple point of B and above the triple point of A. The points d, e, and f show the compositions of coexisting vapor, liquid, and solid, respectively, for the B liquidus. Clearly, the composition of the vapor in equilibrium with B and liquid must be the same as for the liquid in equilibrium with vapor at precisely this point; similarly for the liquid. Thus, the L–V surfaces must have two points in common with the B–L–V equilibrium, i.e., the points d and e. The line e–m shows the intersection of the isothermal plane with the liquid composition surface of the B–L equilibrium. The solid composition surface intersection with the isothermal plane lies along the axis B–f, since the solid composition is unchanging with pressure. The remaining figures require no further discussion, since they represent duplications or obvious extensions of the above.

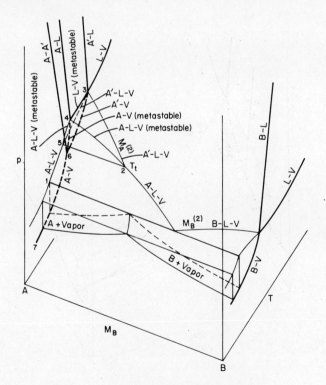

FIG. 13. A system showing a phase change.

A similar projection to that of Fig. 11 on the T–M_x plane of Fig. 10 would look much like Fig. 11 except that each tie line would now represent a constant pressure ($p_A^0 + p_B$ or $p_B^0 + p_A$) at each tie-line temperature.

The variations in the appearance of the T–p–M_x diagram of state due to different assumptions as to the unary triple-point temperatures and pressures are matters of degree rather than principle and will not be considered further. The reader is for the moment, left the task of considering cases such as that where one of the components is basically involatile. Here, one of the unary participants exhibits S–V and L–V curves that lie at essentially zero pressure through the experimental range of interest. The total pressure in the system effectively, then, becomes equal to that of the volatile solid or liquid; a univariant curve such as A–B–V lies displaced close to the p–T plane for A and the isothermal univariant S–V curves degenerate into lines of constant pressure terminating at the sublimation curve of the volatile component on one side of the diagram.

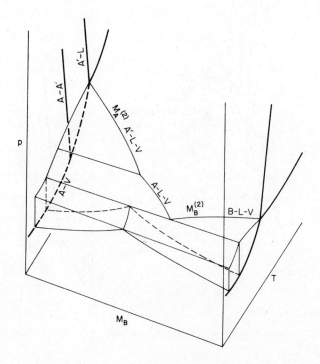

FIG. 14. A simplified view of Fig. 13 in which metastable extensions are left out.

C. Phase Changes—A Second Quadruple Point

If one of the end members undergoes a solid–solid phase transformation, its S–V curve will exhibit a discontinuity. This will result in a corresponding discontinuity in its p–T–M_x diagram with the generation of a second set of quadruple points. The appearance of the lowest pressure boundaries of such a system is given in Fig. 13. For purposes of simplicity, the liquid–vapor regions are not shown, nor are the vapor composition curves in the liquidus regions. The curve 2–3 is the liquidus for which solid A′ (the high temperature phase) is in equilibrium with liquid and vapor. The curve 2–L_Q is the liquidus for the solid A phase. The line 2–4 shows the metastable extension of this solid A–liquid curve to the metastable triple point of the A phase at point 4. This metastable curve lies at a higher pressure at each temperature than does the stable curve 2–3 at temperatures above T_t (point 2). The curve 1–5 shows the total pressure of the solid A–L–V system. The dashed curve 5–4 shows the metastable extension of this total pressure to the metastable melting point A–L–V at point 4. The equilibrium total pressure curve above point 5 is shown by the dashed

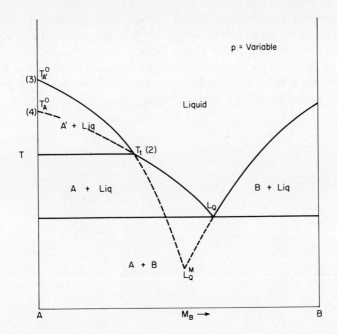

FIG. 15. Metastable extensions of liquidus curves in which a phase transformation occurs. A projection of Fig. 13 on the T–M_B plane.

line 5–3 which terminates at the triple point A′–L–V. The curve 7–6 is the sublimation curve of the A phase and it is shown continuing metastably to point 4, the triple point of the A phase. The curve 6–3 is the sublimation curve of the A′ phase. Finally, the line 2–5 shows the temperature and pressure of the new quadruple point A–A′–L–V, while point 2 itself is the liquid composition at this quadruple point.

Figure 14 shows the system of Fig. 13 in which, to enable easier visualization, metastable extensions have not been drawn. The projection of Fig. 13 on the T–M_X basal plane has the appearance of Fig. 15. It is to be noted that the projection of the metastable curve 2–4 of Fig. 13 lies below that of the stable curve 2–3 of Fig. 13. Similarly, the metastable curve 2–$L_Q{}^M$ (not shown in Fig. 13), which represents the intersection of the A′ liquidus with the B liquidus, lies below the stable liquidus in this temperature interval. On T–M_X representations it is always the case that metastable curves lie at lower temperatures than do the stable curves for the temperature interval in question. This is a consequence of the metastable pressures being greater than the stable ones.

The case just treated is a prelude to consideration of incongruent melting phenomena in binary systems. In the instance of phase transformations, tie lines to both the A and A′ liquidi terminate at pure A and it is evident that only a single binary diagram is present. In the following chapter we will see that a slightly more complex case arises if, instead of one of the end-member phases being unstable above a certain temperature, a solid intermediate compound formed in the primitive system is unstable above a certain temperature.

25

Eutectic Interactions Continued—Intermediate Compound Instability—Incongruently Melting Compounds

A. Introduction

In the same way as a solid unary substance may exhibit instability at a particular temperature and undergo a phase transformation, a binary substance may exhibit one or more phase transformations while continuing to behave in pseudounary fashion. In addition, however, a solid binary substance may exhibit a type of instability in which it decomposes below its melting point to form two phases, each having a different composition than the pure binary substance. Most frequently such decompositions result in the formation of a new solid in equilibrium with a liquid. This type of temperature instability is referred to as incongruent melting. It is characterized by the fact that even when starting with the pure solid binary substance in question, upon reaching the temperature of incongruent melting, the unary character of the material is lost. In other words, upon decomposing, the coexisting phases are not of similar composition.

Since the incongruent melting occurs below the hypothetical melting point of the starting substance, a liquidus, compound temperature maximum is not realized. Because of this, it is impossible for a eutectic intersection to be generated with each of the end members immediately adjacent to this intermediate compound composition.

Since the liquidus equations describing the melting of solid phases in a binary interaction vary between mole fractions of 0 and 1 in the temperature interval, zero to the melting points, and, as a consequence, always exhibit a eutectic intersection, the absence of a compound peak in an incongruently melting material prevents it from participating in subsystem interactions. This represents a corollary to our definition of a compound as it relates to the participation of compounds in subsystem interactions. In order for a binary or higher order solid of given stoichiometry to be suitable for treatment as a component, it must exhibit pseudounary behavior in the entire range of experimental interest. In other words, in solid–liquid phase diagrams, the substance must exhibit a triple point of its own in which, upon melting, a liquid of the same composition as the melting solid results. In this chapter the description of incongruently melting systems will be perused in detail in terms of two- and three-dimensional representations. The applicability of idealized liquidus models previously developed will be described as will be the proper application of the lever arm principle.

B. The Intermediate Compound A_xB_y Melts Incongruently

In Fig. 1 the condensed phase binary system A–B is constructed showing the formation of an intermediate compound A_xB_y. This substance melts incongruently at a temperature T_i, decomposing completely in this process to form pure solid A and liquid. Since A_xB_y does not remain stable until its hypothetical melting point $T^0_{A_xB_y}$, and, as a matter of fact, disproportionates into solid A and a liquid, it cannot participate in a eutectic equilibrium with solid A. It cannot, therefore, be treated as an end member of the subsystem $A–A_xB_y$. If we recall that, in defining the ideal binary liquidus of a specified species, the type of other participating species is subordinate to their concentrations, we are in a position to decide on an appropriate application of liquidus equations. Since A does not equilibrate with A_xB_y excepting, as we shall see, at a unique temperature, the A liquidus is determined by its composition in the system A–B. The A_xB_y incomplete liquidus is determined by the incomplete subsystem A_xB_y–B and similarly for the B liquidus. In other words, assuming A_xB_y or some other species derived from it to be a representative liquid phase spacies below T_i, the composition of B in the B liquidus field depends on the A_xB_y composition and similarly for A_xB_y in its liquidus field, its composition depends solely on the composition of

B. A_xB_y MELTS INCONGRUENTLY

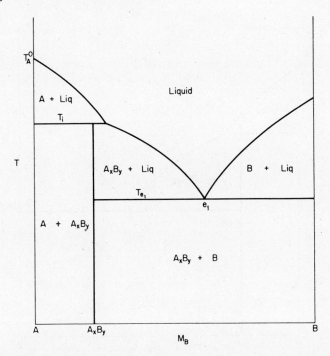

FIG. 1. Field representation of the system A–B showing the incongruently melting compound A_xB_y.

B derived species. Consequently, given $\Delta H^0_{A_xB_y}$ and its hypothetical melting point $T^0_{A_xB_y}$ and similar information for A and B we may generate the entire phase diagram shown in Fig. 1 by constructing the liquidi for the assumed systems A–B and A_xB_y–B. For convenience, this is accomplished by considering the primitive system A–B and the subsystem A_xB_y–B after performing the appropriate transforms. Once these data are acquired, it is most convenient to specify the entire system in terms of the primitive system A–B as is done in Fig. 1.

Before continuing further, it is interesting to note the superficial resemblance between Fig. 1 and Fig. 15 of the preceding chapter. In both instances, the A liquidus exhibits a discontinuity associated with a point of invariance. The notable differences between the two diagrams are that the eutectic tie line T_{e_1} extends over the entire diagram in Fig. 15 of Chapter 24 while in Fig. 1 of this chapter it is interrupted by the A_xB_y isopleth, the latter being absent from Fig. 15 of Chapter 24.

If the composition A_xB_y of Fig. 2 is heated from a temperature below T_i, it will decompose entirely upon reaching this temperature. In the decomposition process, a liquid of composition 2 and a solid of composition

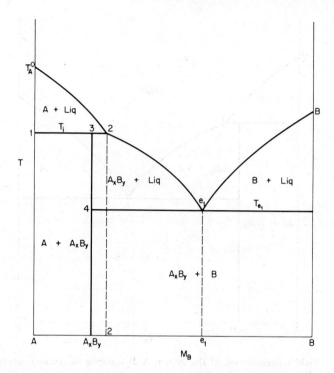

FIG. 2. The binary system A–B showing the formation of the incongruently melting compound A_xB_y.

1 will form. At the temperature T_i, three condensed phases, i.e., solid A, solid A_xB_y, and liquid, coexist. The system is, therefore, isobarically invariant. Until one of these phases disappears, the system possesses zero degrees of freedom, the compositions of the coexisting phases and the temperature being unique. Once all of the solid A_xB_y has completely decomposed, the system consists of solid A and liquid, a univariant condition. Upon further heating, the liquid composition follows the curve $2-T_A^0$ and the solid exhibits the constant composition defined by the vertical line $A-T_A^0$. In the solid A–liquid field bounded by $1-2-T_A^0$, use of the lever arm principle directly on Fig. 2 (as scaled) provides answers in terms of the primitive system A–B. In other words, the solid-to-liquid ratio gives the molecular ratio in terms of A and B independent of their mode of aggregation in the solids formed. Similarly the ratio, solid/solid–liquid represents the fraction of the total number of component moles $m_A + m_B$ that is present as solid.

In the composition interval $A-A_xB_y$, below the temperature T_i, the solids A and A_xB_y coexist. Upon heating any mixture in this

B. A_xB_y MELTS INCONGRUENTLY

composition interval, no change occurs until T_i is reached. Whereupon all of the solid A_xB_y again decomposes to yield additional solid A plus a new phase, namely, liquid. The liquid again has the composition 2 at T_i. Further heating causes the liquid composition to follow the univariant curve $2-T_A^0$. In the two-solid-phase region $A-A_xB_y$, direct application of the lever arm principle to Fig. 2 again provides us with answers based on counting A and B as separate components. If we are interested in defining such ratios, taking into account the stoichiometries of the solid phases, it is convenient to transform the scale into terms $A-A_xB_y$. The mole ratio A to the total number of moles will then relate to the case where we differentiate between the A assumed derived from the component A and that from the defined component A_xB_y. As already pointed out, the liquidus $2-T_A^0$ is calculable based on consideration of the system A–B. In other words, M_A only approaches zero as pure B is approached at $T = 0$. The B liquidus, however, vanishes at A_xB_y and $T = 0$. This is shown in Fig. 3 where the metastable extensions of the A and B liquidus curves are shown intersecting in the metastable eutectic e_m at T_{e_2}.

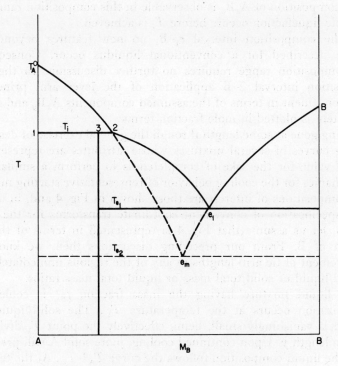

Fig. 3. Metastable extensions of the A and B liquids curves.

If an all solid composition in the interval A_xB_y–2 of Fig. 2 is heated, it will, upon reaching the temperature T_{e_1}, melt to give a mixture of pure solid A_xB_y plus liquid. Here the behavior of A_xB_y is pseudounary and, indeed, the curve 2–e_1 is describable in terms of the component A_xB_y. It is most convenient in this two-phase field to apply the lever arm principle in terms of the incomplete subsystem A_xB_y–B. For the composition interval A_xB_y–2 and the temperature interval T_{e_1}–T_i, the composition of the liquid lies along the incomplete liquidus curve e_1–2. When T_i is reached, solid A_xB_y again completely decomposes to yield solid A and liquid and upon further heating, the liquid composition follows the A liquidus curve.

In the composition interval 2–e_1, heating a mixture of solid A_xB_y and B results in a melting at T_{e_1} with attendant disappearance of the B phase as in any well-behaved binary subsystem. The liquid composition moves along the curve e_1–2 and, depending on the starting composition, final melting occurs between T_{e_1} and T_i. As indicated, in the composition interval 2–e_1, the system behaves in all respects as the subsystem A_xB_y–B, describable in terms of the components A_xB_y and B. No decomposition of A_xB_y is observable in this composition range since complete liquefaction occurs before T_i is achieved.

In the composition interval e_1–B, no new features beyond those already described for a conventional liquidus occur. Consequently, this composition range requires no further discussion. In the entire composition interval 2–B application of the lever arm principle is most convenient in terms of the assumed components A_xB_y and B when the system is plotted in mole fraction terms.

Having gone at some length through the general exercise of describing heating curves of several mixtures whose attributes are representative, it is of value for the sake of completeness to perform a similar series of mechanics for the cooling behavior of representative starting mixtures. The compositions of interest are those shown in Fig. 4 and, in order to avoid specification of convenient coordinate transforms for the several regions, let us assume that Fig. 4 is represented in terms of the mass fraction of B. From our preceding discussions then, we know that measurement of tie arm lengths in any of the regions immediately gives us solid/liquid or solid/total mass or liquid/total mass ratios.

If a liquid mixture having the mass fraction $f_B{}^a$ is cooled, first crystallization occurs at the temperature T_1. The solid/liquid mass fraction is vanishingly small, being effectively the point T_1 divided by the arm length y. Upon continued cooling, more solid A will precipitate while the liquid composition follows the curve $T_A{}^0$–T_i. At the temperature T_y, the solid/liquid mass ratio increases to x/y. When the tempera-

B. A_xB_y MELTS INCONGRUENTLY

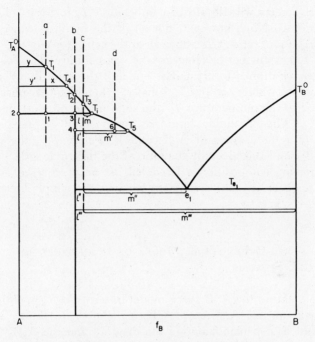

FIG. 4. The system A–B defined in mass fractions.

ture T_i is reached in further cooling, the solid/liquid mass ratio is given by the length ratios $|1 - T_i|/|1 - 2|$ and the liquid composition by the point T_i. At this temperature, solid A and liquid react to form a mixture of solid A and A_xB_y. Once all of the liquid has been consumed, the mass ratio A/A_xB_y is given by the lengths $|1 - 3|/|1 - 2|$. For the composition in question, not only has all of the liquid been used to form the solid A_xB_y, but a quantity of solid A represented by the arm segment $3-T_i$ has also been consumed in forming solid A_xB_y. At all temperatures below T_i, the mass ratio A/A_xB_y established at T_i prevails.

If a sample having the composition b is cooled, first crystallization occurs at T_2, with the liquid composition upon further cooling following the liquidus segment T_2-T_i. When the temperature T_i is reached the mass ratio solid A/liquid is given by $|T_i - 3|/|2 - 3|$. At this temperature the liquid reacts completely to form A_xB_y, i.e., note that the ratio [point 3/(2 − 3)] = 0 showing disappearance of all solid A.

Starting with the composition c, the status within the A liquidus field is evident from the preceding discussion. When T_i is reached, the situation becomes different than before. At this point, solid A completely reacts with liquid of composition T_i to form solid A_xB_y of composition 3. Now, however, not all of the liquid is consumed. The solid A_xB_y/liquid ratio at T_i, once all of the solid A is consumed, is given by m/l. Upon further cooling to the temperature T_5, for example, solid A_xB_y continues to crystallize out and the solid/liquid mass ratio increases to a value $|\ m'\ |/|\ l'\ |$. Finally, when the two-phase mixture has cooled to the temperature T_{e_1} the solid/liquid ratio attains the value $|\ m''\ |/|\ l''\ |$. The remaining liquid of composition e_1 precipitates as a mixture of pure crystals of A_xB_y and B. Below T_{e_1}, the mass ratio solid A_xB_y/solid B is given by the ratio of the tie arm lengths $|\ m'''\ |/|\ l'''\ |$.

If a starting composition d is cooled, no occurrences of an unusual nature are observed, and the system behaves as a simple binary eutectic one in all respects.

C. The Three-Dimensional Model of Incongruently Melting Systems

Figure 5 shows the A–B space model in terms of T, p, and M_B. The unary diagrams for A, B, and A_xB_y are shown on three T–p planes. The A and B T–p planes bound the diagram and the A_xB_y T–p plane lies in the interior of the model. The solid–liquid unary curves for A and B are depicted as having positive slope, while that of A_xB_y exhibits a negative slope. Thus, while the melting points of A and B increase with increasing pressure, that of A_xB_y decreases with increasing pressure. This latter fact is important, as we shall see shortly, since it represents the necessary condition for observance of a congruently melting A_xB_y phase under appropriate conditions.

In Fig. 5, the line 1–I_Q represents the temperature of incongruency T_i and the pressure of incongruency p_i. At this temperature, solids A, A_xB_y, liquid, and vapor coexist and the system is invariant. The liquid composition of this invariant point is I_Q. The metastable extension of the liquidus for A_xB_y, designated $M^{(2)}_{A_xB_y}$, to temperatures above T_i lies at higher pressures than those for A in this same interval (the liquidus I_Q–2). The curve 4–3 shows the total pressure for the stable and metastable portions of the A_xB_y liquidus ($P_t = p^0_{A_xB_y} + p_B$). The curve 1–2 shows the total pressure for only the stable portion of the A liquidus. If the A liquidus is extended to temperatures below T_i, its pressure at each temperature would lie above the sums $p_A{}^0 + p^0_{A_xB_y}$ and $p^0_{A_xB_y} + p_B$.

If the total pressure on the system is raised sufficiently, the melting point of A_xB_y will be depressed below the critical decomposition

C. p–T–M_X REPRESENTATION OF INCONGRUENCY

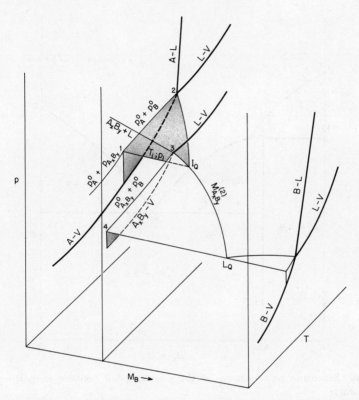

FIG. 5. The three-dimensional model showing incongruent melting in the system A–B.

temperature T_i. The compound A_xB_y will then become congruently melting, and the primitive system A–B will comprise two complete subsystems, i.e., A–A_xB_y and A_xB_y–B.

Figure 6 depicts the case where the applied constant pressure lowers the melting point of A_xB_y precisely to the value T_i. Figure 7 shows the case where the applied pressure is great enough to depress the melting point of A_xB_y below T_i. Both Figs. 6 and 7 represent isobaric slices through the condensed-phase portions of the space model, it being assumed that the pressures necessary to lower $T^0_{A_xB_y}$ below T_i are greater than those at the triple points of both A and B. Figure 8 is a reiteration of Fig. 1 in which the metastable extension of the A_xB_y liquidus to its melting point is depicted under the equilibrium pressure of the system A–B.

Since the effect of pressure on the melting points of A and B is to increase them, the melting points of A and B in Fig. 6 are increased to T_A' and T_B', respectively. The melting point of the compound A_xB_y,

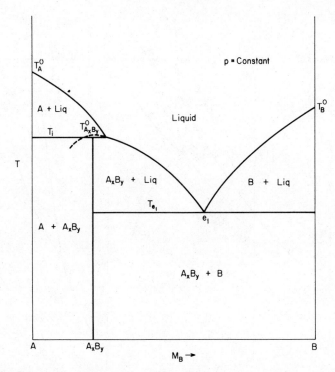

FIG. 6. Schematic representation of the singular point between congruency and incongruency.

on the other hand, is decreased to just the critical temperature T_i. The A liquidus starts at the new melting point of A, T_A' and terminates precisely at pure A_xB_y. This unique point is termed a singular point between congruency and incongruency and the temperature is precisely the critical temperature of decomposition. The A_xB_y liquidus starts at the melting point of A_xB_y and terminates in the eutectic intersection e_1' at T'_{e_1}. Thus, we see that the effect of pressure has not only resulted in the singular point emergence of a stably melting A_xB_y phase, but has also changed the A and B liquidi and the A_xB_y–B eutectic composition and temperature. In the process of generating the singular point, the melting point of A_xB_y has been depressed until it just coincides with its decomposition temperature.

When the total pressure is further increased, the melting point of A_xB_y is decreased further while the melting points of A and B are increased further as in Fig. 7. This now results in the complete emergence of the A_xB_y compound maximum. Now the A liquidus is determined by the subsystem A–A_xB_y and the B liquidus by the subsystem A_xB_y–B.

D. PRESSURE EFFECTS ON INCONGRUENT SYSTEMS

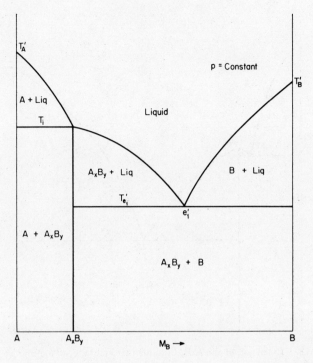

FIG. 7. Schematic representation of the emergence of the A_xB_y compound peak.

Again the eutectic composition and temperature are different from the cases depicted in Figs. 8 and 6. Since the decomposition of solid A_xB_y does not occur below the temperature T_i, it is evident that in Fig. 7 no anomalies should arise in either of the A_xB_y liquidus fields and the diagrams should be well-behaved.

Depending on whether the slopes of the unary S–L curves for pure A and B are negative, positive, or infinite, the effect of pressure on their melting points, and consequently their liquidi, will be similar, or opposite to the case we have just considered.

D. THE EFFECT OF PRESSURE WHEN THE S–L SLOPE OF THE INTERMEDIATE INCONGRUENTLY MELTING COMPOUND IS POSITIVE

In the case just considered, the application of pressure resulted in the emergence of the A_xB_y compound peak. Two other general cases are possible. In the first case, the incongruently melting A_xB_y S–L slope is positive rather than negative. In the second, the solid A_xB_y exhibits

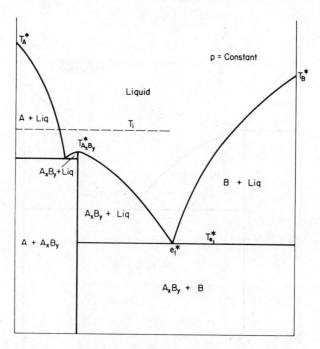

FIG. 8. Extension of the A_xB_y liquidus metastably to its melting point.

a compound maximum under the equilibrium pressure of the system, but also exhibits a positive S–L slope and has an instability temperature above its normal melting point. From Fig. 5 it is seen that if the S–L curve of pure A_xB_y is positive, an increase in pressure would result in an increasing submergence of the A_xB_y liquidus, since the temperature T_i would be achieved at increasingly greater differentials from the A_xB_y melting point. If the S–L curve is very positive, then it is not improbable that the A_xB_y liquidus may be fully submerged. Such a sequence of events starting with the system under its equilibrium pressure and increasing the pressure are depicted in Fig. 9a–e.

In Fig. 9a the system is shown under its equilibrium pressure. As the pressure is increased, Fig. 9b, the melting point of A_xB_y increases and the difference $T_{A_xB_y} - T_i$ increases. The A liquidus as a consequence encompasses a greater portion of the A_xB_y liquidus field. With further increasing pressure, Fig. 9c, the submergence of the A_xB_y liquidus becomes still greater and finally in Fig. 9d, the A_xB_y decomposition temperature occurs at a lower value than that of the A–B eutectic. In the solid field, however, A_xB_y is shown as still possessing a range of existence, the temperature T_i lying slightly under $T_{e_1}^{**}$. When solid A_xB_y

D. PRESSURE EFFECTS ON INCONGRUENT SYSTEMS

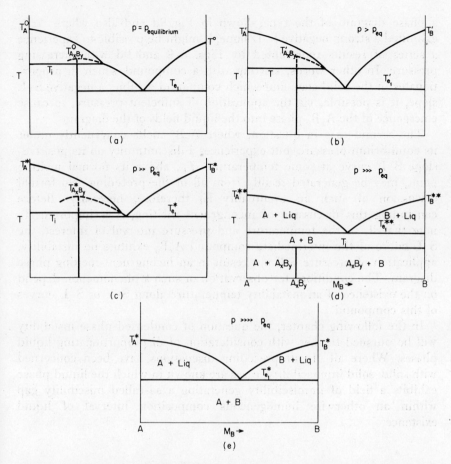

FIG. 9. Effect of increasing pressure on an incongruent melting when the incongruently melting compound exhibits a positive S–L slope.

reaches the temperature T_i, it decomposes to yield a mixture of solid A and solid B.

Finally, in Fig. 9e, the case is shown where T_i lies at a temperature below the experimental range of interest or detectability at the total pressure applied, and the presence of the compound $A_x B_y$ is never observed. Returning momentarily to Fig. 9d, all considerations previously considered for $A_x B_y$ in its incongruently melting cases are applicable where appropriate, excepting that all tie lies intersect vertical lines so that compositional variations among coexisting phases do not occur in the temperature interval in question.

In the event that $A_x B_y$ under its system equilibrium pressure exhibited a phase diagram of the type shown in Fig. 9d and also where A_xB_y exhibited a strong negative S–L slope, it might be possible to experience a series of results represented by Figs. 6–8 and 9d with increasing pressure. In other words, starting with a compound which disproportionates in the solid state and which compound exhibits a negative S–L slope, it is possible, via the application of sufficient pressure, to cause emergence of the A_xB_y phase into the liquid fields of the diagram.

The second case in question, where A_xB_y melts congruently under its equilibrium pressure, but experiences a discontinuity on its positive-slope S–L curve at some temperature, T_i, above its normal melting point, may be generated readily from all of the preceding. No formal discussion of such an eventuality is, therefore, necessary. Before concluding this discussion of incongruent melting, it is important to note that if, in the temperature and pressure interval of interest, the S–L curve of the intermediate compound A_xB_y exhibits no instability, application of pressure will not result in an incongruent-melting phase diagram. The conditions for observation of such a phenomenon depend on the existence of an instability temperature along S–V or S–L curves of this compound.

In the following chapter, the question of condensed-phase instability will be pursued further with consideration of disproportionating liquid phases. Where all of our preceding discussions have been concerned with solid–solid immiscibility, cases are known in which the liquid phase exhibits a field of immiscibility generating a so-called miscibility gap within an otherwise homogeneous composition interval of liquid existence.

26

Liquid Immiscibility in Eutectic Systems

A. INTRODUCTION

To this point, our discussions of binary systems have been confined to cases where the liquid phase was homogeneous and where the solid phases, on the other hand, were completely immiscible in one another. In this chapter we shall consider systems of the eutectic type in which, in addition to solid–solid immiscibility, there exists a temperature–composition interval in which two liquid phases coexist. Such an interval of liquid immiscibility is termed a *miscibility gap* (MG). This gap may be confined to the liquid region alone, or may intersect the liquidus of one of the components, thereby creating an invariant situation in which a solid, two liquids, and a vapor are in equilibrium.

A frequently held viewpoint is that the occurrence of inflected liquidi, of the type described in Chapter 15, is an indication of a tendency toward liquid separation. As we have seen, however, inflected curves are predictable on the basis of entropy of fusion considerations alone. Furthermore, as we have seen in Chapter 21, such inflected behavior also seems to be called for when extensive liquid-phase species interactions occur. Whether either or both of these phenomena lead to liquid-phase separation is a moot point since neither leads to a region of invariance but only to points of inflection. At present, theoretical description of liquid phase MG behavior is lacking, and our discussions must, therefore, be descriptive in nature. Logically, however, a case may be developed which argues strongly against inflected liquidus behavior being indicative of a tendency toward liquid-phase separation.

B. The Miscibility Gap Is Confined to the Liquid Regions

Figure 1 depicts a simple eutectic system showing the appearance of a miscibility gap at temperatures above the liquidi. The figure is

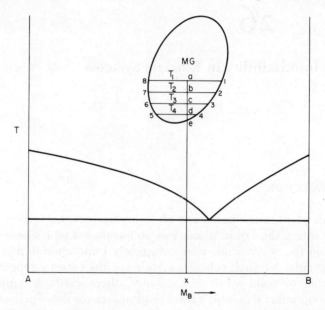

Fig. 1. A eutectic system showing a region of immiscibility in the liquid regions.

representative of a system either under its equilibrium pressure or at constant total pressure. At any temperature within the miscibility gap region, the liquid splits into so-called conjugate solutions, the compositions of which are given by determining the tie-line intersection with the MG boundaries. For example, at the temperature T_1, all starting compositions within the range of compositions 8–1 when heated to T_1 will form two immiscible liquids. The composition of one of these is represented by the point 8 and the composition of the other by the point 1. If we begin with the composition x and heat to the temperature T_1, the moleculur ratio of the components in conjugate

C. T–p–M_X REPRESENTATION OF THE MISCIBILITY GAP

solution α (that having the composition denoted by the point 8) to that in the conjugate solution β (that having the composition denoted by the point 1) is $|a - 1|/|8 - a|$. In cooling to T_2, the composition of conjugate solution α follows the curve 8–7, while the composition of conjugate solution β follows the curve 1–2. At point e, the ratio of molecules in csα/csβ is vanishingly small. In other words, at point e, one of the conjugate solutions has all but vanished and the liquid phase is almost homogeneous. Below point e, the liquid phase does, indeed, become homogeneous. The maximum temperature in the miscibility gap is sometimes referred to as the critical solution temperature, and the system concentration at this point is sometimes referred to as the consolute point.

C. The T–p–M_X Representation of the Eutectic System Showing Liquid Phase Immiscibility

The three-dimensional representation of the system shown in Fig. 1 is presented in Fig. 2. For simplicity, solid and liquid volumes have not been constructed and excepting for the dashed line 1–V_t–2, the vapor composition lines are not shown. In the region denoted liquid–vapor, the isotherms represent the variation of total pressure with composition. It is seen that, in the volume enclosed by the miscibility gap, the pressure is invariant in the three-phase equilibrium L_A, L_B, vapor. It is seen also that above a critical pressure C_0 the MG vanishes. In general also, there is no relationship between the consolute point on a T–M_X projection and the three-dimensional consolute point C_0. Figure 2 has been simplified further in Fig. 3, and an isobaric slice through the MG is shown. In addition, the bounding planes of the S–L regions are outlined.

In Fig. 4a–f a series of isotherms corresponding to those labeled in Fig. 2 are shown. Each of these isothermal planes provides a pressure–composition diagram in which liquid, solid, and vapor compositions are shown. Figure 4a is an isotherm at a temperature just above the maximum temperature of the gap. Here the two phases, liquid–vapor coexist and the system is isothermally univariant over the entire composition range. Figure 4a has been previously described in Chapter 24. Figure 4b represents the case where the isothermal slice barely intersects the highest temperature region of the three dimensional gap. Starting at $p_A^\dagger$ or $p_B^\dagger$, the vapor pressures of pure liquid A or B, respectively, the L–V curves are not unlike those of Fig. 4a. In the gap composition interval, however, the system is isothermally invariant since two liquids and a vapor coexist. The liquid compositions coexisting at the three-

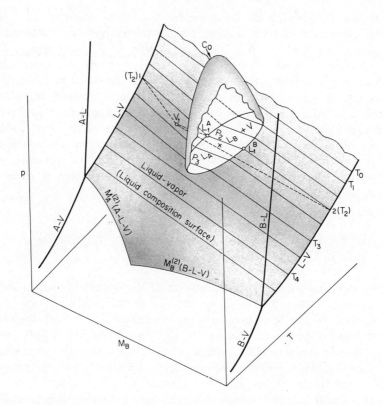

Fig. 2. The liquid MG T–p–M_X diagram showing the liquid composition and total pressure curves only (except for the L–V isothermal slice T_2).

phase points are given by the compositions L_t^A and L_t^B. The composition of the vapor at the same unique three-phase total pressure is given by the point V_t. Obviously, the vapor composition curves $p_A^\dagger$–V_t and $p_B^\dagger$–V_t must intersect at the vapor composition triple point V_t. Consequently, neither of these curves is describable via the Raoult equation since starting at $p_A^\dagger$, the total pressure decreases more rapidly than in the ideal case and the total pressure curve starting at $p_B^\dagger$ increases less rapidly than in the ideal case.

At the temperature T_2, the isothermal plane intersects more deeply into the gap, Fig. 4c, and the composition interval of the gap is greater. We also see the maximum pressure (with no vapor present) up to which the gap exists at this temperature. In Fig. 4d, at a still lower temperature T_3, the isothermal slice catches the gap at about its lowest temperature and between this temperature and the liquidi the isotherms again

C. T–p–M_X REPRESENTATION OF THE MISCIBILITY GAP

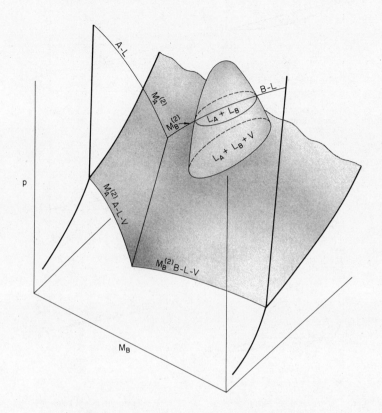

FIG. 3. The T–p–M_X diagram of Fig. 2 showing an isobaric slice through MG.

appear as in Fig. 4a, namely, Fig. 4e. Finally, at the temperature T_5, an isotherm intersects the A liquidus. In the liquid composition interval $p_B{}^\dagger$ to L_t the curves look like Fig. 4a and e. When solid A precipitates at L_t, the system becomes isothermally invariant again. In this region of isothermal invariance, the compositions of the vapor, liquid, and solid at the total constant pressure P_t (where $P_t = p_A{}^0 + p_B$ and $p_A{}^0$ is the vapor pressure of pure solid A at T_5, and p_B is the vapor pressure of B as determined by Raoult's law for the liquid composition on the liquidus) are given by the points V_t, L_t, and S_t, respectively. The dashed lines extending from V_t and L_t to $p_A{}^\dagger$ show the metastable extension of the S–V curve to the metastable pressure $p_A{}^\dagger$ of pure supercooled liquid A. If we assume that the solid–liquid curves are ideal, then it is evident from our previous treatments that the L–V curves must also be ideal.

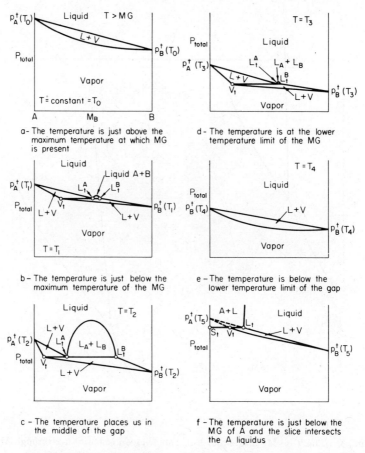

FIG. 4. Isothermal planes through the $T-p-M_X$ diagram of a system showing a MG confined to the liquid regions only.

Only in this way can the point L_t be related to the pressures $p_A{}^†$ and $p_A{}^0$ by the Raoult relationship (see Chapter 13, Section C, and Chapter 24). The construction of isobaric sections through Fig. 2 to give $T-M_X$ graphs is left as an exercise for the reader.

D. The Intersection of a Miscibility Gap with the Liquidus Regions

If a miscibility gap intersects a liquidus curve, then the maximum number of phases in evidence will be four, namely a solid, two liquids, and a vapor. The system will be invariant. Under a constant pressure

D. THE MISCIBILITY GAP INTERSECTS A LIQUIDUS

constraint, the system will be isobarically invariant and the phases, solid and two liquids will coexist. It is clear that, thermodynamically, the preceding fact eliminates any possibility of a gap intersecting both liquidi, since this would lead to a negative number of degrees of freedom.

The appearance of a system showing intersection of the gap with the liquidus of the component A is shown in Fig. 5. Figure 5 is representative

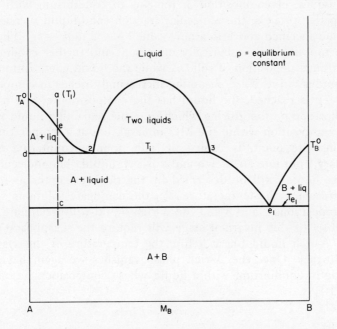

FIG. 5. A eutectic system showing a miscibility gap intersecting a liquidus.

either of a projection of the liquid and solid composition lines onto the basal $T-M_x$ plane of the three-dimensional model, or of an isobaric section through the three-dimensional model.

If a mixture of composition a is cooled from the temperature T_1, it will become saturated with respect to solid A when point e is reached. At this point the system becomes isobarically univariant and the composition of the liquid is a function of temperature. Further cooling results in continued crystallization of solid A, while the liquid composition follows the liquidus 1–2, becoming increasingly rich in the component B in the process. When the temperature reaches T_i, the molecular fraction solid A/liquid is given by the tie arm lengths $\mid b - 2 \mid / \mid d - b \mid$. At this temperature, the liquid achieves the composition 2 and decomposes to form a liquid much richer in the component B (having the composition 3). In this process, the original liquid must divest itself of component A and does this by depositing more solid A. When T_i is reached and the B-rich liquid forms, the system comprising solid A, A-rich liquid, and B-rich liquid is isobarically invariant. The temperature cannot change until one of the phases has vanished. The phase that vanishes is the A-rich liquid phase. This process continues until a single liquid, of composition 3, remains in equilibrium with solid A. During this process, the molecular fraction solid/liquid increases from the value specified above to a new value $|\, b - 3\, |/|\, d - b\, |$. The system at this time is again isobarically univariant and further cooling results in continued deposition of solid A while the liquid composition follows the liquidus curve $3-e_1$, becoming increasingly richer in component B. When T_{e_1} is reached, the liquid has the composition e_1 and a eutectic crystallization occurs, during which time the remaining liquid vanishes.

If a composition within the MG interval, but at a temperature above the consolute point, is cooled, it will, upon encountering the MG boundary, split into an A-rich and a B-rich liquid. The molecular ratios of these liquids will be described by the tie-arm lengths as in Fig. 1. Upon reaching the temperature T_i, the conjugate solutions will have the compositions 2 and 3 and the A-rich liquid will precipitate solid A. The onset of this invariant stage will require the disappearance of all of the A-rich liquid phase before the temperature of the system will drop further. Once the A-rich phase vanishes we again have a solid A phase in equilibrium with a liquid whose composition lies along the curve $3-e_1$.

E. The $T-p-M_X$ Representation
   of the MG Intersecting a Liquidus

In Fig. 6, the $T-p-M_X$ solid model, from which Fig. 5 is taken, is constructed. This model is considerably more involved than the one shown in Figs. 2 and 3. In addition to the univariant line L_Q-L_t that represents the variation of the eutectic liquid composition with total pressure from its invariant termination at L_Q (the liquid composition at the quadruple point) through its succession of isobaric points, the univariant loop $L_Q{}^A-1-L_Q{}^B$ is also present. $L_Q{}^A-1-L_Q{}^B$ shows the surface of the miscibility gap that intersects the liquid–solid surface. This closed loop terminates at the quadruple points $L_Q{}^A$ and $L_Q{}^B$ which show the liquid phase compositions when solid A, two liquids, and a vapor coexist under the equilibrium pressure of the system. To minimize confusion in obtaining the perspective of Fig. 6, the curves showing vapor composi-

E. T–p–M_X DIAGRAM OF THE MG

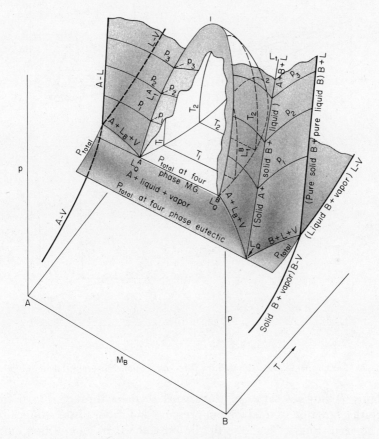

FIG. 6. The T–p–M_X representation of the MG intersecting the liquids of A.

tion have not been constructed. The curve L_Q^A–2–L_Q^B shows the coexistence curve, two liquids–vapor and is analogous to the MG of Fig. 2 except that it terminates in a constant pressure line L_Q^A–L_Q^B rather than exhibiting the somewhat circular shape of Fig. 2. In essence, Fig. 6 only exhibits a part of the three-dimensional MG of Figs. 2 and 3, the lower temperature portion of the gap being cut off by its intersection with the S–L surfaces. The lines p_1, p_2, and p_3 in Fig. 6 show the intersection of isobaric planes with the liquidi and MG at different pressures. The lines T_1 and T_2 show isothermal plane intersections with the gap. The point 1 is the highest pressure at which the gap occurs and clearly does not coincide with the equilibrium pressure consolute point 2.

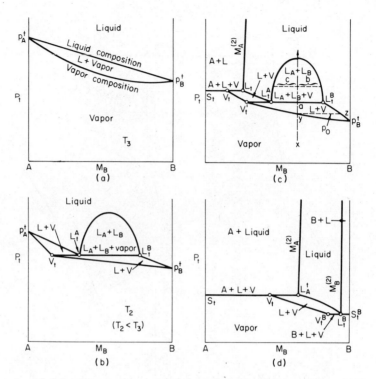

FIG. 7. Isothermal sections through the three-dimensional diagram shown in Fig. 6.

Figure 7 shows a series of isothermal sections through Fig. 6 from which the variation in total system pressure and vapor phase composition may be seen. Figure 7a shows the liquid and vapor composition curves at temperatures above the maximum temperature to which the gap extends. As usual p_A^t and p_B^t represent the pressures of pure liquids A and B, respectively. In Fig. 7b is represented an isothermal slice through the T–p–M_x diagram at a temperature within the gap but above the maximum temperature of the A liquidus (the triple point of A is assumed to be at a higher temperature than the triple point of B). V_t is the composition of the vapor in the three phase, invariant isothermal equilibrium: L_A, L_B, and vapor. The points L_t^A and L_t^B represent the composition of the liquid in equilibrium with vapor at this triple point. The point A represents the three-dimensional consolute point and shows the maximum pressure (in the absence of vapor, of course) to which the gap extends.

In Fig. 7c, the isothermal slice is taken at a temperature which is below the melting point of A, but above the melting point of B. The

E. T–p–M_X DIAGRAM OF THE MG

slice then cuts through the A liquidus. The three-phase equilibrium solid A, liquid, and vapor is isothermally invariant. The compositions of solid, vapor, and liquid in this isothermally invariant equilibrium are given by the points S_t, V_t, and L_t, respectively. The variation of the liquid composition with pressure in the univariant equilibrium solid A–liquid is given by the line denoted as $M_A^{(2)}$. In the composition region between the three-phase liquidus S–L–V and the miscibility gap, there exists a two-phase equilibrium L–V. The liquid composition in this region as a function of p is given by the line L_t–$L_t{}^A$ and the vapor composition by the line V_t–$V_t{}'$. At the gap, the three-phase equilibrium occurs, consisting of vapor, A-rich liquid, L_A, and B-rich liquid, L_B. This represents an isothermally invariant region and the compositions of vapor, L_A, and L_B are given by the points $V_t{}'$–$L_t{}^A$ and $L_t{}^B$, respectively. The constant pressure of the system at this invariant point is denoted by the line $V_t{}'$–$L_t{}^A$–$L_t{}^B$. Past the MG (toward B) the system is again isothermally univariant and consists of liquid and vapor. The vapor composition is given by the line $V_t{}'$–$p_B{}^\dagger$ and the liquid composition by the line $L_t{}^B$–$p_B{}^\dagger$. In this region, the total pressure varies between $p_B{}^\dagger$ (the pressure of pure liquid B) and the triple point pressure for the equilibrium L_A–L_B–V. The closed loop $L_A + L_B$ shows the variation of composition of the univariant two phase equilibrium $L_A + L_B$ at pressures above the triple point pressure.

If we begin with a system whose composition is denoted by x as in Fig. 7c and compress the vapor via a piston, liquid of composition z will condense when the pressure reaches the value p_0. The vapor composition at this point will be given by the point y. The ratio of liquid-to-vapor is, of course, vanishingly small at this point and is given by the tie arms point $y/|y - z|$. As the pressure is increased, the liquid composition moves up along the curve z–$L_t{}^B$ while the vapor composition moves up along the curve y–$V_t{}'$. The ratio of liquid-to-vapor increases and is given by the ratio of tie-arm lengths as usual. When the total pressure achieves the value of the triple point V–L_A–L_B, the composition of the three phases is, of course, V_t, $L_t{}^A$, and $L_t{}^B$ and the pressure cannot increase until the vapor phase vanishes. The ratio $L_t{}^A/L_t{}^B$, when all of the vapor has disappeared, is given by the ratio $|a - L_t{}^B|/|a - L_t{}^A|$. With further increase in pressure, the L_A to L_B ratio is given by the ratio of tie arms such as b/c. Finally, with sufficient increase in pressure, the system becomes homogeneous. From the preceding discussion, the variation with pressure of any starting composition within or outside of the MG composition interval may readily be deduced.

Figure 7d represents an isothermal slice for a temperature above the eutectic but below the melting points of solid A or solid B. Since the slice intersects both liquidus fields, two sets of invariant triple point compositions must be specified. S_t, V_t, and $L_t{}^A$ represent the compositions of solid, liquid, and vapor in the A liquidus three-phase equilibrium. $V_t{}^B$, $L_t{}^B$, and $S_t{}^B$ represent the vapor, liquid, and solid compositions in the three-phase equilibrium for the B liquidus. The dependency of the A liquidus on pressure is shown by the line marked $M_A^{(2)}$, while the dependency of the B liquidus on pressure is $M_B^{(2)}$. Between the A and B liquidi, the two-phase region L–V is intersected. The liquid composition as a function of p is given by the line $L_t{}^A$–$L_t{}^B$ and the vapor composition by the line V_t–$V_t{}^B$.

In the existing literature, intuitive arguments have been offered in which the MG is visualized as varying in stages depending upon its degree of submergence within the solid confines of the phase diagram. That is, starting in stage 1, with the MG lying below the eutectic temperature of Fig. 5, for example, a sequence of stages is visualized in which the gap moves to higher temperatures finally erupting into the liquid regions.

In the solid regions then, the MG is held responsible for the formation of solid conjugate solutions. From Fig. 6, however, such a sequence of events is seen to be improbable. One would expect, for example, that if a system were to exhibit a small liquid phase MG on a temperature–composition diagram, a considerable MG would be evident in the solid regions since only the tip of the gap would have erupted. In the system Cd–Se, for example, the author has found that despite the existence of a confined liquid phase gap, solid solubility of the end members in each other is vanishingly small. One would have expected, based on the intuitive argument cited, to find conjugate solid solutions in which considerable solid solubility was evident.

There is one further argument against the "eruptive type of MG pictorialization." With reference to Fig. 6, it is seen that if the gap is extended to lower temperatures, namely, if the loop $L_Q{}^A$–2–$L_Q{}^B$ is completed metastably, the equilibrium, solid A–L_A–L_B–V will exhibit a vapor pressure higher than that for the equilibrium system which has a vapor phase as one of its parts. For such an occurrence, it is necessary that the gap must be unobserved in solid–vapor regions, for example, when it is observed in liquid–vapor regions. This is clearly not the case since systems are known in which extensive solid solubility is observed simultaneously with liquid immiscibility. Similarly, systems are known with extensive solid solubility and no evidence of liquid immiscibility. Finally, systems are known that exhibit little solid solubility, extensive liquid solubility, and a small liquid MG.

The T–M_X diagrams as a function of increasing pressure are shown

E. T–p–M_x DIAGRAM OF THE MG

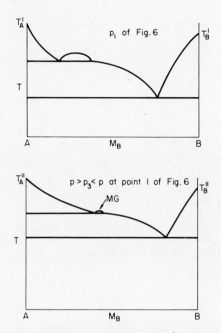

FIG. 8. Effect of increasing pressure on the MG.

in Fig. 8 (see Fig. 6). The compositional shell of the MG narrows with increasing pressure. At a pressure precisely that of the three-dimensional consolute point (point 1 of Fig. 6), the contact of the A liquidus with the gap is tangential giving rise to a point of inflection in the A liquidus. Certainly then this appears to correlate with the argument that points of inflection are indicative of a tendency toward unmixing. A little thought shows this correlation is probably coincidental. First, it is only at the unique pressure at which tangential contact occurs that an inflected liquidus is expected. Second, many inflected systems are known. It would be a strange coincidence if each of these inflected systems (studied for the most part at 1 atm constant total pressure) were examined at precisely the pressure necessary to exhibit tangential contact of the liquidus with the consolute point and that all consolute points fortuitously occurred at the same pressure.

This chapter on MG behavior is a necessary prelude to entertaining discussions on all manner of solid solubility, and is intended to provide some insight into the nature of coexistent solutions as well as to indicate techniques of constructing systems exhibiting limited solubility of end members in one another.

27

Systems Exhibiting Solid Solubility—
Ascending Solid Solutions

A. Introduction

It is evident that the setting up of a boundary of complete solid phase immiscibility represents an idealized version of nature that, to a lesser or greater degree, depending upon the system in question, cannot be realized in practice. Frequently, however, unless one resorts to extremely sophisticated experimental techniques to detect solid solution (solids exhibiting variable composition), systems, to all intents and purposes, of the type previously discussed do exist. On the opposite side of the spectrum there are many known systems where the end members appear to be completely soluble in one another in the solid state. These cases exhibit complete miscibility of end members in both the liquid and solid phases and may be termed continuous solid solution interactions.

A notable difference between systems exhibiting complete solid phase immiscibility, on the one hand, and complete solid phase miscibility, on the other hand, is that the crystal structures of the end members in the temperature interval of interest are, in general, different in the former instance, and must be the same in the latter. It has been observed

A. INTRODUCTION

experimentally that if two end members possess precisely the same crystal structure but exhibit lattice constants that differ by 10–15% or more, continuous solid solution is not likely to occur. If, however, the crystal structures are the same, and in addition, the lattice constants do not differ appreciably, the interaction may be such as to generate a continuous series of solid solutions over the entire composition interval 0–1. In the simplest of such cases, the lattice constant, if the system comprises cubic end members, is purported to vary continuously and linearly with composition (Vegard's law).[1]

The criterion of equivalent crystal structures of end members, for systems exhibiting continuous solid solution, is an obvious one if we realize that only a single solid phase is present over the entire compositional extent. Thus, while the lattice constant may vary as a function of composition, the structure will not, and at any particular starting composition only a single solid is observable. The mechanism involved with changing composition is such that the atoms or molecules of the end member B substitute directly for the atoms or molecules of the end member A as we change the composition from 0 to 1 mole fraction B. Since the "substitutional" solid solution generated is not (in the ideal case) ordered with respect to which atoms of A are replaced in the lattice, the solid solution is obviously disordered. In other words, while the atoms or molecules of A or B are homogeneously distributed on a macroscopic level they need not be homogeneously distributed on a microscopic level. This represents another distinction between solid solution and solid compound phases. The latter, to a good first approximation, exhibit distinct crystallographic ordering of the components, that is, each crystallographic location acts as a preferential site for one or the other of the species derived from one or the other of the components or from the reaction between them.

In this chapter we will consider the simplest case of solid solubility which represents the diametrically opposite case to that of complete immiscibility discussed in the preceding chapters. This simplest solid solution system exhibits melting point variation of the solid phase that is monotonic with composition, lying between the melting points of the end members. It is always the case in these ascending solid solutions, as they are termed, that the liquid in equilibrium with a solid solution of specified composition is richer in that end member exhibiting the lower melting point and that the solid as a consequence is richer in that member exhibiting the higher melting point.

As before, the term liquidus represents either the temperature of

---

[1] See the discussion under Chapter 27 in the Appendix.

first crystallization or alternately the temperature of last melting of a given starting system composition. The term solidus also has the definition previously employed. However, in the ascending solid solution, since never more than a single solid phase coexists with liquid, the solidus cannot represent a condition of isobaric invariance. Here, like the liquidus, the solidus is univariant.

In keeping with the pedagogical approach previously employed, the first case to be treated in idealized fashion is that in which component and species mole fractions are identical. Thus, the component A is present in the liquid and solid only as the species A and the component B is present in both phases as the species B.

B. The Analytical Description of Univariance for Ascending Solid Solutions

Let us consider the interaction A–B as before. As with the simplest eutectic interaction treated, no homogeneous equilibrium is involved. The heterogeneous equilibria involved are described by

$$A^{(1)} \rightleftarrows A^{(2)} \tag{1}$$

$$B^{(1)} \rightleftarrows B^{(2)} \tag{2}$$

where $A^{(1)}$ is the species A present in the solid solution (ss), and $A^{(2)}$ is the species A present in the liquid solution, $B^{(1)}$ is the species B present in the solid solution, and $B^{(2)}$ is the species B present in the liquid solution. Since the solid solution, upon melting, releases both A and B into the liquid phase, it is to be anticipated that the univariant equations to be generated shall contain enthalpies of fusion of both solid A and solid B. The equilibrium constant defining the process described in Eq. (1) is given by

$$K_A = \frac{N_A^{(2)}}{N_A^{(1)}} \bigg/ \frac{N_A^{(2)0}}{N_A^{(1)0}} = \exp\left[-\frac{\Delta H_A^0}{R}\left(\frac{1}{T} - \frac{1}{T_A^0}\right)\right] = e^{-A} \tag{3}$$

and that for the process described in Eq. (2) by

$$K_B = \frac{N_B^{(2)}}{N_B^{(1)}} \bigg/ \frac{N_B^{(2)0}}{N_B^{(1)0}} = \exp\left[-\frac{\Delta H_B^0}{R}\left(\frac{1}{T} - \frac{1}{T_B^0}\right)\right] = e^{-B}. \tag{4}$$

The superscript 0 terms refer to the pure phases and the values $N_X^{(2,1)0}$ each equal one. In previous treatments, the superscript (1) terms referred to pure solids, so that the species concentrations $N^{(1)}$ could be set equal to one. Where the concentration of the solid is not a constant,

B. IDEAL SOLID SOLUTION EQUATIONS

as is the case in a solid solution system, the above simplification cannot be made. Equations (3) and (4), therefore, contain the four unknowns $N_A^{(1)}$, $N_A^{(2)}$, $N_B^{(1)}$, and $N_B^{(2)}$. The two-phase two-component equilibrium in question is univariant and the degree of freedom, as before, is chosen to be the temperature. To provide the two additional independent equations [in addition to Eqs. (3) and (4)] necessary to enable simultaneous solution for the four unknowns, we make use of the constraining equations

$$N_A^{(2)} = 1 - N_B^{(2)} \tag{5}$$

and

$$N_A^{(1)} = 1 - N_B^{(1)}. \tag{6}$$

Finally, before proceeding, let us set forth the component mole fraction conservation conditions

$$m_A^{(2)} = n_A^{(2)}, \tag{7}$$

$$M_A^{(2)} = N_A^{(2)}, \tag{8}$$

$$m_B^{(2)} = n_B^{(2)}, \tag{9}$$

$$M_B^{(2)} = N_B^{(2)}. \tag{10}$$

There is an exactly equivalent set for the solid phase.

In Eq. (3), we may substitute for $N_A^{(1)}$ the relationship Eq. (6) to give

$$N_A^{(2)} = [1 - N_B^{(1)}] e^{-A}. \tag{11}$$

From Eq. (4), we see that

$$N_B^{(1)} = N_B^{(2)}/e^{-B}. \tag{12}$$

This value of $N_B^{(1)}$ may be substituted in Eq. (11) to give

$$N_A^{(2)} = [1 - (N_B^{(2)}/e^{-B})] e^{-A}. \tag{13}$$

We may now invoke Eq. (5) and substitute for $N_B^{(2)}$ in Eq. (13) to give

$$N_A^{(2)} = [1 - [(1 - N_A^{(2)})/e^{-B}]] e^{-A}. \tag{14}$$

Solving for $N_A^{(2)}$ yields

$$N_A^{(2)} = \frac{e^{-A}[e^{-B} - 1]}{e^{-B} - e^{-A}}. \tag{15}$$

Expanding the terms A and B in the exponentials gives finally as the variation of the concentration of A in the liquid with T

$$N_A^{(2)} = \frac{\exp\left[-\frac{\Delta H_A^0}{R}\left(\frac{1}{T} - \frac{1}{T_A^0}\right)\right]\left\{\exp\left[-\frac{\Delta H_B^0}{R}\left(\frac{1}{T} - \frac{1}{T_B^0}\right)\right] - 1\right\}}{\exp\left[-\frac{\Delta H_B^0}{R}\left(\frac{1}{T} - \frac{1}{T_B^0}\right)\right] - \exp\left[-\frac{\Delta H_A^0}{R}\left(\frac{1}{T} - \frac{1}{T_A^0}\right)\right]}. \qquad (16)$$

Invoking Eq. (5), the solution for $N_B^{(2)}$ is given by

$$N_B^{(2)} = \frac{\exp\left[-\frac{\Delta H_B^0}{R}\left(\frac{1}{T} - \frac{1}{T_B^0}\right)\right]\left\{\exp\left[-\frac{\Delta H_A^0}{R}\left(\frac{1}{T} - \frac{1}{T_A^0}\right)\right] - 1\right\}}{\exp\left[-\frac{\Delta H_A^0}{R}\left(\frac{1}{T} - \frac{1}{T_A^0}\right)\right] - \exp\left[-\frac{\Delta H_B^0}{R}\left(\frac{1}{T} - \frac{1}{T_B^0}\right)\right]}. \qquad (17)$$

Knowing $N_A^{(2)}$ and $N_B^{(2)}$ as a function of T we may now substitute these values into Eqs. (3) and (4) to obtain the temperature dependencies of $N_A^{(1)}$ and $N_B^{(1)}$, respectively:

$$N_A^{(1)} = \frac{\exp\left[-\frac{\Delta H_B^0}{R}\left(\frac{1}{T} - \frac{1}{T_B^0}\right)\right] - 1}{\exp\left[-\frac{\Delta H_B^0}{R}\left(\frac{1}{T} - \frac{1}{T_B^0}\right)\right] - \exp\left[-\frac{\Delta H_A^0}{R}\left(\frac{1}{T} - \frac{1}{T_A^0}\right)\right]}, \qquad (18)$$

$$N_B^{(1)} = \frac{\exp\left[-\frac{\Delta H_A^0}{R}\left(\frac{1}{T} - \frac{1}{T_A^0}\right)\right] - 1}{\exp\left[-\frac{\Delta H_A^0}{R}\left(\frac{1}{T} - \frac{1}{T_A^0}\right)\right] - \exp\left[\frac{\Delta H_B^0}{R}\left(\frac{1}{T} - \frac{1}{T_B^0}\right)\right]}. \qquad (19)$$

It is easily found that the sums $N_A^{(2)} + N_B^{(2)}$ and $N_A^{(1)} + N_B^{(1)}$ in Eqs. (16)–(19) each equal one, consistent with Eqs. (5) and (6). Examination of Eqs. (16)–(19) further reveals that none of the relationships is linear. Since the relationships $N_A^{(2)}$ and $N_A^{(1)}$ are different, it is evident also that the liquidus and solidus curves do not coincide. Furthermore, looking at the extremes when $T = T_A^0$ or T_B^0 and $T_A^0 \leqslant T \leqslant T_B^0$ or $T_B^0 \leqslant T \leqslant T_A^0$, we see that $N_B = 1$ when $T = T_B^0$, $N_B = 0$ when $T = T_A^0$, and likewise for N_A. From Eqs. (3) and (4) it is seen that with $T_A^0 \leqslant T \leqslant T_B^0$ and $\Delta H > 0$, the ratio $N_A^{(2)}/N_A^{(1)}$ is always greater than 1 and the ratio $N_B^{(2)}/N_B^{(1)} < 1$. With $T_A^0 \geqslant T \geqslant T_B^0$ the reverse is the case. This demonstrates that the liquid is richer than the solid in the lower melting end member. Conversely, it demonstrates that the solid is richer in the higher melting end member.

It has been stated above that the sums $N_A^{(1)} + N_B^{(1)}$ or $N_A^{(2)}$ and $N_B^{(2)}$, as required, equal one. If it can be demonstrated that each of the terms

C. SOLID SOLUTION PHASE DIAGRAMS

$N_A^{(1)}$, $N_B^{(1)}$, $N_A^{(2)}$, and $N_B^{(2)}$ are, in addition, positive, it is proved that each of them, separately, lies in the interval 0–1.

As an example, consider Eq. (18). Assume that $T_A^0 \leqslant T \leqslant T_B^0$. It is seen that both the numerator and denominator are negative. Consequently, $N_A^{(1)}$ is positive. Conversely, for the same set of conditions, the numerator and denominator of Eq. (19) are positive so that $N_B^{(1)}$ is positive. In Eq. (16) both numerator and denominator are negative so $N_A^{(2)}$ is positive. Conversely, in Eq. (17), the numerator and denominator are positive requiring $N_B^{(2)}$ to be positive. Equivalent results are obtained for the case $T_B^0 \leqslant T \leqslant T_A^0$. Thus we have demonstrated that if T lies in the interval between the melting points of A and B, physically meaningful answers are obtained, i.e., each of the mole fractions lies in the interval 0–1 and the sum of species mole fractions in either the liquid or solid equals one.

C. THE GRAPHICAL REPRESENTATION OF ASCENDING SOLID SOLUTIONS

Leaving to a later chapter presentation of calculated data for hypothetical ascending solid solutions, where the values ΔH_X^0 and T_X^0 are varied, we will in this section, present an interpretive description of the binary ascending solid-solution phase diagram. A schematic represen-

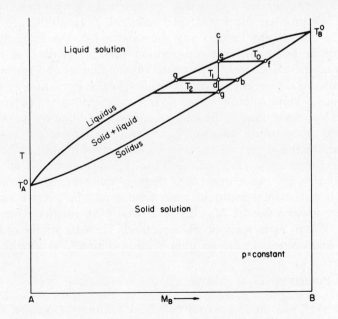

FIG. 1. Schematic representation of an ascending solid solution.

tation for a general form of such a system in terms of temperature and composition is given in Fig. 1. Three distinct regions are evident, an isobarically bivariant liquid region, an isobarically univariant solid–liquid region, and an isobarically bivariant solid solution region. At all temperatures above the liquidus curve, the system is molten and at all temperatures below the solidus, the system is solid and homogeneous. As required, the liquidus and solidus are both monotonic with respect to T. A tie line constructed isothermally between the liquidus and solidus, i.e., the tie line a–b, defines the compositions of coexisting liquid and solid at a specified temperature. As demonstrated mathematically in Section B, the liquid is seen to be richer in that component exhibiting the lower melting point. Conversely, the solid is always richer in that component exhibiting the higher melting point.

A notable difference between the eutectic diagrams previously discussed and the solid solution depicted in Fig. 1 is that tie line intersections in the latter never occur with vertical lines.

If a sample having a composition c is cooled, first traces of solid crystallize at the temperature T_0. The crystallized solid is a homogeneous solid solution of composition f and the solid-to-liquid ratio at this point is essentially zero. If the sample is further cooled to the temperature T_1, solid solution continues to crystallize while the liquid composition moves along the curve e–a toward a. If equilibrium is maintained during this cooling process, the initial solid solution, f, reequilibrates in the system so that at any point along the solidus f–b, the solid phase is homogeneous and exhibits a composition variation with T along f–b. In practice it is frequently difficult to obtain this reequilibration of a solid phase while cooling, and this results in the formation of a non-homogeneous solid solution phase. While the liquidus temperature for all starting compositions in the interval a–b will be different, it is evident that at any temperature such as T_1, the compositions of the coexisting phases will be the same, only the relative amounts of the phases differing. For the constructed case, the solid-to-liquid mole ratio at T_1, is $|a - d|/|d - b|$. At a somewhat higher value of M_B within the interval a–b, the solid-to-liquid mole ratio would be greater and at a somewhat lower value of M_B this ratio would be smaller when T_1 is reached. When T_2 is reached, the system is almost entirely solid and finally a homogeneous solid solution of composition C is obtained.

D. The Purification of Materials

As we shall see later, a variety of solid solution forms are known. In the limit, of course, all eutectic systems exhibit some degree of solid

D. THE PURIFICATION OF MATERIALS

solubility of end members. When dealing with an essentially pure substance, the behavior of impurities is frequently such that, to a first approximation, the conglomeration of impurities appear to behave as if they represented a single impurity.[2] Furthermore, if the phase diagram, "primary substance—conglomeration of impurities," is examined, it almost always has the appearance of an ascending solid solution near the primary substance axis. This has enabled the application of a technique known as zone refining to the purification of the primary substance to an extremely high degree. The principle by which the continuous-process zone-refining approach operates is best described with reference to an older, step-type operation, used extensively in purification of water soluble materials, organic mixtures, and metals. This latter technique, which is still in use, but which is being displaced by the more efficient zone refining technique (where feasible), is termed fractional crystallization.

In Fig. 2 a solid solution diagram of the type constructed in Fig. 1, where we have expanded the scale around the pure B phase, is shown. Assume we start with a solid phase containing approximately 0.98 mole fraction of B and desire to purify it. If we melt this phase and cool to T_1, the composition of the first solid is greater than 0.99 mole fraction B, i.e., the point b. However, the quantity of solid that is present at T_1 is vanishingly small making it impractical to cool only to this temperature. If we cool to T_2, the solid-to-liquid ratio becomes $|c-e|/|e-d|$ and if we cool to T_3, it increases to $|f-h|/|h-g|$. However, note that in lowering the temperature to T_3, the enhanced purity of the solid is less than at T_2 which is less than at T_1. If we cool to T_3, then separate the solid and liquid phases, we can reheat the solid of composition g to above the liquidus and repeat the cooling process. The g composition phase may be termed the "first cut." If we cool this composition to T_4, the solid fraction composition is again enriched relative to B and now has the composition i. Each time we repeat the process of separating solid from liquid, we lose some of the desired material to the liquid phase. Consequently, one attempts to cool each time to a temperature at which the solid-to-liquid ratio is large. The number of times the fractionation is effected depends on how pure a final product is desired and how much one is prepared to lose. In an industrial approach to the problem one would retain the liquid fractions and combine them. These could then be subjected to a further step-wise fractional crystallization.

A dynamic approach to the fractional crystallization method in which,

---

[2] This behavior is roughly analogous to ideal gas behavior in that the number of molecules present rather than their chemical composition is the determining factor.

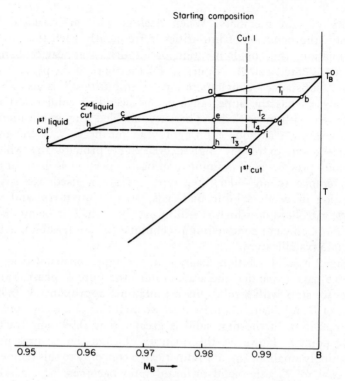

FIG. 2. Expanded system of that shown in Fig. 1.

because of the continuous nature of the process, the loss of desired material is minimal, is the technique of zone refining alluded to above. Because it is somewhat inefficient, in that equilibrium is not obtained in individual stages, it is applied often in a so-called multiple pass process and is capable of yielding a desired elemental or pseudounary substance in remarkable purities, i.e., containing 1 part in 10^9 or less of impurities. The method functions as follows.

Suppose the material to be purified is placed into a suitable container, as in Fig. 3. This container is located in a furnace in which either a vacuum can be maintained or an ultrapure inert gas ambient environment may be established.

Starting at the end A, a narrow heater is traversed slowly toward point B. Immediately under the heater the solid melts forming a liquid zone, whence the term zone refining. Consistent with practical requirements, the zone is slowly passed through the distance A—B. If the impurities behave as the lower melting component, they are swept forward in the molten zone. Immediately in back of the traversing

D. THE PURIFICATION OF MATERIALS

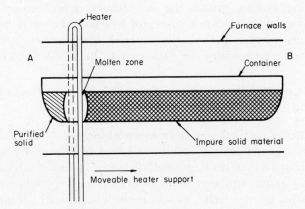

FIG. 3. Schematic representation of a zone melting apparatus.

zone (toward A) the solid recrystallizes and is richer in the higher melting component. An important consideration is the so-called segregation coefficient. This is the ratio of the unwanted material in the solid relative to the liquid. This ratio is determined by the divergence of the solidus and liquidus. It is obvious that this quantity should be less than one. As the pure component B axis is approached, this coefficient increases since the liquidus and solidus curves converge, and the process is less efficient. On the other hand, as the molten zone acquires more and more impurity, the effective overall composition of the next solid to melt plus the existent melt is richer in the lower melting impurity. This tends to increase the segregation coefficient since the liquidus and solidus curves are more divergent. Counteracting this is the inability to achieve less than the equilibrium segregation value of the original composition for any impurity enriched "existent zone."[3] Consequently, the general approach is to repeat the process after the first pass is completed. In order to achieve maximum purification, techniques such as floating zones are used where applicable. Here the impure material in the form of an ingot is vertically suspended and affixed at both ends, obviating the need for a container. As the zone sweeps vertically past the ingot, the two solid halves of the ingot are held together by the surface tension of the liquid. While resistance heaters are usable in a

---

[3] By this we mean that while the ratio, impurity in solid/impurity in liquid is decreasing as the liquid becomes richer in impurity, the absolute concentration of impurity in the solid under equilibrium conditions is increasing. This happens as a consequence of the fact that as the net overall impurity concentration in the molten zone plus adjacent solid increases, the liquidus and solidus curves of Fig. 2 both move to the left.

zone refining system, greater use is made of radio frequency heated systems. Since these require a susceptor to which they may be coupled, it is often necessary to preheat the material using focused light or some other means, except where the substance to be purified is a conductor.

E. The General Experimental Approach to the Resolution of Solid Solution Diagrams

If a melt of a particular starting composition is cooled, it will, at the point of first crystallization, exhibit a latent heat anomaly which will persist during further cooling until all traces of liquid have vanished. On a T–M_x graph, this first point of crystallization would be recorded as the liquidus point for that composition. Theoretically, if equilibrium could be maintained through the entire crystallization process, the point of disappearance of a latent heat anomaly might be recorded as the solidus for that composition. Unfortunately, it is, in practice, a difficult task to maintain equilibrium through the entire solid–liquid interval. This fact necessitates evaluation of the solidus curve using specially prepared samples that have been melted, quenched to freeze in a state of homogeneity, and then heat treated to enable crystal growth of the quenched particles to a size that will exhibit thermodynamic behavior. The process of quenching involves a sudden rapid cooling from an elevated temperature so that crystalline grain size is kept small. The system may then be composed of particles in the so-called glassy or amorphous state. Such a state may or may not be representative of the bulk material depending upon whether it is merely amorphous to X rays (i.e., is of a small enough particle size to prevent obtaining interference or reinforcement of the X rays), or whether it possesses a structure having little or no long-range order. If such a material is held for a sufficient length of time at temperatures as close to the solidus as possible without melting, it will either increase in grain size if it was composed of small but crystalline particles, or it will convert from the glassy state to the crystalline state. This growth and/or conversion process proceeds via solid state diffusion processes or a combination of solid and vapor phase processes. In any event, once crystallinity has been achieved, heating cycles may then be conducted. When the solidus is reached, a latent heat anomaly occurs and the temperature associated with this anomaly, whether determined visually by observation of melting, or physically by detection of the latent heat anomaly, is recorded as the solidus temperature for the composition in question. Physical methods for evaluating the liquidus and solidus will be discussed in Chapter 36, which is devoted to experimental techniques.

F. THE SPACE MODEL FOR SOLID SOLUTION SYSTEMS

The three-dimensional representation of idealized systems exhibiting complete miscibility in solid, liquid, and vapor phases is considerably simpler than that of eutectic and eutectic compound-forming systems. In the temperature range in which solid and vapor alone coexist, the system is bivariant requiring specification of two of the three available parameters to fix the value of the third. Alternatively, such a region is isothermally univariant or isobarically univariant, etc. If we describe a solid–vapor equilibrium on a p–M_X graph (an isothermal slice through the space model), two curves will be generated, one showing the composition of the solid (the total pressure curve), and the other showing the composition of the vapor at this total system pressure.

As we shall see, in the temperature range in which the three-phase equilibrium, solid solution–liquid–vapor is possible, there must also be depicted two two-phase equilibria in a constant temperature graph.

Above the temperature interval of S–L–V coexistence, the system is again two phase, i.e., liquid–vapor, and quite similar in appearance to the low temperature, solid–vapor region.

A representation of the three-phase univariant portion of the diagram is best given via projection since, at any one temperature, for example, the system is isothermally invariant. The compositions of the three coexisting phases would then be represented by points in an isotherm, these points all lying at the same total pressure.

Figure 4 represents an isotherm through the solid solution–vapor portion of the solid model. $p_A{}^0$ and $p_B{}^0$ are the vapor pressures of pure solid A and pure solid B at the temperature of the section T_1. The top curve, $M_B^{(1)}$ shows the composition of the solid solution, the bottom curve, that of the vapor in equilibrium with the solid. In an ideal system, it is the case that the total vapor pressure varies linearly from $p_A{}^0$ to $p_B{}^0$ according to

$$P_t = N_A^{(1)} p_A{}^0 + N_B^{(1)} p_B{}^0 = N_A^{(1)} p_A{}^0 + (1 - N_A^{(1)}) p_B{}^0. \tag{20}$$

The vapor composition $M_B^{(3)}$ is again proportional to the partial pressures of the end members, i.e.,

$$M_B^{(3)} = p_B/(p_A + p_B) = N_B^{(1)} p_B{}^0 / (N_A^{(1)} p_A{}^0 + N_B^{(1)} p_B{}^0). \tag{21}$$

As in the ideal liquid–vapor case, the vapor is richer in that component exhibiting the higher vapor pressure at the temperature in question. The discussion of tie lines, etc., for Fig. 4 follows the paths previously explored and need not be repeated here.

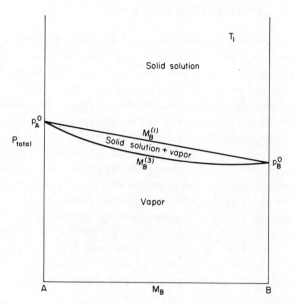

Fig. 4. An isotherm through the solid solution–vapor region.

An isobaric slice through the solid solution–vapor region is presented in Fig. 5. The temperatures T_A^s and T_B^s are those at which the sublimation pressures of the pure solids are equal to the total pressure. These temperatures then are the "normal sublimation temperatures" for the pressure in question. The curves $M_B^{(1)}$ and $M_B^{(3)}$ show at what temperature

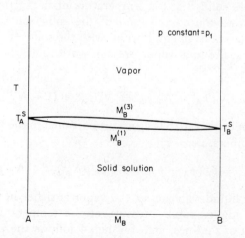

Fig. 5. An isobaric section through the solid solution–vapor region of the system A–B.

F. THE SPACE MODEL FOR SOLID SOLUTION SYSTEMS

a solid of a specific composition exhibits a pressure equal to the pressure of the isobaric slice, and isothermal tie lines show the compositions of the coexisting phases.

The analytical expressions describing the univariance of solid solution–vapor equilibrium are derived in precisely the same manner as used in Section B for the specification of liquidus and solidus curves. The only changes are that the terms $N_A^{(2)}$ and $N_B^{(2)}$ of Section B are replaced by terms $N_A^{(3)}$ and $N_B^{(3)}$, and the fusion enthalpies ΔH_A^0 and ΔH_B^0 are replaced by the sublimation enthalpy terms ΔH_A^s and ΔH_B^s. The terms T_A^0 and T_B^0 are replaced by the terms T_A^s and T_B^s, which are isobarically invariant, since they are functions of the pressure only. Again, tie-line phenomena and a description of what occurs as temperature is varied need not be pursued for Fig. 5.

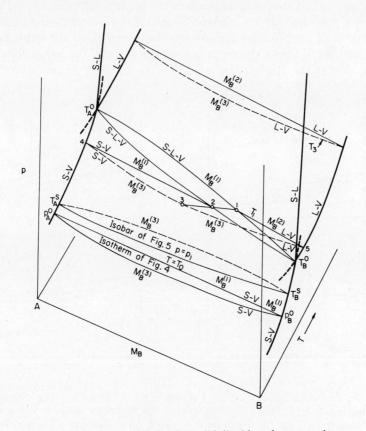

FIG. 6. The space model showing solid, liquid, and vapor regions.

Before continuing with the description of isobaric and isothermal sections, it is helpful, to present that portion of the space model in which vapor is one of the coexisting phases. This is shown in Fig. 6. The lens-shaped shaded area extending from $T_A{}^0$, the triple point of pure A, to $T_B{}^0$, the triple point of pure B, shows the univariant region solid solution–liquid–vapor, and specifically, the curves $M_B^{(1)}$ and $M_B^{(2)}$. Since this region is univariant, it can be seen only in projection. If an isotherm is sectioned at a temperature T_1, for example, this isotherm will cut the S–L–V region in three points only (1, 2, and 3 in Fig. 6). The three points will all lie at the same total pressure since with the temperature defined, the system is isothermally invariant so long as three phases coexist. Also note that such an isothermal slice will cut the solid–vapor surfaces (marked $M_B^{(1)}$ and $M_B^{(3)}$ and the liquid–vapor surfaces (marked $M_B^{(2)}$ and $M_B^{(3)}$) generating two incomplete intersecting sets of univariant curves in addition to the triple points described above. The S–V surfaces that are intersected are those connecting the sublimation curve of pure A and the metastable extension of the sublimation curve of pure B above the triple point of the latter. We have arbitrarily assumed that the triple point of A lies at a higher temperature than

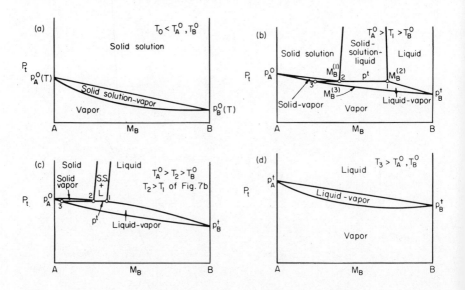

FIG. 7. Isotherms through the space model of Fig. 6. (a) An isotherm through the two-phase region S–V; (b) an isotherm through the two-phase region S–V, the three-phase region S–L–V, and the two-phase region L–V; (c) similar isotherm to Fig. 7b with increasing T; and (d) an isotherm through the two-phase region L–V.

F. THE SPACE MODEL FOR SOLID SOLUTION SYSTEMS

the triple point of B. The L–V surfaces that are intersected are those connecting the vaporization curve of pure B with the metastable extension of the vaporization curve of liquid A below the triple point temperature of A.

A series of isotherms starting in the solid–vapor portion of the diagram and continuing to higher temperatures through the three-phase region is shown in Fig. 7a–d. Figure 7a is the same as Fig. 4, and shows the isotherm T_0 of Fig. 6. Figure 7b shows the isotherm T_1 and Fig. 7c shows a higher temperature isotherm (not shown in Fig. 6). Figure 7d shows the isotherm T_3 of Fig. 6.

It is to be noted that any isotherm in the temperature interval $T_A^0 - T_B^0$ will always cut one three-phase region and two two-phase regions. One of the two-phase regions is a S–V and the other a L–V equilibrium. The pressure region above the former is a solid region, and the region

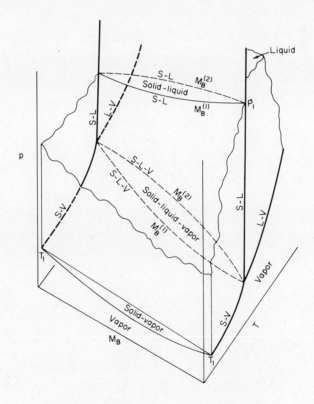

FIG. 8. Solid portions of the space model.

above the latter is a liquid region. Between the two, the volume is occupied by solid solution and liquid. As T_A^0 is approached, this solid solution–liquid region has a smaller compositional interval and when T_A^0 is reached the interval vanishes.

The points 1, 2, and 3 of Figs. 7b and 7c and of Fig. 6 show the composition of liquid, solid, and vapor, respectively, at the three-phase pressure p^t. The description of the L–V surfaces follows exactly that given for the S–V surfaces. Isotherms are described by the Raoult equation and isobars by the equations derived in Section B. Figure 8 shows the condensed-phase regions of the space model lying above the condensed vapor-phase regions.

28

Solid Solutions Continued—Graphical Representation of Idealized Systems and Some Comment on Real Systems

A. Introduction

In the preceding chapter, a general description of the ascending solid solution type of interaction was offered. In addition, analytical descriptions of these systems were derived assuming that species and component mole numbers were equal and that solubility was continuous in both the liquid and solid state. Thus, where $M_A^{(x)} = N_A^{(x)}$ and $M_B^{(x)} = N_B^{(x)}$, the values $M_A^{(x)}$ and $M_B^{(x)}$ as a function of T are given by

$$M_A^{(2)} = \frac{\exp\left[-\frac{\Delta H_A^0}{R}\left(\frac{1}{T} - \frac{1}{T_A^0}\right)\right]\left\{\exp\left[-\frac{\Delta H_B^0}{R}\left(\frac{1}{T} - \frac{1}{T_B^0}\right)\right] - 1\right\}}{\exp\left[-\frac{\Delta H_B^0}{R}\left(\frac{1}{T} - \frac{1}{T_B^0}\right)\right] - \exp\left[-\frac{\Delta H_A^0}{R}\left(\frac{1}{T} - \frac{1}{T_A^0}\right)\right]}, \quad (1)$$

$$M_B^{(2)} = \frac{\exp\left[-\frac{\Delta H_B^0}{R}\left(\frac{1}{T} - \frac{1}{T_B^0}\right)\right]\left\{\exp\left[-\frac{\Delta H_A^0}{R}\left(\frac{1}{T} - \frac{1}{T_A^0}\right)\right] - 1\right\}}{\exp\left[-\frac{\Delta H_A^0}{R}\left(\frac{1}{T} - \frac{1}{T_A^0}\right)\right] - \exp\left[-\frac{\Delta H_B^0}{R}\left(\frac{1}{T} - \frac{1}{T_B^0}\right)\right]}, \quad (2)$$

$$M_A^{(1)} = \frac{\exp\left[-\frac{\Delta H_B^0}{R}\left(\frac{1}{T} - \frac{1}{T_B^0}\right)\right] - 1}{\exp\left[-\frac{\Delta H_B^0}{R}\left(\frac{1}{T} - \frac{1}{T_B^0}\right)\right] - \exp\left[-\frac{\Delta H_A^0}{R}\left(\frac{1}{T} - \frac{1}{T_A^0}\right)\right]}, \quad (3)$$

$$M_B^{(1)} = \frac{\exp\left[-\frac{\Delta H_A^0}{R}\left(\frac{1}{T} - \frac{1}{T_A^0}\right)\right] - 1}{\exp\left[-\frac{\Delta H_A^0}{R}\left(\frac{1}{T} - \frac{1}{T_A^0}\right)\right] - \exp\left[-\frac{\Delta H_B^0}{R}\left(\frac{1}{T} - \frac{1}{T_B^0}\right)\right]}. \quad (4)$$

Using these relationships, arbitrary values for ΔH_A^0 and ΔH_B^0, the latent molar heats of fusion of A and B, and T_A^0 and T_B^0, the melting points of A and B on the absolute temperature scale, were introduced to obtain solutions for the liquidus and solidus. Since the liquidus and solidus curves are each continuous, solutions were required only for $N_A^{(2)}$ and $N_A^{(1)}$ or for the set $N_B^{(2)}$ and $N_B^{(1)}$. The cases treated were:

I. $T_A^0 = 1000°K$; $T_B^0 = 1500°K$
$\Delta H_A^0 = 2000$ cal/mole; $\Delta H_B^0 = 2000$ cal/mole

II. $T_A^0 = 1000°K$; $T_B^0 = 1500°K$
$\Delta H_A^0 = 5000$ cal/mole; $\Delta H_B^0 = 5000$ cal/mole

III. $T_A^0 = 1000°K$; $T_B^0 = 1500°K$
$\Delta H_A^0 = 10,000$ cal/mole; $\Delta H_B^0 = 10,000$ cal/mole

IV. $T_A^0 = 1000°K$; $T_B^0 = 1500°K$
$\Delta H_A^0 = 20,000$ cal/mole; $\Delta H_B^0 = 20,000$ cal/mole

V. $T_A^0 = 1000°K$; $T_B^0 = 1500°K$
$\Delta H_A^0 = 5000$ cal/mole; $\Delta H_B^0 = 10,000$ cal/mole

VI. $T_A^0 = 1000°K$; $T_B^0 = 1500°K$
$\Delta H_A^0 = 5000$ cal/mole; $\Delta H_B^0 = 15,000$ cal/mole

VII–XII. Repeat of problems I–VI except that $T_A^0 = 1000°K$; $T_B^0 = 1100°K$

XIII. Problem V is repeated with $T_A^0 = 1000°K$; $T_B^0 = 1100°K$
$\Delta H_A^0 = 10,000$ cal/mole and $\Delta H_B^0 = 5000$ cal/mole

XIV. Problem VI is repeated with $T_A^0 = 1,000°K$; $T_B^0 = 1100°K$
$\Delta H_A^0 = 15,000$ cal/mole and $\Delta H_B^0 = 5000$ cal/mole

B. Temperature–Composition Diagrams for Cases I–XIV

The cases I–IV inclusive form part of a set in which the heats of fusion of the end members for each case are assumed to vary while

B. $T-M_X$ DIAGRAMS FOR CASES I–XIV

the melting points of the end members remain constant. The $T-M_X$ diagrams are shown in Figs. 1–4. With low ΔH_X^0 values as in Fig. 1, it is observed that both the liquidus and solidus are convex with respect to the composition axis. As the ΔH_X^0 values increase, the separation between the liquidus and solidus increases. The solidus becomes even more convex relative to the composition axis while the liquidus tends to become concave relative to this axis, creating a somewhat distorted lens shaped envelope (see Figs. 2–4).

In Figs. 5 and 6, examples V and VI, we have the case where the ΔH_A^0 of the end members are different. Here we see that the curvature of the concave liquidus is greater than that of the solidus. Furthermore, with an increasingly greater latent heat of the higher melting end member, there is a tendency for the solidus, in the vicinity of the higher melting end member, to become concave resulting in an inflected solidus curve.

Figures 7–10, inclusive, represent a similar sequence to that shown in Figs. 1–4 except that the melting points of the end members are assumed to be closer together (100°K difference). The trend in liquidus and solidus contours of Figs. 7–10 is precisely the same as in Figs. 1–4 except that the separation of curves is less in the former. This indicates that the separation is more a function of melting points than of latent heats, although the latter also contribute. We see also that the curves tend to be more symmetrical when the melting points of the end members are closer in value. Again, with lower ΔH_X^0 values, the tendency toward convexity of the liquidi curves becomes more pronounced.

In Figs. 11 and 12 representing examples XI and XII, we have the analogous cases to those depicted in Figs. 5 and 6, but with a smaller separation in melting point. Now, the tendency observed in Figs. 5 and 6 for concave solidus curves to form becomes fully developed and both Figs. 11 and 12 exhibit both concave solidi and liquidi relative to the composition axis. This indicates that with similar melting points but dissimilar latent heats, double concave curves are more likely to occur.

Finally in Figs. 13 and 14, the situation in problems XI and XII is reversed so that the higher melting end member has the lower latent heat of fusion. This causes a corresponding reversal in the liquidus and solidus curves of Figs. 11 and 12 such that instead of both being concave to the composition axis, they are both convex. Summarizing then, the following trends may be specified.

Biconvex (lens shaped envelopes) liquidus–solidus curves, which are considered the normal case for ascending solid solutions, are most likely when the latent heats of fusion are similar and have values greater than

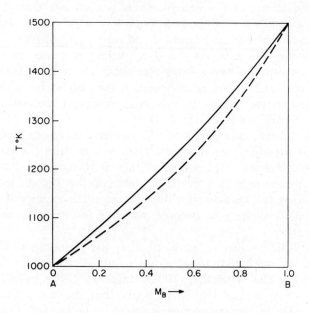

FIG. 1. $\Delta H_A = 2000$ cal/mole; $\Delta H_B = 2000$ cal/mole.

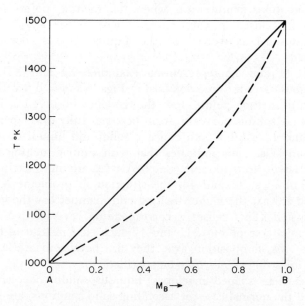

FIG. 2. $\Delta H_A = 5000$ cal/mole; $\Delta H_B = 5000$ cal/mole.

B. T–M_X DIAGRAMS FOR CASES I–XIV

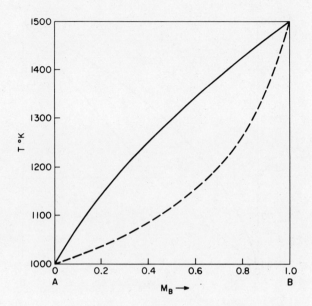

FIG. 3. $\Delta H_A = 10,000$ cal/mole; $\Delta H_B = 10,000$ cal/mole.

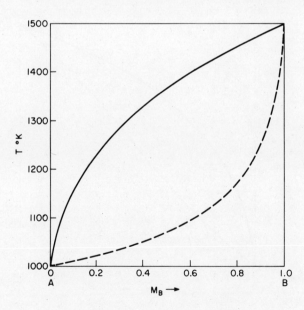

FIG. 4. $\Delta H_A = 20,000$ cal/mole; $\Delta H_B = 20,000$ cal/mole.

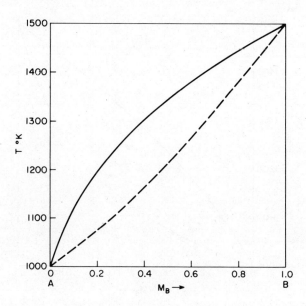

Fig. 5. $\Delta H_A = 5000$ cal/mole; $\Delta H_B = 10{,}000$ cal/mole.

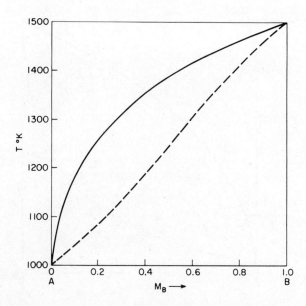

Fig. 6. $\Delta H_A = 5000$ cal/mole; $\Delta H_B = 15{,}000$ cal/mole.

B. T–M_X DIAGRAMS FOR CASES I–XIV

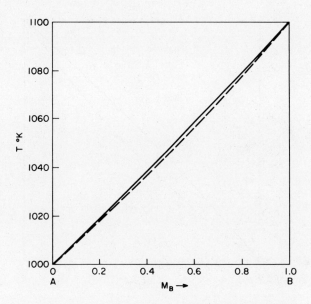

Fig. 7. $\Delta H_A = 2000$ cal/mole; $\Delta H_B = 2000$ cal/mole.

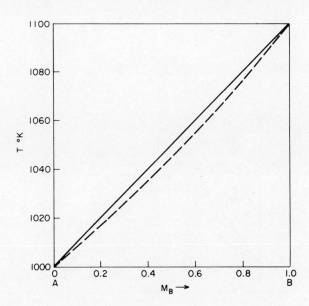

Fig. 8. $\Delta H_A = 5000$ cal/mole; $\Delta H_B = 5000$ cal/mole.

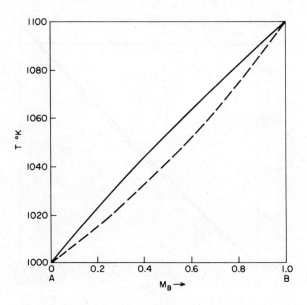

Fig. 9. $\Delta H_A = 10{,}000$ cal/mole; $\Delta H_B = 10{,}000$ cal/mole.

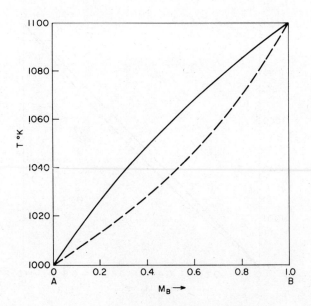

Fig. 10. $\Delta H_A = 20{,}000$ cal/mole; $\Delta H_B = 20{,}000$ cal/mole.

B. $T-M_x$ DIAGRAMS FOR CASES I–XIV

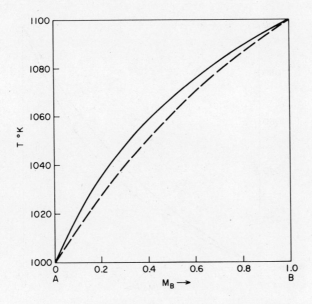

FIG. 11. $\Delta H_A = 5000$ cal/mole; $\Delta H_B = 10,000$ cal/mole.

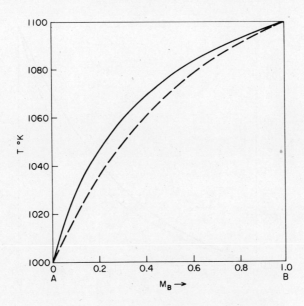

FIG. 12. $\Delta H_A = 5000$ cal/mole; $\Delta H_B = 15,000$ cal/mole.

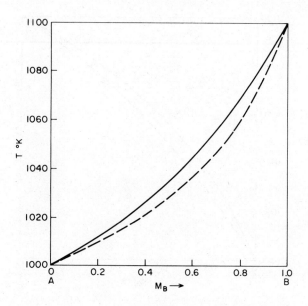

Fig. 13. $\Delta H_A = 10{,}000$ cal/mole; $\Delta H_B = 5000$ cal/mole.

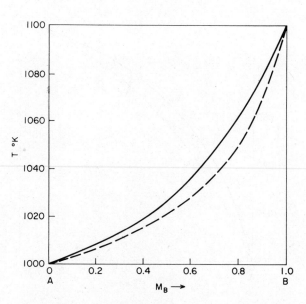

Fig. 14. $\Delta H_A = 15{,}000$ cal/mole; $\Delta H_B = 5000$ cal/mole.

B. T–M_x DIAGRAMS FOR CASES I–XIV

2000 cal/mole. As the difference in melting points between the end members decreases (again with similar latent heats of fusion) biconvex curves appear at fairly low latent heats and become more pronounced with increasing latent heats. As the difference in melting points increases, the tendency for doubly convex curves becomes greater in systems where both end members have similar latent heats.

When the difference in latent heats between end members becomes appreciable and the lower melting end member possesses the lower latent heat of fusion, two effects may be noted. If the melting points are also quite different, it appears that a biconvex (lens shaped) situation will persist with the solidus showing a tendency to inflect, and tending, therefore, to become concave relative to the composition axis in the process. The liquidus becomes even more concave relative to this axis. If the melting points are not too different, then the system exhibits doubly concave curves, the degree of concavity increasing with greater divergence of the end member latent heats of fusion.

When the difference between latent heats becomes appreciable, with the lower melting end member possessing the greater latent heat, a complete reversal occurs. With the melting points similar for both

TABLE I
Ascending Binary Systems[a]

System	Solidus contour	Liquidus contour	T_A^0 (°K)	T_B^0 (°K)	ΔH_A^0 (cal/mole)	ΔH_B^0 (cal/mole)
Ag–Au	Concave	Concave	1234	1336	2690	3050
Ag–Pd	Concave	Concave		1825	2690	4000
Au–Pd	Concave	Concave				
Pt–Rh	Concave	Concave	2042	2239	5200	—
Ba–Sr	Convex	Convex	985	1043	1830	2100
Cd–Mg	Convex	Convex	594	922	1530	2100
Cu–Pd	Convex	Convex	1356		3100	
Au–Pt	Inflected	Inflected				
Cu–Pt	Inflected	Inflected				
Se–Te	Inflected	Inflected	488	723	9000	8360
Si–Sb	Convex	Concave	544	903	2000	6500
Gd–La	Convex	Concave	—	1138	—	2500
Ge–Si	Convex	Concave	1213	1685	7700	11,100
Ti–V	Convex	Concave	1933	1403	4500	3000

[a] The data presented in this table are taken from the work of Hansen (1958) and Elliott (1965). For the complete references, see p. 535 in the Appendix.

end members, for example, the liquidus and solidus both become convex relative to the composition axis (doubly convex curves).

Finally, it should be noted that if one attempts to introduce values of T lying outside the interval $T_A{}^0-T_B{}^0$ into Eqs. (1)–(4), physically meaningless answers result, i.e., $N_A{}^x$ and $N_B{}^x$ do not both lie in the interval 0–1 with their sum equal to one.

C. Real Systems

A survey of the literature dealing with binary interactions (not including pseudobinary systems) is only partly rewarding when looking for experimental examples that exhibit the contour types discussed in Section B. First, the reported data for both the liquidi and solidi contain sufficient experimental uncertainty to obscure the precise contour character. This is particularly true of solidus data. Second, latent heat data are frequently not known with sufficient accuracy to warrant extensive comparison of real and theoretical data, particularly when this uncertainty is coupled with the experimental data uncertainties.

In Table I, the reported status of several ascending binary system, together with latent heat and melting point data where available, is presented for examination without further comment.

29

Solid Solutions Continued—Complete Liquid Phase Dissociation and Minimum Type Solid Solutions

A. INTRODUCTION

In Chapter 27, we saw that the simplest type of solid solution interaction occurs when it is assumed that species and component mole numbers are the same. Such an assumption leads to equations which, when depicted graphically in a temperature–composition diagram, (Chapter 28), provide us with pictorial representations of the well-known ascending type of solid solution. Furthermore, in Chapter 28 we saw that the same ideal equations yield liquidus and solidus contours that one would have intuitively attributed to nonideality. In all instances, however, the curves were of the ascending type.

Another frequently encountered solid solution type is one which exhibits a minimum in both its liquidus and solidus, the two curves touching tangentially at this minimum. Such a solid solution system for obvious reasons is generally termed a minimum type solid solution and is depicted schematically in Fig. 1. As might have been anticipated, solubility curves having characteristics of the type shown in Fig. 1 have been termed nonideal and are indicative of a miscibility gap lying submerged somewhere beneath the solidus. Also because of a superficial

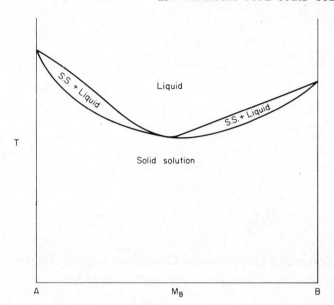

FIG. 1. A schematic representation of a minimum type solid solution.

resemblance to eutectic systems exhibiting limited solid solubility (a topic yet to be considered) sequential approaches have been advanced showing a gradually rising miscibility gap that finally erupts through the minimum type solid–liquid field thereby generating the limited solid solubility eutectic diagram. While it is the case that solid-phase miscibility gaps are observed in solid solution systems it should be noted that, by definition, a miscibility gap is only possible in a solution type system. It is also not unreasonable to expect that mutual solid miscibility should tend to decrease with decreasing temperature. Consequently, the observation of exsolution behavior in solid solution systems generally, might well be anticipated intuitively.

More significantly perhaps, it should be noted that in a minimum type solid solution, the end members have always been found to be isostructural (isomorphic), that is, they exhibit the same crystal structure. Thus, in each of the solid solution–liquid fields shown in Fig. 1, it is the same solid phase that coexists with liquid. In eutectic systems, of course, there are generally two distinct solid phases present. Also, in the minimum type solid solution, both the solidus and liquidus are continuous, touching at only a single point. In eutectic systems, the liquidi are separate curves, intersecting at a point. In other words, metastable extensions of the minimum solid solution liquidus to lower temperatures is not possible.

C. COMPLETE DISSOCIATION IN SOLID SOLUTION SYSTEMS

These apparent inconsistencies in the intuitive arguments that have been advanced to explain minimum type solid solutions lead one to search for a different, more palatable explanation of minimum type behavior.

B. THE QUALITATIVE ASPECTS OF FIG. 1

If a solid solution (Fig. 1), having a composition lying in the interval bounded by pure A and the composition of the minimum, is heated, it will begin to melt upon reaching the solidus. If we visualize the diagram as consisting of two ascending solid solutions touching at the minimum, the compositions of coexisting solid and liquid as a function of T are determined in precisely the same way as in a bonafide ascending-type system. If a sample of the minimum composition is heated, it will melt completely at the minimum temperature. At all binary compositions including the minimum composition, the system is isobarically univariant. This brief description should suffice as a guide to graphical evaluation of such systems. It is now of interest to attempt a quantitative description of behavior such as that depicted in Fig. 1.

C. COMPLETE DISSOCIATION IN SOLID SOLUTION SYSTEMS

Two possible treatments may be visualized in considering the effects of complete dissociation on the liquidus and solidus contours of a solid solution system. We may, for example, consider the system AC–BC, which is pseudobinary, and assume that the melting of the end members proceeds via

$$AC^{(1)} \rightleftarrows A^{(2)} + C^{(2)}, \tag{1}$$

$$BC^{(1)} \rightleftarrows BC^{(2)}. \tag{2}$$

As a consequence of Eqs. (1) and (2) it is seen that despite the presence of a common part in each pseudounary end member (the part C), no common species effect is present upon melting since BC is assumed to melt without dissociating.

Alternatively, we may consider a truly binary interaction (which concept is equally applicable to the pseudobinary case) where one (or both) of the end members is present as a dimer in the solid phase and as a monomer in the liquid phase. The description of the melting of such a system is given by

$$A_2^{(1)} \rightleftarrows 2A^{(2)}, \tag{3}$$

$$B^{(1)} \rightleftarrows B^{(2)}. \tag{4}$$

It is evident that each of these boundary case systems may be perturbed by (1) partial dissociation of the dissociating end member or by (2) partial or complete dissociation of both end members with and without common species effects, etc. We will restrict our treatments to the complete dissociation cases represented by Eqs. (1)–(4). Before doing so, however, let us attempt to develop the dimer–monomer concept, somewhat more fully.

Up to this point we have more or less tacitly avoided an introduction of the concept of a solid phase species. In simple eutectic systems, where only pure solid phases were assumed present, this concept was not a necessary part of our treatment since the mole fraction of the precipitating end member in the solid was unity at all system concentrations independently of how we chose to designate it symbolically. In the treatment of ascending solid solutions, however, the solid phase was not made up of a pure end member except at component mole fractions of unity. Consequently, at all intermediate concentrations, species quantities $n_X^{(a)}$ and concentrations $N_X^{(a)}$ had to be assumed. What was done implicitly was to assume that one could make a one-to-one equivalence between the component designation A or B and its apparent state of aggregation in the solid phase. Thus, if the component were designated by the symbol A, it would be assumed that in the solid, the concept of a species having the stoichiometry A could be invoked. Since solids, in general, are continuous three-dimensional arrays, it is, perhaps, difficult to see why any particular species notation should have any more validity than any other species notation or for that matter why one should even be able to invoke a solid phase species concept. The rationale employed by the author is as follows.

If in the solid solution A–B it is the case that precipitation of the A liquid-phase species involves precipitation of single A atoms from the liquid, the solid is assumed to be composed of A species plus species from the other end member. In the disordered solid solution lattice then, we should expect to find individual atoms of A surrounded by atoms of species derived from the other end member. If on the other hand, the precipitation of A involves two individual A liquid-phase species combining at the solid surface, the solid phase species derived from the end member is assumed to have the stoichiometry A_2. In this case, it is assumed that in the disordered lattice one would not find single A atoms surrounded by atoms from the second end member. Instead, the minimum number of A atoms found separately in the structure at adjacent lattice sites will be two. In this instance the component A is present in the solid as A_2 species.

Alternatively, we may think of the above via an assumed melting process. If individual A atoms may break free from the lattice, then the solid-state species derived from the component A is assumed to have the stoichiometry A, and species counting is done in terms of this stoichiometry. On the other hand, if melting of the solid involves breaking free of A_2 dimers, which then dissociate completely upon entering the liquid phase, the solid state aggregation of the component A is assumed to have the stoichiometry A_2. We may arbitrarily designate the component stoichiometry by either its species stoichiometry in the solid or liquid phases. Conventionally, and purely arbitrarily, we have chosen to assign to the component the stoichiometry corresponding to the minimum molecular weight.

For cases where the solid phase consists of species derived from both components, it is seen that we could (if we chose) include in our treatments hypothetical cases where, in addition to partial liquid phase dissociation, association, or what have you, we could also assume similar perturbations in the solid solution. We will not attempt such treatments since they do not offer additional insight and add considerable complexity. The interested reader may find such exercices rewarding.

D. Case I—Dimer–Monomer Behavior

Let us consider the system AC–BC whose behavior is pseudobinary. Let us assume that the components AC and BC are completely miscible in both solid and liquid phases. In Case I, it shall further be assumed that the component AC is present in the solid as the dimer species $(AC)_2$ and melts according to

$$(AC)_2^{(1)} \rightleftarrows 2AC^{(2)}, \tag{5}$$

via complete dissociation of the dimer. This is precisely the same case shown in Eq. (3) for a unary substance and can be considered for either the pseudounary or unary species. The example given in Eq. (1), on the other hand, is possible only for the pseudounary substance. Because we are afforded greater flexibility in developing our arguments by considering pseudobinary interactions, we will consider the system AC–BC in all discussions. It is to be recognized that in certain instances [as, for example, the one shown in Eq. (5)], the treatments are applicable to binary as well as to pseudobinary systems, while in others they are applicable only to pseudobinary interactions, there being no counterpart in the binary regime.

The component BC is, for the sake of simplicity, assumed to melt as shown in Eq. (2), it being present in both the solid and liquid as the species BC. The equilibrium constants describing the phenomena represented by Eq. (5) and (2), respectively, are given by

$$K_{AC}/K_{AC}^0 = \exp\left[-\frac{\Delta H_{AC}^0}{R}\left(\frac{1}{T} - \frac{1}{T_{AC}^0}\right)\right] = e^{-AC}, \tag{6}$$

$$K_{BC}/K_{BC}^0 = \exp\left[-\frac{\Delta H_{BC}^0}{R}\left(\frac{1}{T} - \frac{1}{T_{BC}^0}\right)\right] = e^{-BC}. \tag{7}$$

In anticipation of the degree of freedom to be chosen, the equilibrium constants are defined as a function of temperature. As usual, the terms T_X^0 are the melting points and the terms K_X^0 are the equilibrium constants at the melting points. The terms K_X represent the equilibrium constants at any other temperature. The terms ΔH_X^0 represent the standard molar enthalpies of fusion of the solid whose species stoichiometry is X.

In terms of concentrations of species N_X, Eqs. (6) and (7) may be written

$$\frac{(N_{AC}^{(2)})^2}{N_{(AC)_2}^{(1)}} \bigg/ \frac{(N_{AC}^{0(2)})^2}{N_{(AC)_2}^{0(1)}} = e^{-AC}, \tag{8}$$

$$\frac{N_{BC}^{(2)}}{N_{BC}^{(1)}} \bigg/ \frac{N_{BC}^{0(2)}}{N_{BC}^{0(1)}} = e^{-BC}. \tag{9}$$

In these equations, the terms $N_{AC}^{0(2)}$, $N_{(AC)_2}^{0(1)}$, $N_{BC}^{0(2)}$, and $N_{BC}^{0(1)}$ are each equal to 1. Since the solid phases at compositions intermediate between the end members are made use of species derived from both components, the terms $N_{(AC)_2}^{(1)}$ and $N_{BC}^{(1)}$ must appear in the equilibrium relationships.

Consequently, we have for the temperature dependencies of the equilibrium constants describing Eqs. (5) and (2);

$$(N_{AC}^{(2)})^2/N_{(AC)_2}^{(1)} = e^{-AC}, \tag{10}$$

$$N_{BC}^{(2)}/N_{BC}^{(1)} = e^{-BC}. \tag{11}$$

The system of two components coexisting in two phases is isobarically univariant. Since four species unknowns exist, as can be seen from Eqs. (10) and (11), four independent relationships are required for an analytical solution. Choosing the temperature as the degree of freedom as defined by Eqs. (10) and (11), and the constraints

$$N_{AC}^{(2)} + N_{BC}^{(2)} = 1, \tag{12}$$

$$N_{(AC)_2}^{(1)} + N_{BC}^{(1)} = 1, \tag{13}$$

D. CASE I—DIMER–MONOMER BEHAVIOR

the four necessary independent relationships are at hand. In order to depict graphically the results to be obtained, we require transforms from the species quantities N_X to the component quantities M_X. The conservation equations relating component and species concentrations may be arrived at as follows;

$$M_{AC}^{(2)} = m_{AC}^{(2)}/(m_{AC}^{(2)} + m_{BC}^{(2)}), \tag{14}$$

$$M_{BC}^{(2)} = m_{BC}^{(2)}/(m_{AC}^{(2)} + m_{BC}^{(2)}), \tag{15}$$

$$M_{AC}^{(1)} = m_{AC}^{(1)}/(m_{AC}^{(1)} + m_{BC}^{(1)}), \tag{16}$$

$$M_{BC}^{(1)} = m_{BC}^{(1)}/(m_{AC}^{(1)} + m_{BC}^{(1)}), \tag{17}$$

where

$$m_{AC}^{(2)} = n_{AC}^{(2)} = N_{AC}^{(2)} n_t^{(2)}, \tag{18}$$

$$m_{BC}^{(2)} = n_{BC}^{(2)} = N_{BC}^{(2)} n_t^{(2)}, \tag{19}$$

$$m_{AC}^{(1)} = 2n_{(AC)_2}^{(1)} = 2N_{(AC)_2}^{(1)} n_t^{(1)}, \tag{20}$$

$$m_{BC}^{(1)} = n_{BC}^{(1)} = N_{BC}^{(1)} n_t^{(1)}. \tag{21}$$

Therefore,

$$M_{AC}^{(2)} = N_{AC}^{(2)}/(N_{AC}^{(2)} + N_{BC}^{(2)}), \tag{22}$$

$$M_{BC}^{(2)} = N_{BC}^{(2)}/(N_{AC}^{(2)} + N_{BC}^{(2)}), \tag{23}$$

$$M_{AC}^{(1)} = 2N_{(AC)_2}^{(1)}/(2N_{(AC)_2}^{(1)} + N_{BC}^{(1)}), \tag{24}$$

$$M_{BC}^{(1)} = N_{BC}^{(1)}/(2N_{(AC)_2}^{(1)} + N_{BC}^{(1)}). \tag{25}$$

In view of the constraint imposed by Eq. (12), it is seen that the denominators of Eqs. (22) and (23) are each equal to 1. Furthermore, from Eq. (13) it is seen that the denominators of Eqs. (24) and (25) are equal to $1 + N_{(AC)_2}^{(1)}$. Substituting these values in Eqs. (22)–(25), we obtain, for the bridges between component and species terms, the conservation equation

$$M_{AC}^{(2)} = N_{AC}^{(2)}, \tag{26}$$

$$M_{BC}^{(2)} = N_{BC}^{(2)}, \tag{27}$$

$$M_{AC}^{(1)} = 2N_{(AC)_2}^{(1)}/[1 + N_{(AC)_2}^{(1)}], \tag{28}$$

$$M_{BC}^{(1)} = N_{BC}^{(1)}/[1 + N_{(AC)_2}^{(1)}]. \tag{29}$$

From Eq. (10) we have

$$(N_{AC}^{(2)})^2 = N_{(AC)_2}^{(1)} e^{-AC} \tag{30}$$

Invoking Eq. (13), we may substitute for $N_{(AC)_2}^{(1)}$ in Eq. (30) to give

$$(N_{AC}^{(2)})^2 = (1 - N_{BC}^{(1)}) e^{-AC} \tag{31}$$

Invoking Eq. (11), we may now substitute for $N_{BC}^{(1)}$ in Eq. (31) to give

$$(N_{AC}^{(2)})^2 = [(e^{-BC} - N_{BC}^{(2)})/e^{-BC}] e^{-AC} \tag{32}$$

Finally, for the term $N_{BC}^{(2)}$ in Eq. (32) we may substitute its equivalent value in Eq. (12) to give an equation in $N_{AC}^{(2)}$;

$$(N_{AC}^{(2)})^2 e^{-BC} - N_{AC}^{(2)} e^{-AC} - e^{-AC}(e^{-BC} - 1) = 0 \tag{33}$$

Solving this quadratic for $N_{AC}^{(2)}$ gives

$$N_{AC}^{(2)} = \frac{e^{-AC} \pm \{e^{-2AC} + 4e^{-(AC+BC)}[e^{-BC} - 1]\}^{1/2}}{2e^{-BC}} \tag{34}$$

Substituting for $N_{AC}^{(2)}$ in Eq. (12) we may solve for $N_{BC}^{(2)}$;

$$N_{BC}^{(2)} = \frac{2e^{-BC} - e^{-AC} \mp \{e^{-2AC} + 4e^{-(AC+BC)}[e^{-BC} - 1]\}^{1/2}}{2e^{-BC}} \tag{35}$$

Substituting now for $N_{BC}^{(2)}$ in Eq. (11) we can solve for $N_{BC}^{(1)}$;

$$N_{BC}^{(1)} = \frac{2e^{-BC} - e^{-AC} \mp \{e^{-2AC} + 4e^{-(AC+BC)}[e^{-BC} - 1]\}^{1/2}}{2e^{-2BC}} \tag{36}$$

Substituting the value of $N_{BC}^{(1)}$ from Eq. (36) into Eq. (13) enables us to obtain the solution for $N_{(AC)_2}^{(1)}$;

$$N_{(AC)_2}^{(1)} = \frac{2e^{-BC}[e^{-BC} - 1] + e^{-AC} \pm \{e^{-2AC} + 4e^{-(AC+BC)}[e^{-BC} - 1]\}^{1/2}}{2e^{-2BC}} \tag{37}$$

Substituting the appropriate terms from Eqs. (34)–(37) into the Eqs. (26)–(29) provides us finally with the desired temperature dependencies of the components AC and BC in the liquid and solid phases, which are;

$$M_{AC}^{(2)} = \frac{e^{-AC} \pm \{e^{-2AC} + 4e^{-(AC+BC)}[e^{-BC} - 1]\}^{1/2}}{2e^{-BC}}, \tag{38}$$

$$M_{BC}^{(2)} = \frac{2e^{-BC} - e^{-AC} \mp \{e^{-2AC} + 4e^{-(AC+BC)}[e^{-BC} - 1]\}^{1/2}}{2e^{-BC}}, \tag{39}$$

$$M_{\text{AC}}^{(1)} = \frac{2\{2e^{-BC}[e^{-BC} - 1] + e^{-AC} \pm [e^{-2AC} + 4e^{-(AC+BC)}(e^{-BC} - 1)]^{1/2}\}}{2e^{-BC}[2e^{-BC} - 1] + e^{-AC} \pm \{e^{-2AC} + 4e^{-(AC+BC)}[e^{-BC} - 1]\}^{1/2}}, \quad (40)$$

$$M_{\text{BC}}^{(1)} = \frac{2e^{-BC} - e^{-AC} \mp \{e^{-2AC} + 4e^{-(AC+BC)}[e^{-BC} - 1]\}^{1/2}}{2e^{-BC}[2e^{-BC} - 1] + e^{-AC} \pm \{e^{-2AC} + 4e^{-(AC+BC)}[e^{-BC} - 1]\}^{1/2}}. \quad (41)$$

Addition of Eqs. (38) and (39) and of Eqs. (40) and (41) satisfies the requirements that

$$M_{\text{AC}}^{(2)} + M_{\text{BC}}^{(2)} = 1, \quad (42)$$

and

$$M_{\text{AC}}^{(1)} + M_{\text{BC}}^{(1)} = 1. \quad (43)$$

E. CASE II

The alternative mechanism, by which twice as many species may be derived from a component present in the liquid phase as compared to the number derived from it in the solid phase, is that in which complete dissociation attends the melting as in Eq. (1). For simplicity, it will again be assumed that the second component BC is represented in both the solid and liquid phases by the same species stoichiometry as in Eq. (2).

The temperature dependencies of the equilibrium constants for Eqs. (1) and (2), respectively, are given by

$$K_{\text{AC}}/K_{\text{AC}}^0 = \exp\left[-\frac{\Delta H_{\text{AC}}^0}{R}\left(\frac{1}{T} - \frac{1}{T_{\text{AC}}^0}\right)\right] = e^{-AC} \quad (44)$$

$$K_{\text{BC}}/K_{\text{BC}}^0 = \exp\left[-\frac{\Delta H_{\text{BC}}^0}{R}\left(\frac{1}{T} - \frac{1}{T_{\text{BC}}^0}\right)\right] = e^{-BC} \quad (45)$$

In expanded form, Eqs. (44) and (45) are given by

$$\frac{[N_{\text{A}}^{(2)}][N_{\text{C}}^{(2)}]/[N_{\text{AC}}^{(1)}]}{[N_{\text{A}}^{0(2)}][N_{\text{C}}^{0(2)}]/[N_{\text{AC}}^{0(1)}]} = e^{-AC}, \quad (46)$$

$$\frac{N_{\text{BC}}^{(2)}/N_{\text{BC}}^{(1)}}{N_{\text{BC}}^{0(2)}/N_{\text{BC}}^{0(1)}} = e^{-BC}. \quad (47)$$

Since AC is assumed to be completely dissociated in the liquid phase, it is a consequence that

$$N_{\text{A}}^{(2)} = N_{\text{C}}^{(2)}, \quad (48)$$

and at pure AC

$$N_A^{0(2)} = N_C^{0(2)} = 0.5. \tag{49}$$

Since the terms $N_{BC}^{0(2)}$ and $N_{BC}^{0(1)}$ as well as $N_{AB}^{0(1)}$, each equal unity, Eqs. (46) and (47) may be rewritten as

$$(N_A^{(2)})^2/N_{AC}^{(1)} = 0.25 e^{-AC}, \tag{50}$$

$$N_{BC}^{(2)}/N_{BC}^{(1)} = e^{-BC}. \tag{51}$$

For the four unknowns present in Eqs. (50) and (51), four independent relationships are required for simultaneous solution of the isobarically univariant system. Equations (50) and (51) together with the constraints,

$$N_A^{(2)} + N_C^{(2)} + N_{BC}^{(2)} = 2N_A^{(2)} + N_{BC}^{(2)} = 1, \tag{52}$$

$$N_{AC}^{(1)} + N_{BC}^{(1)} = 1. \tag{53}$$

provide the independent set.

The component conservation equations which are useful for the graphical description of the results to be obtained may be deduced in a manner similar to that employed in the preceding section starting with Eqs. (14)–(17) inclusive and the continuing sequence

$$m_{AC}^{(2)} = n_A^{(2)} = N_A^{(2)} n_t^{(2)}, \tag{54}$$

$$m_{BC}^{(2)} = n_{BC}^{(2)} = N_{BC}^{(2)} n_t^{(2)}, \tag{55}$$

$$m_{AC}^{(1)} = n_{AC}^{(1)} = N_{AC}^{(1)} n_t^{(1)}, \tag{56}$$

$$m_{BC}^{(1)} = n_{BC}^{(1)} = N_{BC}^{(1)} n_t^{(1)}. \tag{57}$$

Substituting these equivalencies in the appropriate places in Eqs. (14)–(17) leads to

$$M_{AC}^{(2)} = N_A^{(2)}/(N_A^{(2)} + N_{BC}^{(2)}), \tag{58}$$

$$M_{BC}^{(2)} = N_{BC}^{(2)}/(N_A^{(2)} + N_{BC}^{(2)}), \tag{59}$$

$$M_{AC}^{(1)} = N_{AC}^{(1)}/(N_{AC}^{(1)} + N_{BC}^{(1)}) = N_{AC}^{(1)} \quad \text{via Eq. (53)}, \tag{60}$$

$$M_{BC}^{(1)} = N_{BC}^{(1)}/(N_{AC}^{(1)} + N_{BC}^{(1)}) = N_{BC}^{(1)} \quad \text{via Eq. (53)}. \tag{61}$$

Substituting in Eqs. (58) and (59) the value of the sum $N_A^{(2)} + N_{BC}^{(2)}$ from Eq. (52) we obtain

$$M_{AC}^{(2)} = N_A^{(2)}/(1 - N_A^{(2)}), \tag{62}$$

E. CASE II

and

$$M_{BC}^{(2)} = N_{BC}^{(2)}/(1 - N_A^{(2)}). \tag{63}$$

Equations (60)–(63) represent the sought after bridges between component and species terms.

From Eq. (50) it is seen that

$$[N_A^{(2)}]^2 = 0.25 N_{AC}^{(1)} e^{-AC}. \tag{64}$$

Invoking Eq. (53) we may substitute for $N_{AC}^{(1)}$ in Eq. (64) to obtain

$$[N_A^{(2)}]^2 = 0.25[1 - N_{BC}^{(1)}] e^{-AC}. \tag{65}$$

$N_{BC}^{(1)}$ in Eq. (65) may be replaced by its value determined from Eq. (51) to give

$$[N_A^{(2)}]^2 = 0.25[1 - (N_{BC}^{(2)}/e^{-BC})] e^{-AC}. \tag{66}$$

Finally we may substitute for $N_{BC}^{(2)}$ in Eq. (66) its value from Eq. (52) to give a quadratic in $N_A^{(2)}$;

$$[N_A^{(2)}]^2 e^{-BC} - 0.5 e^{-AC} N_A^{(2)} + 0.25 e^{-AC}[1 - e^{-BC}] = 0. \tag{67}$$

The solution for $N_A^{(2)}$ is given by

$$N_A^{(2)} = \frac{0.5 e^{-AC} \pm \{0.25 e^{-2AC} - e^{-(AC+BC)}[1 - e^{-BC}]\}^{1/2}}{2 e^{-BC}}. \tag{68}$$

From Eqs. (52) and (68) we may solve for $N_{BC}^{(2)}$;

$$N_{BC}^{(2)} = \frac{e^{-BC} - 0.5 e^{-AC} \mp \{0.25 e^{-2AC} - e^{-(AC+BC)}[1 - e^{-BC}]\}^{1/2}}{e^{-BC}}. \tag{69}$$

From Eqs. (69) and (51) we may derive the value for $N_{BC}^{(1)}$ as

$$N_{BC}^{(1)} = \frac{e^{-BC} - 0.5 e^{-AC} \mp \{0.25 e^{-2AC} - e^{-(AC+BC)}[1 - e^{-BC}]\}^{1/2}}{e^{-2BC}} \tag{70}$$

and from Eqs. (70) and (53) we may derive the value of $N_{AC}^{(1)}$

$$N_{AC}^{(1)} = \frac{e^{-BC}[e^{-BC} - 1] + 0.5 e^{-AC} \pm \{0.25 e^{-2AC} - e^{-(AC+BC)}[1 - e^{-BC}]\}^{1/2}}{e^{-2BC}}. \tag{71}$$

From Eqs. (60) and (71) we see that

$$M_{AC}^{(1)} = \frac{e^{-BC}[e^{-BC} - 1] + 0.5 e^{-AC} \pm \{0.25 e^{-2AC} - e^{-(AC+BC)}[1 - e^{-BC}]\}^{1/2}}{e^{-2BC}}, \tag{72}$$

and from Eqs. (61) and (70) we see that

$$M_{BC}^{(1)} = \frac{e^{-BC} - 0.5e^{-AC} \mp \{0.25e^{-2AC} - e^{-(AC+BC)}[1 - e^{-BC}]\}^{1/2}}{e^{-2BC}}. \quad (73)$$

From Eqs. (68) and (62) it is found that

$$M_{AC}^{(2)} = \frac{0.5e^{-AC} \pm \{0.25e^{-2AC} - e^{-(AC+BC)}[1 - e^{-BC}]\}^{1/2}}{2e^{-BC} - 0.5e^{-AC} \mp \{0.25e^{-2AC} - e^{-(AC+BC)}[1 - e^{-BC}]\}^{1/2}}, \quad (74)$$

and from Eqs. (69) and (63)

$$M_{BC}^{(2)} = \frac{2[e^{-BC} - 0.5e^{-AC} \mp \{0.25e^{-2AC} - e^{-(AC+BC)}[1 - e^{-BC}]\}^{1/2}]}{2e^{-BC} - 0.5e^{-AC} \mp \{0.25e^{-2AC} - e^{-(AC+BC)}[1 - e^{-BC}]\}^{1/2}}. \quad (75)$$

The sums obtained by addition of Eqs. (72) and (73) as well as of Eqs. (74) and (75) satisfy the requirements of Eqs. (43) and (42), respectively.

F. The Case Where Both Components Are Dimers in the Solid

An obvious extension of the cases presented in Section D is one in which both components are present in the solid as dimers and in the liquid as monomers. We have then for the melting equations

$$(AC)_2^{(1)} \rightleftarrows 2AC^{(2)}, \quad (76)$$

$$(BC)_2^{(1)} \rightleftarrows 2BC^{(2)}, \quad (77)$$

and

$$K_{AC}/K_{AC}^0 = e^{-AC}, \quad (78)$$

$$K_{BC}/K_{BC}^0 = e^{-BC}. \quad (79)$$

The conservation conditions for species are defined by

$$N_{AC}^{(2)} + N_{BC}^{(2)} = 1, \quad (80)$$

$$N_{(AC)_2}^{(1)} + N_{(BC)_2}^{(1)} = 1. \quad (81)$$

Since the component mole conservation equations now have the form

$$m_{AC}^{(2)} = n_{AC}^{(2)} = N_{AC}^{(2)} n_t^{(2)}, \quad (82)$$

$$m_{BC}^{(2)} = n_{BC}^{(2)} = N_{BC}^{(2)} n_t^{(1)}, \quad (83)$$

$$m_{AC}^{(1)} = 2n_{(AC)_2}^{(1)} = 2N_{(AC)_2}^{(1)} n_t^{(1)}, \quad (84)$$

F. BOTH COMPONENTS ARE DIMERS IN THE SOLID

and

$$m_{BC}^{(1)} = 2n_{(BC)_2}^{(1)} = 2N_{(BC)_2}^{(1)} n_t^{(1)}, \tag{85}$$

it is the case that

$$M_{AC}^{(2)} = N_{AC}^{(2)}, \tag{86}$$

$$M_{BC}^{(2)} = N_{BC}^{(2)}, \tag{87}$$

$$M_{AC}^{(1)} = N_{(AC)_2}^{(1)}, \tag{88}$$

$$M_{BC}^{(1)} = N_{(BC)_2}^{(1)}. \tag{89}$$

Using a procedure identical to that presented in Section D we may solve for $M_{AC}^{(2)}$ and $M_{AC}^{(1)}$. The solutions for $M_{BC}^{(2)}$ and $M_{BC}^{(1)}$ are obtained from Eqs. (80) and (81), and need not be given in order to define the phase diagrams.

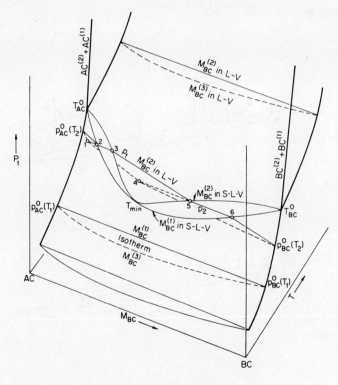

FIG. 2. The space model showing S–L, S–L–V, and L–V regions.

The desired results are given by

$$M_{AC}^{(2)} = N_{AC}^{(2)} = \frac{e^{-AC} \pm \{e^{-(AC+2BC)} + e^{-(2AC+BC)} - e^{-(AC+BC)}\}^{1/2}}{e^{-AC} + e^{-BC}}, \qquad (90)$$

$$M_{AC}^{(1)} = N_{(AC)_2}^{(1)} = \frac{\begin{bmatrix} e^{-AC} \pm 2\{e^{-(AC+2BC)} + e^{-(2AC+BC)} - e^{-(AC+BC)}\}^{1/2} \\ + e^{-BC}[e^{-BC} + e^{-AC} - 1] \end{bmatrix}}{(e^{-AC} + e^{-BC})^2}. \qquad (91)$$

G. Space Models

In the next chapter, the theoretical results of the two preceding sections will be employed to generate hypothetical cases of minimum type solid

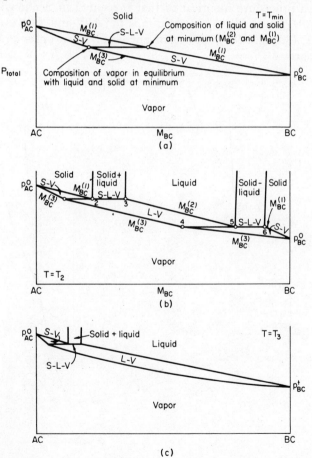

FIG. 3. Several isotherms through the S–L–V portion of the space model of Fig. 2.

G. SPACE MODELS

solution diagrams. These will be so chosen as to enable comparison with the experimentally determined system Na_2CO_3–K_2CO_3. In the present section we will consider two generalizations of the p–T–M_X diagrams for minimum type systems. The first of these assumes that even though the solid–liquid equilibrium exhibits dimer–monomer or complete dissociation behavior, the equilibrium S–V (or L–V) is of the simple ascending type. The second approach assumes that the S–V equilibrium is similar in nature to the S–L equilibrium, involving as it were a two-fold increase in species count of one of the components in the phase of higher heat content.

Figure 2 shows a simplified version of the p–T–M_{BC} diagram for the

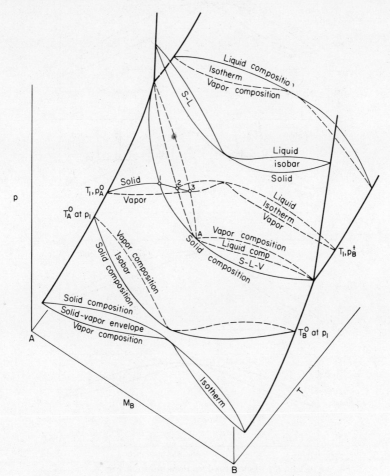

FIG. 4. Minima are formed in S–V, S–L, and L–V interactions.

minimum solid solution system. The L–V and S–V surfaces are identical with those defined for the ascending solid solution case treated in Chapter 27, and the all liquid and all solid bivariant portions of the space model are obvious as is the double envelope tied at each of its ends to the S–L curves of the end members. The area that is different from Fig. 6 of Chapter 27 involves the S–L–V regions of the model.

Starting at T_{AC}^0, there is shown in the shaded construction, the S–L equilibrium under its equilibrium partial pressure. At all compositions on the solid or liquid curve, the pressure for the three-phase coexistence S–L–V varies. At any particular composition, however, it is a consequence

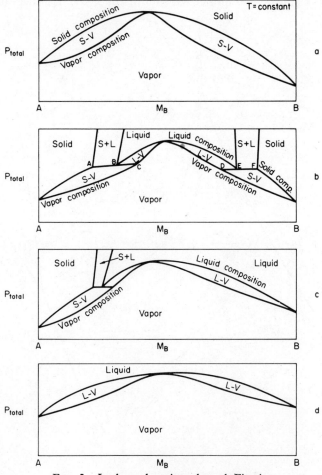

Fig. 5. Isothermal sections through Fig. 4.

G. SPACE MODELS

of the phase rule that the pressure is immediately fixed. For example, if the composition of the system is given by point 3, then when a liquid of the composition of point 3 is in equilibrium with solid and vapor, the solid has the composition of point 2, and the vapor has the composition point 1. At this composition the total pressure in the three-phase equilibrium is p_1.

Similarly, if a liquid in equilibrium with solid and vapor has the composition of point 5, the solid in equilibrium with it has the composition of point 6, while the vapor has the composition of point 4. The tie line 4–5–6 connecting the compositions of vapor, liquid, and solid, respectively, lies at the total pressure p_2.

In the composition interval points 3–5, the isothermally univariant system at temperature T_2 is shown to be composed of liquid and vapor phases. The curve of points 3–5 shows the variation of the total pressure and liquid composition, while the curve of points 1–4 shows the variation of vapor composition with variation of liquid composition at this temperature.

In the composition interval pure AC–point 2, the system is composed of solid and vapor at the temperature T_2 with the curve $p_{AC}^0(T_2)$–point 2 showing the composition of the solid and the total pressure, while the curve $p_{AC}^0(T_2)$–point 1 shows the composition of vapor in equilibrium with a solid of specified composition. A similar region exists on the opposite side of the diagram.

Figure 3 shows several isothermal sections through S–L–V portions of Fig. 2 which need not be discussed further since their significance is readily deduced by analogy with such isothermal sections in preceding chapters.

Figure 4 shows a three-dimensional construction where minima are present in T–M_X representations of L–V, S–L, and S–V regions of the diagram. Figure 5 shows isothermal sections through the space model depicted in Fig. 4. These are straightforward and require no further discussion.

30

Hypothetical Examples Using the Equations of Chapter 29

A. Introduction

In the preceding chapter, equations were derived for solution of solid–liquid univariant equilibria in which the following assumptions were made: (1) One of the end members was present as a dimer in the solid and as a monomer in the liquid while the other end member was present as a monomer in both phases; (2) one of the end members dissociated completely in the liquid phase while the other underwent no species stoichiometry change in undergoing its phase change; or (3) both of the end members were present as dimers in the solid and as monomers in the liquid phases.

Logical extensions of these treatments would involve assumptions such as partial monomerization or dissociation upon melting. This, in fact, was the approach employed in treating eutectic interactions. In the present instance, introduction of such assumptions into our arguments is left as an exercise for the interested reader. Another logical extension of the treatments applied to solution equilibria involves assumptions opposite to those invoked in the preceding chapter, i.e., monomers are present in the solid and dimers in the liquid, etc. Such assumptions would lead to equations which exhibit maxima rather than minima. The number

B. SOME HYPOTHETICAL EXAMPLES

of such cases which have been verified experimentally is few indeed, and the development of analytical treatments based on this model or a modification of it is again left to the reader.

In the present chapter, we shall present computer derived data for the cases presented in the preceding chapter in order to show qualitatively the effects on the solidus and liquidus curves due to invoking different assumptions. The final section of this chapter will be concerned with the system Na_2CO_3–K_2CO_3, which is of the minimum solid solution type. The experimental data for this system will be compared with those obtained theoretically.

B. Some Hypothetical Examples

Using the assumptions of one dimer in the solid, two dimers in the solid, and complete dissociation of a pseudounary end member in the liquid, phase diagrams were computer derived for several ΔH_X^0 combinations. In all instances the melting points for the system AC–BC were chosen as 1127 and 1174°K, which represent the values for Na_2CO_3 and K_2CO_3, respectively. In order to enable better comparison of the derived data, liquidus and solidus data are plotted separately.

Figure 1 comprises eight liquidus curves. Curves 1–4 show the cases where either AC or BC is present as a dimer in the solid while the other end member is present as a solid monomer. Curves 5–8 show the cases where either AC or BC dissociates completely in the melting process while the other end member melts without incident.

In curves 1 and 2 it is assumed that the higher melting end member BC undergoes the dimer → monomer transformation upon melting while the lower melting end member AC, undergoes no species change. For a constant value of ΔH_{BC}^0, two different values of ΔH_{AC}^0 are shown, i.e., 6700 and 8000 cal/mole. It is seen that the temperature of the minimum increases with increasing ΔH^0 of AC. The minimum location is affected only slightly, moving slightly toward pure BC as ΔH_{AC}^0 is increased.

In curves 2 and 3, the situation of curves 1 and 2 is reversed, i.e., it is assumed that AC, the lower melting end member, undergoes the dimer–monomer transformation upon melting while BC, the higher melting end member, undergoes no species change upon melting. Again, two values of ΔH_{AC}^0, 6700 and 8000 cal/mole, are compared with a constant value of ΔH_{BC}^0, 6600 cal/mole. Again, with increasing ΔH_{AC}^0, the temperature of the minimum is shifted to higher temperatures, but the location of the minimum is hardly affected. Note that in the four cases generated T_{min} varies over some 17°K while the M_{AC} minimum varies only over a 3% composition interval.

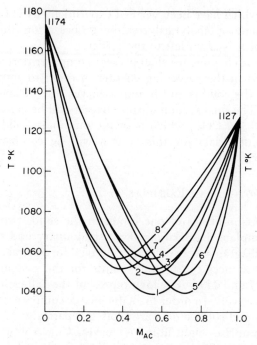

Fig. 1. Liquidus curves for minimum type systems.

Liquidus	Solid dimer	Cal/mole ΔH	Solid monomer	Cal/mole ΔH	Minimum $T°K$	M_{AC}
1	BC	6600	AC	6700	1039	0.55
2	BC	6600	AC	8000	1048	0.53
3	AC	6700	BC	6600	1051	0.56
4	AC	8000	BC	6600	1056	0.54
	Diss. end member	Cal/mole ΔH	Nondiss. end member	Cal/mole ΔH	Minimum $T°K$	M_{AC}
5	BC	6600	AC	6700	1040	0.72
6	BC	6600	AC	8000	1048	0.70
7	AC	6700	BC	6600	1052	0.39
8	AC	8000	BC	6600	1057	0.37

This leads us to conclude that when only one of the end members undergoes a dimer–monomer transformation in the melting process, an uncertainty in the heat of fusion will cause a significant variation in the minimum temperature, but not in the minimum composition. The second

B. SOME HYPOTHETICAL EXAMPLES

effect which perturbs the minimum temperature, but not the minimum composition, appears to be related to the melting points of the end members. If the higher melting end member undergoes the dimer–monomer transformation, the minimum will lie at a lower temperature than if the lower melting end member undergoes this effect.

Curves 5 and 6 represent the cases where the higher melting end member BC dissociates completely into two fragments upon melting, while the lower melting end member AC melts normally. Again for a single value of ΔH^0_{BC}, 6600 cal/mole, two values of ΔH^0_{AC} are examined, i.e., 6700 and 8000 cal/mole. As in cases 1–4, an increase in ΔH^0_{AC} causes an increase in T_{min}. The effect on the location of the minimum caused by a ΔH^0 change is, as before, minor.

Curves 7 and 8 represent the cases where the lower melting end member undergoes the dissociative process. The same effects are noted as in cases 5 and 6.

If we compare curves 5 and 6 with 7 and 8, we note that now, as opposed to the results obtained in cases 1–4 choice of which end member undergoes the dissociative process has a significant effect on where the minima are located. Thus, if the higher melting end member is assumed to undergo the dissociation, the minimum is displaced markedly toward the lower melting end member. If the lower melting end member is assumed to undergo the dissociation, the minimum is displaced toward the higher melting end member.

As in cases 1–4, however, the temperature of the minimum is displaced to lower values if the higher melting end member is assumed to undergo an anomaly upon melting.

Figure 2 treats the single case where the higher melting end member alone undergoes the dimer–monomer transformation. The ΔH^0 of the higher melting end member is kept constant while the ΔH^0 of the lower melting end member is varied over a range of some 3 kcal/mole. Figure 2 is a more extensive treatment, in fact, of the cases shown in curves 1 and 2 of Fig. 1. The trend in T_{min} and the M_{AC} minimum is more readily discerned now. For the range ΔH^0_{AC}, 5000–6000 cal/mole, T_{min} varies between 1022 and 1048°K, respectively, while the M_{AC} minimum varies between 0.58 and 0.53, respectively. This minor variation of the minimum concentration with increasing ΔH^0_{AC} is toward pure BC.

In Fig. 3, the liquidus and solidus curves for the boundary cases of Fig. 2 are presented.

For the sequence of ΔH^0_{AC} variations shown in Fig. 2, Fig. 4 shows the effects due to dimer–monomer behavior on the part of both end members. As before, the melting points of the end members are assumed to remain the same and ΔH^0_{BC} has been given the constant value of 6600 cal/mole.

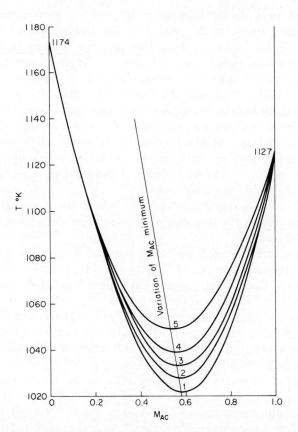

FIG. 2. Effects of ΔH^0_{AC} on the minimum solid solution. $\Delta H^0_{BC} = 6600$ cal/mole; $\Delta H^0_{AC} = $ (1) 5000, (2) 5500, (3) 6000, (4) 6700, (5) 8000 cal/mole. BC undergoes dimer→ monomer change.

It is seen, comparing Fig. 4 with Fig. 2, that while the minimum temperature increases with increasing ΔH^0_{AC}, the minima themselves are at much lower temperatures generally in Fig. 4. Thus curve 1 ($\Delta H^0_{AC} = 5000$) shows a minimum at 903°K and curve 5 ($\Delta H^0_{AC} = 8000$) shows its minimum at 944°K. Despite this large effect of the double dimer cases on $T_{\min}$, it is equally notable that the compositions of the minima are not drastically different from those of Fig. 2. Contrast, for example, Fig. 2, curve 1, where the M_{AC} minimum is about 0.58 with Fig. 4, curve 1, where the M_{AC} minimum is about 0.57. On the other end of the ΔH^0_{AC} spectrum, the M_{AC} minimum of Fig. 2 falls at approximately 0.53 and the M_{AC} minimum of Fig. 4 falls at approximately 0.5.

Finally in Fig. 5, the liquidus and solidus curves for two of the systems

C. THE SYSTEM Na_2CO_3–K_2CO_3

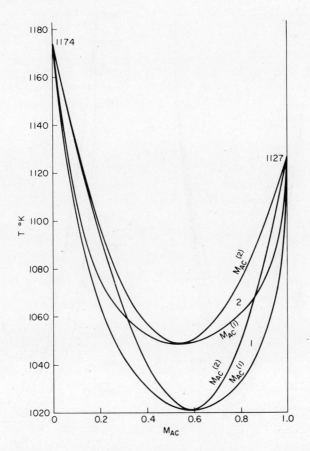

FIG. 3. Liquidus and solidus curves for the boundary cases of Fig. 2; ΔH^0_{BC} = 6600 cal/mole; ΔH^0_{AC} = (1) 5000, (2) 8000 cal/mole.

whose liquidi are shown in curve 1 of Figs. 2 and 4 are superimposed to provide an overall picture of the difference between the single and double dimer–monomer systems.

C. THE SYSTEM Na_2CO_3–K_2CO_3—A COMPARISON BETWEEN EXPERIMENTAL AND THEORETICAL SOLIDUS AND LIQUIDUS CURVES

Figure 6 shows the solid–liquid regions of the pseudobinary system K_2CO_3–Na_2CO_3 as determined by the author during part of an investigation of both the solid and solid–liquid portions of the diagram. The system is of the minimum type with $T_{min} \sim 982°K$ and the $M_{Na_2CO_3}$

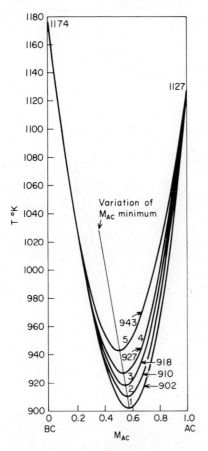
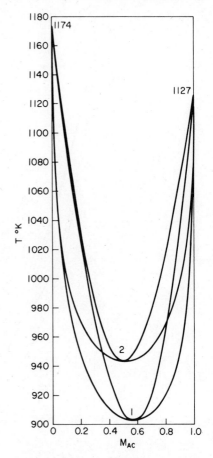

FIG. 4. Liquidus curves for a minimum system in which both end members undergo a dimer–monomer transformation upon melting; $\Delta H^0_{BC} = 6600$ cal/mole; $\Delta H^0_{AC} =$ (1) 5000, (2) 5500, (3) 6000, (4) 6700, (5) 8000 cal/mole.

FIG. 5. Solidus and liquidus curves for the boundary cases shown in Fig. 4; $\Delta H^0_{BC} = 6600$ cal/mole; $\Delta H^0_{AC} =$ (1) 5000, (2) 8000 cal/mole.

minimum approximately 0.58. All of the cases treated in the preceding section have essentially been used as test cases to attempt to define whether the system Na_2CO_3–K_2CO_3 is of the single dimer, double dimer, or dissociative kind of minimum system. A comparison of the experimental results with the theoretically derived curves of the preceding section indicates that the case involving the best data fit is the double dimer one. For example, the dissociative assumption yields minima in extremely poor agreement with the experimentally determined

C. THE SYSTEM Na_2CO_3–K_2CO_3

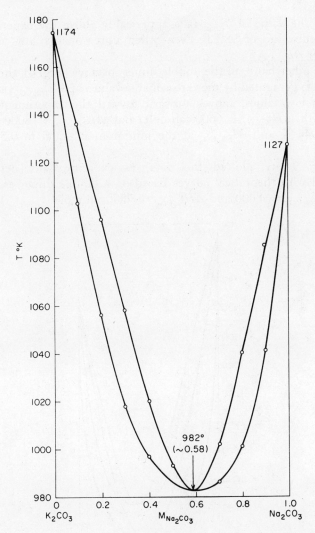

FIG. 6. Experimental solidus and liquidus curves for the system Na_2CO_3–K_2CO_3

results. The single dimer case (whether the dimer–monomer behavior is attributed to either Na_2CO_3 or K_2CO_3), while providing minima in the right composition interval, provides very poor fit relative to the temperature of the minimum. The heats of fusion of Na_2CO_3 and K_2CO_3 are reported in the literature for the sodium compound as varying between 6 and 8 kcal/mole and for the potassium compound as varying between 6.6 and 7.8 kcal/mole. In the single dimer cases it is seen that the value of T_{min} falls appreciably above the experimentally determined value of 982°K even when unreasonably low values are chosen for $\Delta H^0_{Na_2CO_3}$.

On the other hand, in the double dimer instance (Fig. 4), the value of T_{min} that is generated by more reasonable values of $\Delta H^0_{Na_2CO_3}$, starting at obviously low values, moves upward toward the experimental value. Thus, with $\Delta H^0_{K_2CO_3} = 6600$ cal/mole and $\Delta H^0_{Na_2CO_3} = 8000$ cal/mole, $T_{min} = 944°$K and $M_{Na_2CO_3}$ at the minimum is equal to 0.5 (Fig, 4, curve 5).

In Fig. 7 are plotted the data, as experimentally determined, compared with theoretical curves based on a double dimer assumption with $\Delta H^0_{Na_2CO_3} = 8000$ and $\Delta H^0_{K_2CO_3} = 7800$ cal/mole.

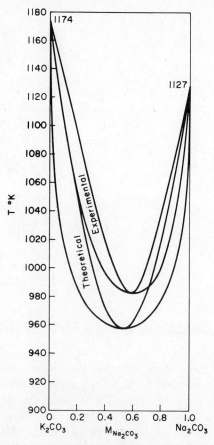

FIG. 7. Comparison of experimental and theoretical solidus and liquidus curves for the system Na_2CO_3–K_2CO_3. $\Delta H^0_{Na_2CO_3} = 8000$ cal/mole; $\Delta H^0_{K_2CO_3} = 7800$ cal/mole.

C. THE SYSTEM Na_2CO_3–K_2CO_3

The agreement between the curves, while not outstanding, is still good enough to merit notice. It should be emphasized that the model employed is quite simple and no attempt at refinement via introduction of liquid phase dissociation or partial monomerization has been attempted. Furthermore, considering the significant effects on the curves of variations in ΔH^0, the uncertainty in reported values of ΔH_{fusion} of the end members may be contributory. Finally, while the hypothetical solidus is a consequence of a purely mathematical argument, the experimental solidus, as in all solid-solution systems, is always subject to question because of the difficulty in obtaining solid-phase equilibrium. This latter is noted because greater discrepancy exists between the hypothetical and experimental solidi, where one might have expected it.

31

Condensed-Vapor Phase Binary Diagrams

A. Introduction

The main concern in preceding chapters on binary interactions has been with solid–liquid equilibria (S–L), for which analytical relationships based on assumed species models have been developed. Discussions concerning solid–vapor (S–V), or liquid–vapor (L–V) equilibria have been offered for the most part within the context of the three-dimensional p–T–M_X space model (see Chapter 24 *et seq.*). In Chapter 13, however, we saw how, based on vapor pressure considerations alone, one could derive the liquidus equations for a simple eutectic system, and in Chapter 24, S–V systems containing immiscible solids were examined to some extent. By and large, however, both the T–M_X and p–M_X relationships in condensed-vapor phase systems have been treated cursorily. The present chapter presents a closer inspection of such systems as a prelude to the next chapter, which considers S–V systems containing several compounds whose vaporization behavior may be termed incongruent, namely, where one of the components volatilizes preferentially. Typical of this type of system are those in which hydrates are formed. The latter consist generally of pseudounary components such as $CuSO_4$, Na_2CO_3, K_2CO_3, and a host of others plus H_2O as the second component. These

B. Boiling and Sublimation Point Diagrams

sulfates, carbonates, etc. behave, over a range of conditions, as pseudo-unary, involatile constituents while the H_2O functions as a pseudounary volatile component.

This and the next chapter then, serve to fill the void left in preceding chapters, and as we shall see lead us more or less naturally into a discussion of limited solid solubility, the topic of Chapter 33.

B. Boiling and Sublimation Point Diagrams (Constant Pressure Diagrams)

If we examine the several derivations offered earlier pertaining to S–L equilibria in eutectic and solid-solution systems, the following may be noted; In all instances, only univariant systems (those possessing a single degree of freedom, e.g., binary two-phase equilibria at either constant total pressure or temperature) were considered. In all instances, the starting point was; that a phase of lower heat content was in equilibrium with a phase of higher heat content, as evidenced by the presence of a latent heat associated with melting or freezing phenomena. While it was generally the case that the phase of lower heat content was specified as solid and the phase of higher heat content was specified as liquid, we now recognize that such a restriction was not implicit in any of the derivations. Had we chosen to change the focus of the preceding chapters, we could have picked as a starting point a solid in equilibrium with a vapor or a liquid in equilibrium with a vapor. The steps in each derivation would have remained the same, and as a matter of fact, the final equations would have been identical in appearance to their S–L counterparts. What would have changed would pertain solely to the definition of terms.

Let us assume for the moment that we are concerned with a S–V equilibrium under a constant total pressure constraint. This equilibrium represents the analog of the S–L equilibrium under the identical constraint. For a particular specified total constant pressure P_t, the melting point of an end member is uniquely defined by its unary S–L curve in the p–T diagram. In a S–V equilibrium this melting point concept must be replaced by the concept of the sublimation point. Specifically, the sublimation point of a component is defined as the temperature at which the vapor pressure of the component achieves some particular value. In a L–V equilibrium the corresponding term would be the boiling point. In a binary interaction, the sublimation or boiling points would represent the temperature at which the binary system develops a pressure equal to some specified value. The representation would be a constant pressure T–M_x slice through the space model. To go along with this, we also introduce the concept of the vaporous curve which defines the variation of sublimation or boiling point with composition in a binary isobaric univariant S–V or L–V interaction, respectively. We see, then, that in isobaric condensed-vapor phase equilibria designated C–V, the vaporous is the counterpart of the liquidus in a S–L equilibrium. In anticipation of the following discussions of isothermal C–V equilibria we may note that in such equilibria, the vaporous will refer to the pressure at which a vapor phase first liberates a condensed phase. Thus, the vaporous represents the saturation, freezing point, or solubility curve of a vapor.

Let us now, as a case in point, consider the familiar liquidus equation previously developed for a eutectic S–L equilibrium;

$$\ln N_X^{(2)} = -\frac{\Delta H_X^0}{R}\left(\frac{1}{T} - \frac{1}{T_X^0}\right). \tag{1}$$

The system in question is one in which the phase of lower heat content is a pure solid substance whose latent heat of fusion is ΔH_X^0 and whose melting point at some constant specified pressure is T_X^0. If we are examining the isobaric S–V analog of the eutectic system, then it consists of a pure solid in equilibrium with its vapor (see Chapter 24, Section B). The term ΔH_X^0 in Eq. (1) is replaced by the term ΔH_X^s, the standard molar enthalpy of sublimation. The term T_X^0 is replaced by the term T_X^s which represents the temperature at which the pure substance X develops the specified value of pressure. For this pure substance this specified pressure shall be denoted by p_X^s and be equal in value to the total pressure P_t at intermediate compositions.

If the system is being examined open to the earth's ambient atmosphere, $P_t \sim 1$ atm, and T_X^s would be referred to as the "normal" sublimation point of X. In general, the normal sublimation point is the temperature at which a system exposed to an arbitrary constant inert gas pressure P_t^i achieves this pressure.

For a pure unary or pseudounary solid or liquid substance, the variation in vapor pressure p_X^c with temperature is given by an equation of the general form

$$\begin{aligned}\ln p_X^c &= -\frac{\Delta H_X^c}{R}\left(\frac{1}{T_X^c} - \frac{1}{T_X^*}\right) + \ln p_X^* \\ &= -\frac{\Delta H_X^c}{RT_X^c} + \left[\frac{\Delta H_X^c}{RT_X^*} + \ln p_X^*\right] = -\frac{\Delta H_X^c}{RT_X^c} + C,\end{aligned} \tag{2}$$

where ΔH_X^c is the standard molar enthalpy of sublimation or condensation of the condensed phase. T_X^c is the temperature at which

B. BOILING AND SUBLIMATION POINT DIAGRAMS

this phase achieves the particular vapor pressure $p_X{}^c$ that we are currently interested in, and $p_X{}^*$ is the vapor pressure of X at some temperature $T_X{}^*$ other than the one at which the desired pressure is achieved. $p_X{}^*$ and $T_X{}^*$ are arbitrary reference values which we happen to know.

Given a value $p_X{}^c$ we can solve for the corresponding sublimation or boiling point temperature $T_X{}^c$ providing we know the value $p_X{}^*$ at some other sublimation or boiling point temperature $T_X{}^*$. Under a constraint of constant total pressure P_t, the value $p_X{}^c$ must, of course, equal P_t. (The value $p_X{}^*$ is also under the same constraint, equal to some value of P_t, i.e., $P_t{}'$.) Solving for $T_X{}^c$ in the second equality in Eq. (2) we obtain

$$T_X{}^c = -\Delta H_X{}^c (R\{\ln P_t - [(\Delta H_X{}^c/RT_X{}^*) + \ln p_X{}^*]\})^{-1}$$
$$= -\Delta H_X{}^c (R\{\ln p_X{}^c - [(\Delta H_X{}^c/RT_X{}^*) + \ln p_X{}^*]\})^{-1}. \quad (3)$$

In exponential form Eq. (2) may be written alternatively as

$$p_X{}^c = P_t = p_X{}^* e^{-x} = P_t' e^{-x}, \quad (4)$$

where P_t' is used as the symbol for the reference case available.

For a particular constant pressure plane through the S–V region of a T–p–M_X diagram, the value for $T_X{}^c$ in Eq. (1), as redefined, may be substituted for by its expanded form given in Eq. (3), i.e.,

$$\ln N_X^{(3)} = -\frac{\Delta H_X{}^s}{R}\left(\frac{1}{T} + \frac{R\{\ln P_t - [(\Delta H_X{}^s/RT_X{}^*) + \ln p_X{}^*]\}}{\Delta H_X{}^s}\right). \quad (5)$$

If we have available a table or graph showing the values $T_X{}^s$ versus $p_X{}^s$, it is clear that the appropriate value of $T_X{}^s$ may be substituted directly into the redefined form of Eq. (1). The value $N_X^{(3)}$ at any particular temperature T may thus be evaluated for the constant pressure section either from Eq. (1) or Eq. (5), whichever is most convenient.

Equation (1) or (5), then, enables us to depict graphically the A and B vaporous curves in a binary isobaric section through the p–T–M_X space model. Such a T–M_X diagram is shown schematically in Fig. 1. We see that, as might have been expected, the diagram is of similar appearance to its S–L counterpart. For each new isobaric section through the space model we would require a knowledge of the applicable sublimation points of the end members. All of the descriptions of S–L equilibria are immediately transferable to Fig. 1. Thus tie lines may be constructed and used to determine solid to vapor ratios, etc.

In precisely the same way as shown for the simple case considered, we may arrive at the analytic expressions for solid-solution systems, and, as in the eutectic case, the T–M_X diagrams are similar in appearance to the

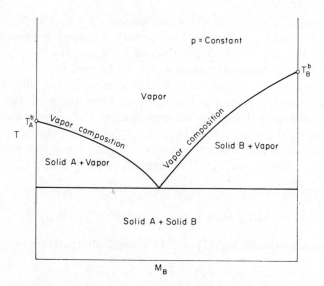

Fig. 1. The diagram S–V in the system A–B at constant total pressure. The boiling point diagram where pure solids are in equilibrium with vapor.

S–L cases. Schematic representations of S–V equilibria for ascending and minimum type solid–solid-solution systems are offered in Figs. 2 and 3, respectively.

Again, without further ado, L–V analogs may be defined and diagrams of the types shown in Figs. 1–3 will result. One minor difference arises in the L–V case in that the immiscible solids of Fig. 1 are replaced by immiscible liquids. (The system $Hg-H_2O$ might be considered representative of such a system.) These latter are liquid immiscibility type systems exhibiting complete liquid phase immiscibility.

C. Constant Temperature Condensed-Vapor Phase Diagrams

We have treated isothermal sections briefly in several places earlier. In Chapter 24, for example, systems involving pure condensed phases in equilibrium with vapor were examined and relationships showing the variation of vapor phase composition with total pressure as well as the variation of the three phase coexistence solid A–solid B–vapor with temperature were derived from first principles. In this section, a more extensive discussion of isothermal diagrams will be offered, and an alternative means of defining vaporous composition curves in the three main types of condensed-vapor phase behavior will be considered.

C. CONSTANT TEMPERATURE DIAGRAMS

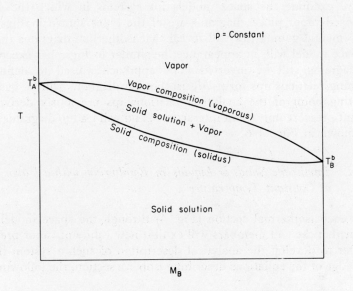

FIG. 2. An ascending S–V equilibrium at constant total pressure.

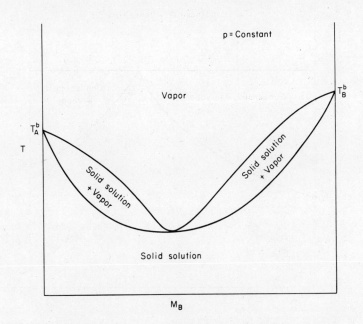

FIG. 3. A minimum type S–V equilibrium at constant total pressure.

If we examine the space models for systems in which the $T–M_x$ condensed-vapor phase diagrams are of the types shown in Figs. 1–3, a little mental gymnastics will reveal that isothermal diagrams through the space model will, in appearance, be similar to Figs. 1–3 except that the diagrams will be inverted. The approaches used to define the governing relationships for $p–M_x$ isothermal diagrams will be based on a starting point of the analytical relationships previously derived for constant pressure univariant interactions, and will lead to diagrams of the type shown in Figs. 4–6.

Case I.   Immiscible Solids or Liquids in Equilibrium with a Vapor Phase at Constant Temperature

For each isothermal section (Fig. 4) through the space model, it is evident that the end members will exhibit new values of vapor pressure. In order to develop the analytical description of such a system from a knowledge of the equations describing isobaric section, the following will

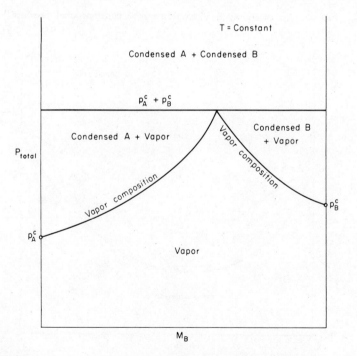

Fig. 4. Schematic representation of a condensed-vapor phase equilibrium at constant temperature where pure condensed phases are in equilibrium with vapor.

C. CONSTANT TEMPERATURE DIAGRAMS

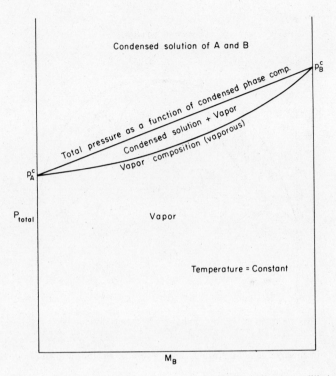

FIG. 5. The isothermal diagram for a condensed-vapor phase equilibrium where the condensed phase is a solution exhibiting equivalent mole and species fractions.

be helpful. If we visualize a series of isobaric planes stacked one upon the other in the space model, we note that each of these planes intersects the p–T curves of the end members at new and unique values of p and T. If through this set of stacked planes we insert an isothermal plane (at a right angle), it can be seen that the latter will cut each of the vaporous T–M_x curves at only a single point and that this point will coincide with a particular total pressure which happens also to be the vapor pressure of each of the end members in the isobaric section. These points of intersection of the isothermal plane with the infinite number of isobaric planes generate the isothermal diagram. Thus, while the isothermal plane cuts through the p–T curves of the end members at only a single point corresponding to the temperature of the isotherm, it cuts through all of the constant total pressure planes. Each of these constant total pressure planes in turn cut through the end member p–T diagrams at a different pressure. Consequently, the isothermal diagram cuts through the equivalent of the end member p–T diagrams since these latter coincide

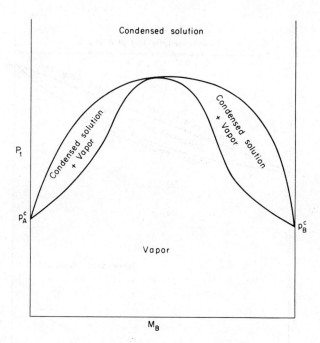

Fig. 6. Isothermal section through a space model exhibiting a minimum in T–M_X curves.

with the corresponding value of the total pressure at any particular value of the latter in an isobaric section.

If we examine Eq. (5) for a condition of constant temperature, T, we see that as a function of P_t it reduces to

$$\ln M_X^{(3)} = \ln N_X^{(3)} = -\ln P_t + C = -\ln P_t + \ln C_1, \qquad (6)$$

since a constant can be set equal to the logarithm of some other constant, Thus,

$$M_X^{(3)} = C_1/P_t. \qquad (7)$$

Equation (7) is identical to Eq. (17) of Chapter 24 if $C_1 = p_X^0$, where p_X^0 is the vapor pressure of pure solid A or B at temperature T. That this is the case is seen from the following.

The value of p_X^c at the temperature of the isotherm T_X^c given the value p_X^c at some other temperature T_X^* is given by Eq. (2). If we expand

C. CONSTANT TEMPERATURE DIAGRAMS

Eq. (5) and invoke Eq. (4), we obtain

$$\begin{aligned}\ln N_X^{(3)} &= -\frac{\Delta H_X{}^c}{RT} - \frac{R\,\Delta H_X{}^c\{\ln P_t - [(\Delta H_X{}^c/RT_X{}^*) + \ln p_X{}^*]\}}{R\,\Delta H_X{}^c}\\ &= -(\Delta H_X{}^c/RT) - \ln P_t + (\Delta H_X{}^c/RT_X{}^*) + \ln p_X{}^*\\ &= -\ln P_t + \ln p_X{}^* - (\Delta H_X{}^c/R)[(T^{-1}) - (T_X{}^*)^{-1}]\\ &= -\ln P_t + \ln p_X{}^c.\end{aligned} \quad (8)$$

At the temperature T, $p_X{}^c$ is a constant so that Eqs. (7) and (8) are seen to be equivalent.

The fact that each value of P_t coincides with some value of both $p_A{}^c$ and $p_B{}^c$ enables us to use a simple approach for the conversion of isobaric to isothermal expressions.

In exponential form Eq. (1) may, for a solid–vapor equilibrium, be written as

$$N_X^{(3)} = \exp\left[-\frac{\Delta H_X{}^c}{R}\left(\frac{1}{T} - \frac{1}{T_X{}^0}\right)\right]. \quad (9)$$

If we multiply and divide the right hand side of Eq. (9) by P_t we obtain

$$N_X^{(3)} = \frac{P_t \exp\left[-\dfrac{\Delta H_X{}^c}{R}\left(\dfrac{1}{T} - \dfrac{1}{T_X{}^0}\right)\right]}{P_t}. \quad (10)$$

Let the term P_t represent the total pressure at some point along the vaporous of an isotherm. This pressure coincides with some value of the pure component pressure $p_X{}^c$ at a different temperature $T_X{}^0$ in an isobar. At the temperature of the isotherm, T, the product

$$P_t \cdot \exp\{-(\Delta H_X{}^c/R)[T^{-1} - (T_X{}^0)^{-1}]\} = p_X^{c'},$$

which is the pressure of pure X at the temperature T of the isotherm.

Thus,

$$N_X^{(3)} = p_X^{c'}/P_t, \quad (11)$$

which is the identical solution arrived at in Chapter 24, Eq. (17).

Case II. Isothermal p–M_X Relationships for Completely Miscible Solids or Liquids in Equilibrium with Vapor

In Chapter 27, the equations defining $M_A^{(1)}$ and $M_A^{(2)}$ in an isobaric ascending solid–liquid equilibrium, rewritten now for the corresponding condensed–vapor phase equilibrium, are given by

$$M_A^{(c)} = \frac{e^{-B} - 1}{e^{-B} - e^{-A}}, \quad (12)$$

$$M_A^{(3)} = \frac{e^{-A}[e^{-B} - 1]}{e^{-B} - e^{-A}} = M_A^{(c)} e^{-A}, \tag{13}$$

where $M_A^{(c)}$ refers to the component mole fraction of A in the condensed phase (liquid or solid),

$$e^{-A} = \exp\left[-\frac{\Delta H_A{}^c}{R}\left(\frac{1}{T} - \frac{1}{T_A{}^c}\right)\right], \quad e^{-B} = \exp\left[-\frac{\Delta H_B{}^c}{R}\left(\frac{1}{T} - \frac{1}{T_B{}^c}\right)\right],$$

where $\Delta H_A{}^c$ and $\Delta H_B{}^c$ are the molar enthalpies of sublimation or vaporization of A and B, respectively, and $T_A{}^c$ and $T_B{}^c$ are the sublimation or vaporization temperatures at the total pressure of the system P_t.

If we multiply numerator and denominator of Eq. (12) by P_t we obtain,

$$M_A^{(c)} = \frac{P_t e^{-B} - P_t}{P_t e^{-B} - P_t e^{-A}}. \tag{14}$$

The values $P_t e^{-A}$ and $P_t e^{-B}$ represent, as in Case I, the values of the vapor pressures of pure A and B at some common temperature T. Thus, we may write

$$M_A^{(c)} = (p_B{}^c - P_t)/(p_B{}^c - p_A{}^c). \tag{15}$$

Solving for P_t as a function of $M_A{}^c$ we obtain

$$P_t = p_A{}^c M_A^{(c)} + p_B{}^c[1 - M_A^{(c)}], \tag{16}$$

which shows the variation of the total pressure of the condensed phase with composition at constant temperature.

This relationship [Eq. (16)] is precisely that obtained assuming the Raoult equation to be descriptive of a solution-vapor equilibrium. This can be seen from the following;

$$P_t = p_A + p_B, \tag{17}$$

$$p_A = N_A^{(c)} p_A{}^c, \tag{18}$$

where $p_A{}^c$ is the vapor pressure of pure A at the temperature in question. For p_B and P_t in expanded form then we have:

$$p_B = N_B^{(c)} p_B{}^c = [1 - N_A^{(c)}] p_B{}^c, \tag{19}$$

$$P_t = N_A^{(c)} p_A{}^c + [1 - N_A^{(c)}] p_B{}^c, \tag{20}$$

The variation in composition of the vapor phase along the vaporous

C. CONSTANT TEMPERATURE DIAGRAMS

with total pressure at constant temperature is obtained from Eq. (13) by multiplying numerator and denominator by P_t to give

$$M_A^{(3)} = N_A^{(3)} = N_A^{(c)}(P_t e^{-A}/P_t) = N_A^{(c)} p_A{}^c/P_t \tag{21}$$

at a specified value T.

The Raoult solution is again the same and utilizing Eq. (18), is seen to be

$$N_A^{(3)} = p_A/(p_A + p_B) = N_A^{(c)} p_A{}^c/P_t. \tag{22}$$

Case III. Isothermal p–M_x Relationships for Completely Miscible Solids or Liquids in Equilibrium with Vapor Involving Dimer–Monomer Behavior

As has been indicated, isothermal sections through the space model provide p–M_x diagrams that are inverted in appearance to those of their T–M_x counterparts. Therefore, if we visualize the three-dimensional space model exhibiting minimum solid solution–vapor behavior on a T–M_x plot, we would expect that a maximum would be generated on a p–M_x plot. The analytical description of the solidus and vaporous curves for a system exhibiting dimer–monomer behavior may be deduced in similar manner to that employed in the preceding sections.

In the S–V system AC–BC where AC is present as the dimer in the solid phase and as the monomer in the vapor phase, the variations of $M_{AC}^{(1)}$ and $M_{AC}^{(3)}$ with temperature are given by

$$M_{AC}^{(1)} = \frac{2\{2e^{-BC}[e^{-BC} - 1] + e^{-AC} \pm [e^{-2AC} + 4e^{-(AC+BC)}(e^{-BC} - 1)]^{1/2}\}}{2e^{-BC}[2e^{-BC} - 1] + e^{-AC} \pm [e^{-2AC} + 4e^{-(AC+BC)}(e^{-BC} - 1)]^{1/2}} \tag{23}$$

$$M_{AC}^{(3)} = \frac{e^{-AC} \pm [e^{-2AC} + 4e^{-(AC+BC)}(e^{-BC} - 1)]^{1/2}}{2e^{-BC}} \tag{24}$$

in accordance with Eqs. (38) and (40) of Chapter 29.

If Eqs. (23) and (24) are multiplied by $P_t{}^2$ and the appropriate substitutions are made, the variation of $M_{AC}^{(1)}$ and $M_{AC}^{(3)}$ with P_t is given by

$$M_{AC}^{(1)} = \frac{\begin{bmatrix} 2P_t[p_{AC}^0 - 2p_{BC}^0] + 4(p_{BC}^0)^2 \\ \pm 2\{P_t^2[(p_{AC}^0)^2 - 4p_{AC}^0 p_{BC}^0] + 4P_t p_{AC}^0 (p_{BC}^0)^2\}^{1/2} \end{bmatrix}}{\begin{bmatrix} P_t[p_{AC}^0 - 2p_{BC}^0] + 4(p_{BC}^0)^2 \\ \pm \{P_t^2[(p_{AC}^0)^2 - 4p_{AC}^0 p_{BC}^0] + 4P_t p_{AC}^0 (p_{BC}^0)^2\}^{1/2} \end{bmatrix}}, \tag{25}$$

$$M_{AC}^{(3)} = \frac{P_t p_{AC}^0 \pm \{P_t^2[(p_{AC}^0)^2 - 4p_{AC}^0 p_{BC}^0] + 4P_t p_{AC}^0 (p_{BC}^0)^2\}^{1/2}}{2P_t p_{BC}^0}. \tag{26}$$

Equations (72) and (74) of Chapter 29 [substituting $M_{AC}^{(3)}$ for the term $M_{AC}^{(2)}$ of Eq. (74)], which define the complete dissociation case for minimum type formation, may be converted in a similar fashion and this is left as an exercise for the reader.

Equations (25) and (26) do not have any obvious derivational counterpart based on the Raoult equation or some trivial modification of it.

D. Some Hypothetical Examples of Maximum Type S–V Equilibria in Isothermal Diagrams

In this section some hypothetical examples of maximum type S–V equilibria based on Eqs. (25) and (26) are presented. Data were obtained by computer analysis.

Figures 7–10 show the effect of maintaining p_{AC}^0 constant at a value of 20 mm Hg while p_{BC}^0 is varied from 10 to 20 mm. The phase diagrams are plotted in terms of P_t (mm Hg) as a function of M_{AC} for the two-phase univariant S–V equilibrium. When p_{AC}^0 is approximately twice the size of p_{BC}^0 (Fig. 7), it is seen that the maximum occurs essentially at pure AC. The extensions of the solidus and vaporous curves into the hypothetical region $M_{AC} > 1$ is shown in order to point out more clearly that the curves are of the maximum type. As p_{BC}^0 approaches the value p_{AC}^0 we see that the maximum shifts into the physically real range where M_{AC} varies between 0 and 1. Thus, in Fig. 8 with $p_{AC}^0 = 20$ and $p_{BC}^0 = 12.5$, the maximum occurs at $M_{AC} \simeq 0.79$. When $p_{BC}^0 = 15$, $M_{AC} = 0.67$ (Fig. 9), and when $p_{BC}^0 = 20$, M_{AC} at the maximum is about 0.51 (Fig. 10). Thus, we may conclude that when the extraordinary component (that exhibiting the dimer–monomer behavior) has a higher vapor pressure in the pure form than the second component, the maximum will be displaced toward higher mole fractions of the extraordinary component. Concomitantly, the total pressure at the maximum increases as the vapor pressure of the "normal" component approaches the vapor pressure of the component showing dimer–monomer behavior. Thus, in Figs. 7–10 we not that as p_{BC}^0 varies from 10 to 20 mm, P_t at the maximum varies from 20 to 26.6 mm, the value for p_{AC}^0 being constant in all instances at 20 mm.

That such a sequence of events does not seem to depend upon the absolute values of p_{AC}^0 and p_{BC}^0, but rather on the ratio of the two values is seen from Figs. 11–19. Here, keeping the value of p_{BC}^0 constant and increasing the value of p_{AC}^0 (the extraordinary component) from values of less than, to values of more than twice that of the pure normal component BC, it is seen that when p_{AC}^0 is twice the value p_{BC}^0 (Fig. 17),

D. EXAMPLES OF MAXIMUM S–V EQUILIBRIA

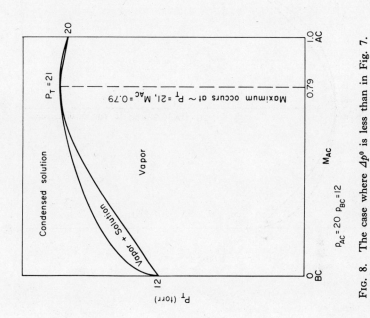

FIG. 8. The case where Δp^0 is less than in Fig. 7.

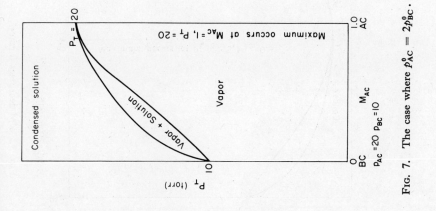

FIG. 7. The case where $p_{AC}^0 = 2p_{BC}^0$.

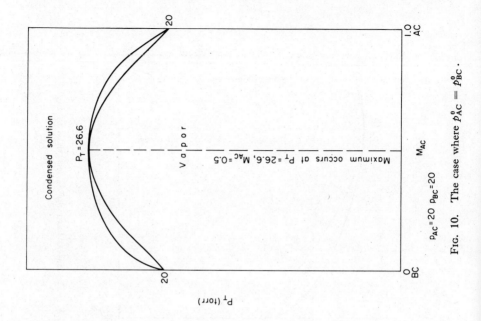

FIG. 10. The case where $p_{AC}^0 = p_{BC}^0$.

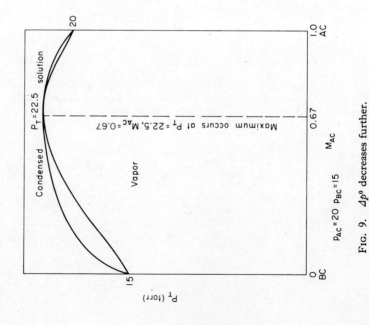

FIG. 9. Δp^0 decreases further.

D. EXAMPLES OF MAXIMUM S–V EQUILIBRIA

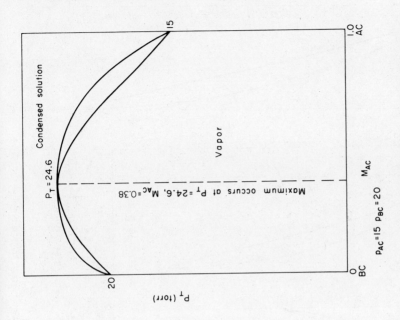

FIG. 12. The case where $p^0_{AC} < p^0_{BC}$, but the difference Δp is less than in Fig. 11.

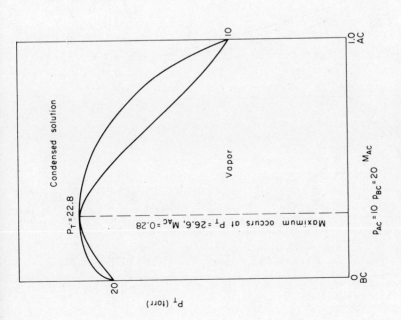

FIG. 11. The case where $p^0_{AC} < p^0_{BC}$.

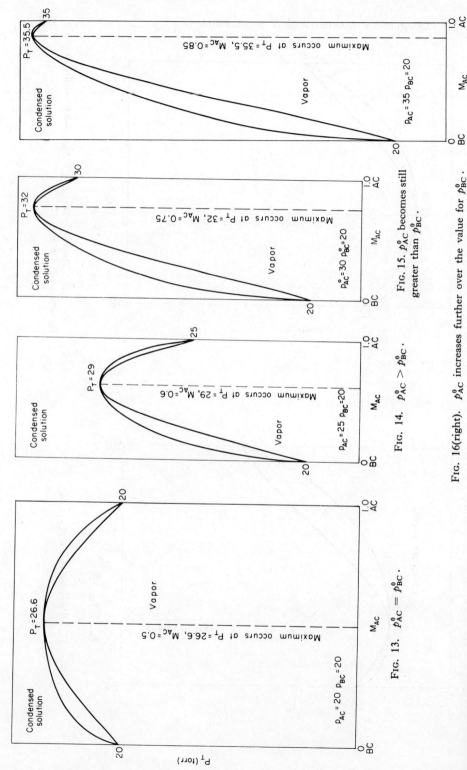

Fig. 13. $p^0_{AC} = p^0_{BC}$.

Fig. 14. $p^0_{AC} > p^0_{BC}$.

Fig. 15. p^0_{AC} becomes still greater than p^0_{BC}.

Fig. 16 (right). p^0_{AC} increases further over the value for p^0_{BC}.

D. EXAMPLES OF MAXIMUM S–V EQUILIBRIA

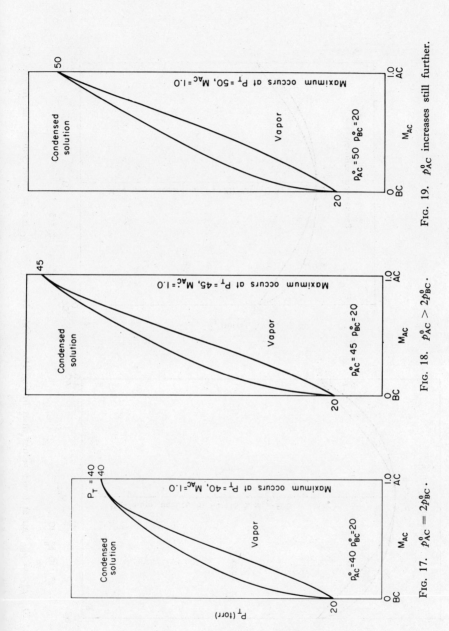

Fig. 17. $p_{AC}^0 = 2p_{BC}^0$.

Fig. 18. $p_{AC}^0 > 2p_{BC}^0$.

Fig. 19. p_{AC}^0 increases still further.

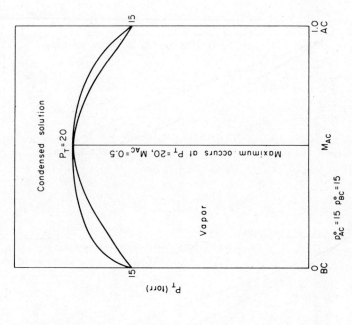

Fig. 21. $p^0_{AC} = p^0_{BC}$ but at lower values than in Fig. 10.

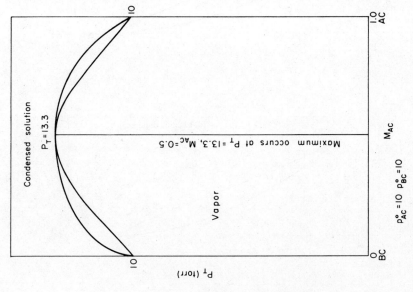

Fig. 20. $p^0_{AC} = p^0_{BC}$ but at lower values than in Fig. 10.

the maximum value of P_t occurs at $M_{AC} = 1$. When $p^0_{AC} > 2p^0_{BC}$ (Figs. 18 and 19), the maximum shifts into the physically unreal region where M_{AC} achieves a value greater than one. With $p^0_{AC} < p^0_{BC}$ (Figs. 11 and 12), the maximum is displaced toward pure BC, the normal component. The maximum value of P_t occurs for the case where $p^0_{AC} = p^0_{BC}$: and furthermore, when $p^0_{AC} = p^0_{BC}$, the value of M_{AC} at the maximum is equal to 0.5.

From Figs. 11 and 12, we also note that, with p^0_{BC} twice the value of p^0_{AC}, the effect on the phase diagram is not nearly so drastic as for the converse. While the maximum is again displaced toward the pure component exhibiting the greater vapor pressure, this maximum is located at values of $M_{BC} < 1$.

It appears then, that while the location of the maximum depends upon which end member exhibits the greater vapor pressure, the effect is much greater when the abnormal end member possesses the greater vapor pressure. In the event that the vapor pressures are equal, the maximum occurs at the middle of the phase diagram. This latter is demonstrated further with reference to Figs. 20 and 21. In these cases, the absolute values of both end members are varied while both end members exhibit the same vapor pressure in the isotherm.

E. Isothermal Distillations

In the ideal ascending solution case defined in Section C, Case II (where species and mole numbers are equal), the variation of the total pressure with composition of the condensed phase is linear, while the variation of gas phase composition with pressure is nonlinear. This leads to a diagram of the type shown in Fig. 5. In practice, however, the condensed-phase composition variation with pressure is not generally linear, being either concave or convex with respect to the composition axis. The concave curves are said to be indicative of so-called "positive deviations" from ideality, while convex liquidus curves are said to exhibit "negative deviations." It is often stated that, when such positive or negative deviations become excessive, maxima or minima, respectively, are generated. While this may or may not be the case, we have seen that maxima may be a natural consequence of dimer-monomer or complete dissociation phenomena. Maximum type S–L curves which would give rise to minimum type S–V or L–V equilibria have not been discussed here, but using the identical approaches previously employed, these may be developed by assuming monomers in the condensed phase and dimers in the vapor phase.

Curves of the type shown in Fig. 5 are sometimes referred to as isothermal distillation curves. If vapor is removed, for example, the liquid becomes richer in the less volatile component while the vapor becomes richer in the more volatile component. A sequence of separation steps in an isothermal distillation of a L–V ascending equilibrium may be visualized better with the aid of Fig. 22. Assuming

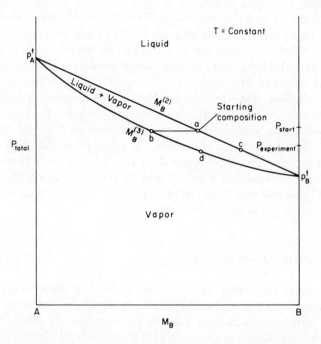

FIG. 22. An isothermal distillation process involving an ascending L–V equilibrium.

a starting composition a, the vapor has a composition b which is richer than the liquid in the component A. If we realize that as soon as a liquid of composition a is permitted to vaporize, this liquid will become richer in the component B, it is evident that if vapor is withdrawn, the liquid composition will move down the curve from b–$p_B^\dagger$. If the total pressure of the system at the start is P_{start}, and we remove vapor at such a rate that the distillation pressure is less than P_{start}, i.e., P_{exp}, the liquid composition will continue to increase in B until the total pressure of the system at equilibrium is P_{exp}. At the same time that the liquid composition is increasing with respect to component B, the vapor composition, while richer in A than the liquid, will not be as rich in A as at point b. The vapor composition will move down along the curve b–d, while the

E. ISOTHERMAL DISTILLATIONS

liquid composition will move along a–c. If the vapor phase is condensed and resubjected to an isothermal distillation while we continue to reduce the pressure on the liquid phase, it is seen that a complete separation of A and B may be effected (in principle at least). We may visualize an iterative process in which different fractions are continuously recombined and redistilled isothermally. This method of purification, while certainly possible, is not of very great practical utility, isobaric distillation processes being preferable.

While a complete separation of components in an ascending system is possible, such a complete separation is not possible in a maximum type L–V or S–V system as can be seen from the following with reference to Fig. 23.

If the starting point is a liquid of composition a, the vapor in equilibrium with it has the composition b. As soon as some vapor has formed, the liquid becomes somewhat deficient in component B and its composition moves up along the curve a–m. If we continuously extract vapor, this process of enrichment of liquid in component B continues.

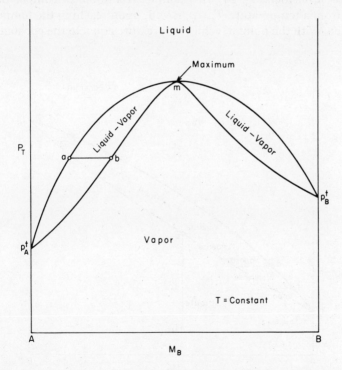

FIG. 23. The isothermal L–V diagram for a system showing a minimum in the T–M_X equilibrium.

Note, however, that when point m is reached, the liquid and vapor achieve the same composition and further volatilization merely decreases the quantity of liquid, but no longer causes a change in its composition. The point m is reached starting with a liquid composition on either side of the maximum. This maximum composition mixture is called an azeotrope and is a constant boiling mixture. For each new temperature of isothermal distillation it is to be anticipated that the composition of the maximum would shift.

F. ISOBARIC DISTILLATIONS

The ascending curve shown in Fig. 2 is for a system at constant total pressure and, as noted, each composition along the condensed-phase curve defines the temperature at which the pressure of the condensed curve achieves a pressure equal to the constant specified value. Diagrams like Fig. 2 may be termed boiling point diagrams or isobaric distillation diagrams. Consider Fig. 24, for example. If a liquid of composition a is heated from a temperature T_{start}, it will, upon reaching the temperature coincident with the point a, achieve a pressure equal to the constant value

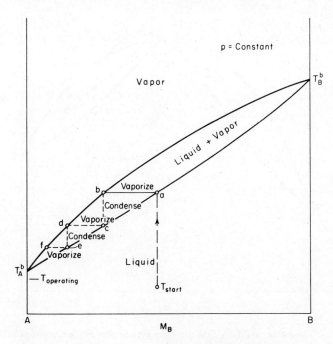

FIG. 24. Boiling point curves for an ascending L–V system.

F. ISOBARIC DISTILLATIONS

specified. The vapor will be richer in that component having the lower boiling point. In this instance, it is the component A and the vapor has the composition b. If we extract vapor, the liquid will become richer in the higher boiling of the two components and its composition will move up along the curve a–T_B^b. At the same time, of course, the composition of the vapor moves up along b–T_B^b. In practice, curves of miscible systems are used only in a quasi-equilibrium fashion, in a process known as fractional distillation.

Fractional distillation depends on the following considerations: In a unary system, it will be recalled that one must differentiate experimentally between the truly unary p–T experiment involving a cylinder, a piston, and only the unary component, and the pseudounary case where the applied constant or varying pressure is achieved via a combination of an inert gas pressure and the vapor pressure of the component. The latter system is actually binary in behavior and dependent on the applied inert gas pressure. Thus, with application of the inert gas pressure, the triple point S_A–L_A–V_A moves up along the S_A–L_L melting point curve and the S_A–V and L_A–V curves are displaced upward as shown schematically in Fig. 25. The solid curve in Fig. 25 is that for the unary system where the vapor phase is made up only of the unary component A. The dashed curve is that for the pseudounary system A–inert gas, where the total pressure above the condensed phase is kept constant via

$$P_t = p_A + p_{\text{inert}}. \tag{27}$$

When p_A achieves a value P_t, then obviously $p_{\text{inert}} = 0$ and the curves intersect.

The binary system analog of the unary system under an inert pressure may be visualized with reference to Fig. 26, as follows. If such a system, open to the atmosphere or to some constant pressure via the use of a controlled vacuum pump, is heated, a temperature will be reached at which the system boils. At this point, the boiling pot or chamber will effectively be represented by Fig. 24, since the vapors of A + B will exclude the inert gas from it. P_t will then be approximated by the sum $p_A + p_B$. If our starting liquid has the composition a of Fig. 24, the vapor will have the composition b. If a liquid at a temperature $T_{\text{operating}}$ at point X on the condenser, which is sufficiently low to condense vapor of composition b, is introduced into the condenser portion, we will have a liquid of approximate composition b in quasi-equilibrium with a vapor. Since the partial pressure of a liquid of composition b is now less than P_t, the vapor-phase total pressure is given again by

$$P_t = p_A + p_B + p_{\text{inert}} \tag{28}$$

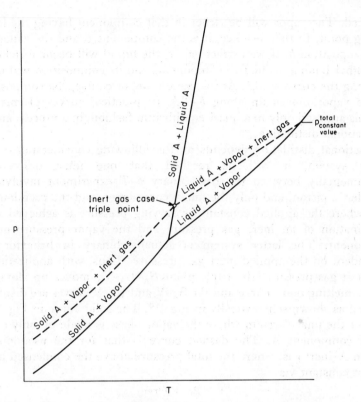

FIG. 25. p–T curve of the component A in the unary and pseudounary experiments.

at T. Furthermore, the curves in Fig. 24 are no longer applicable to the pseudobinary system in the same fashion that the solid curve in Fig. 25 was no longer applicable to the pseudounary system in the inert gas case. If the constant total pressure condition imposed is represented for pure A by the inert gas case triple point noted in Fig. 25, and an equivalent displacement exists on the pure-B unary diagram, the two phase ascending curve analog of Fig. 24 will be displaced upwards. In other words, for a particular binary composition, a higher temperature will be required to achieve the constant total pressure under which the system operates.

Obviously, the process is quite complicated, since the temperature is varying along the length of the condenser, and the mix of inert gas and A and B molecules is also varying. However, if we assume the hypothetical case where T in the condenser is constant and the total constant pressure does not appreciably affect the curves shown in Fig. 24, a continuous series of events can be visualized, where closest to the

F. ISOBARIC DISTILLATIONS

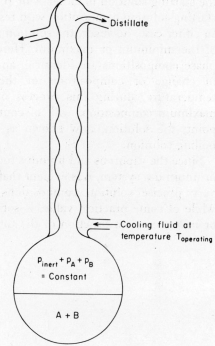

FIG. 26. A pseudobinary fractional distillation apparatus.

bottom of the condenser, the condensed vapor has a composition approximating that at the point b. This vapor is condensed and comes into equilibrium with a vapor of composition d (Fig. 24), which condenses and comes into equilibrium with a vapor of composition f, etc. It can be seen that a liquid of composition approximating pure A distills over at the very top of the condenser.

This process of continuous fractional distillation has extensive industrial application and the design of condenser systems capable of providing a large number of effective boiling pots along its length is well advanced. Each potential separation site in the condenser is known as a boiling plate and in effect each plate behaves as a separate boiling pot where a liquid, ever richer in the lower boiling component, is being distilled.

The isobaric distillation of a condensed-vapor phase system exhibiting either a maximum or a minimum cannot in theory lead to a complete separation of components. If, in such a system, a solution comprising the components A and B is permitted to boil under a constant pressure, the composition of the liquid will (in either type of system) move toward the composition of the minimum or maximum. The vapor will be richer than the starting solution in either A or B depending upon the location of the starting solution composition with respect to the minimum or maximum. In either case, however, the solution will finally achieve the composition of the minimum or maximum. Here the liquid and coexistent vapor-phase compositions are identical, and continued boiling will not result in change of compositions of the coexisting phases. The boiling temperature, during this process of attainment of the minimum or maximum composition, will, of course, vary. Upon attainment of this point, the solution, now known as an azeotrope, becomes a constant boiling solution.

Since the vaporous and liquidus touch only tangentially in a maximum or minimum system, it is evident that this provides a means of achieving very precise solution compositions where needed. This application while of some practical value, is somewhat limited since the minimum or maximum is a function of the total pressure.

32

Condensed-Vapor Phase Binary Diagrams Continued—Incongruently Vaporizing Systems

A. Introduction

Given a set of boundary conditions pertaining to the solid phase in a binary solid–liquid equilibrium, we have seen that one limit is represented by the simple eutectic interaction while the other limit involves continuous solid solubility of the end members. Since the behavior of equilibrium systems is not quantized, we might anticipate a continuum of solid-phase solubility variations from completely immiscible pure solids, to slightly miscible, to completely miscible end members. We will, in fact, focus on this continuum in Chapter 33. In a condensed-vapor phase equilibrium, given a binary or pseudobinary system composed of two immiscible solids in equilibrium with vapor, we can visualize the continuum as starting where both end members exhibit, at a particular temperature, similar vapor pressures, or an opposite extreme where one of the end members is essentially involatile. Such a sequence of systems is the topic of the present chapter, particularly with reference to systems where the boundary involves an involatile end member and a system in which one or more compounds is generated.

B. Vaporous Curves in Isothermal Diagrams

For the binary system exhibiting completely immiscible solid end members, five situations may prevail at temperatures beneath the eutectic. Thus we may have two solids, one or the other solid in equilibrium with vapor, both solids in equilibrium with vapor, or finally, vapor alone. In isothermal sections through the space model, it follows that when only vapor is present the system is bivariant, when two phases are present the system is univariant, and when three phases are present the system is invariant.

In the state of invariance, the total pressure of the system maximizes according to

$$P_{\max} = p_A{}^0 + p_B{}^0, \tag{1}$$

where, as usual, the superscript zero refers to the vapor pressure of the pure solid end members at the temperature in question. The minimum pressure possible in such a system, where at least one of the solids is in equilibrium with vapor, is either the value $p_A{}^0$ or $p_B{}^0$, depending upon which end member exhibits the lower vapor pressure. In the region where pure A coexists with vapor, the pressure of the system may vary between $p_A{}^0$ and $P_{\max}$. Similarly, in the region where pure B coexists with vapor, the pressure may vary between $p_B{}^0$ and $P_{\max}$.

In either of the two phase regions mentioned, it is the case that the total pressure at a particular system composition is composed of a $p_x{}^0$ value plus a $p_y{}^0$ value. Thus, along the A vaporous we know that, in general, the mole fraction of A in the vapor is given by

$$M_A^{(3)} = p_A/(p_A + p_B) = p_A/P_t. \tag{2}$$

However, since pure A is in equilibrium with the vapor along the A vaporous it must be the case that $p_A = p_A{}^0$. Thus,

$$M_A^{(3)} = p_A{}^0/(p_A{}^0 + p_B) = p_A{}^0/P_t. \tag{3}$$

As has been pointed out before, in the interval $p_A{}^0$ to $P_{\max}$, all pressures are possible and it is a simple matter to solve for the value $M_A^{(3)}$ as a function of P_t, thereby defining the equilibrium vaporous for A.

Since the isothermal triple point represents an invariant condition, there must be a single pressure and composition coincident with the occurrence of a triple-point. Uniquely, this simultaneity is given by

$$M_A^{\text{triple}} = p_A{}^0/P_{\max} = p_A{}^0/(p_A{}^0 + p_B{}^0). \tag{4}$$

B. VAPOROUS CURVES IN ISOTHERMAL DIAGRAMS

Since Eq. (1) defines $P_{\max}$, it is evident that the triple point will be displaced toward the end member exhibiting the higher vapor pressure at the temperature of the isotherm. If one of the end members exhibits a vanishingly small vapor pressure, then from Eq. (1) we see that

$$P_{\max} \sim p_X^0, \tag{5}$$

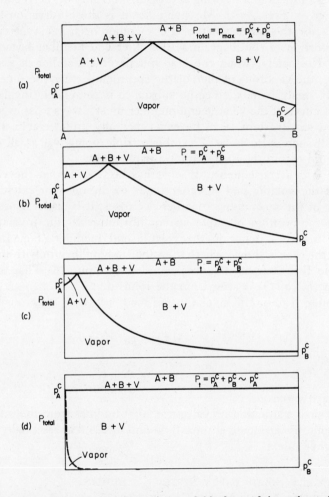

FIG. 1. The disappearance of the two-phase field of one of the end members when the other end member becomes less volatile. (a) Both A and B exhibit appreciable vapor pressures; (b) B exhibits a lower vapor pressure than A; (c) B exhibits a much lower vapor pressure than A; and (d) B exhibits almost no vapor pressure.

where p_x^0 is the end member whose vapor pressure is not vanishingly small and

$$M_A^{\text{triple}} \sim 1 \tag{6}$$

when B is involatile, and

$$M_A^{\text{triple}} \sim 0 \tag{7}$$

when A is involatile.

A series of diagrams showing decreasing volatility of the B component is depicted in Figs. 1a–d. Most significant is the observation that the vaporous for the more volatile end member occupies a decreasingly smaller portion of the diagram as the other end member becomes less volatile. Ultimately, when the less volatile end member is essentially involatile, the vaporous of the volatile end member may become microscopically small. In fact, since no substance is completely involatile the vaporous curve of the volatile end member must always exist even if it is of such a small extent as to be experimentally undetectable. In other words, all such systems must exhibit a triple point just as all eutectic solid–liquid equilibria must exhibit a triple point.

While it is a consequence of what has been said, that the vaporous curve of the volatile end member tends to diminish in extent as the volatility of the second end member decreases, it is also a consequence that the composition of vapor in equilibrium with the involatile end member becomes richer in the volatile end member (Fig. 1a–d). In Fig. 1d, the case is shown where the vapor in equilibrium with involatile pure solid B is composed almost completely of pure A. That this is in accord with Eq. (3) is seen from the following:

Assuming B to be the involatile end member, the B vaporous is given by

$$M_B^{(3)} = [1 - M_A^{(3)}] = p_B^0/P_t \sim 0. \tag{8}$$

Therefore,

$$M_A^{(3)} \sim 1 \tag{9}$$

along the B vaporous.

When such a situation prevails, note that tie lines from the interior of the B vaporous terminate at pure B on one side and at essentially pure A on the other side.

C. Immiscible Solid–Vapor Systems Exhibiting Congruently Vaporizing Compounds

Before we attempt to make use of the conclusions of Section B, particularly that expressed in the last sentence of that section, it is of

C. CONGRUENTLY VAPORIZING COMPOUNDS

interest to examine S–V diagrams in a multicompound system. Let us assume that in the interaction A–B a number of compounds having the stoichiometry A_xB_y are generated. Let us assume further that both of the end members are reasonably volatile and further that each of the compounds generated is pseudounary in behavior. By invoking the constraint that each of the compounds formed is pseudounary in behavior, it is implicit that each one of these compounds, when it volatilizes, provides coexistent condensed phases and vapor phases of the same stoichiometry. By analogy with S–L systems, vaporization of the type specified is termed congruent and the compound is termed congruently vaporizing. A hypothetical completely congruently vaporizing system of the type being discussed is shown schematically in Fig. 2. It is to be noted, as might have been expected, that Fig. 2 comprises a

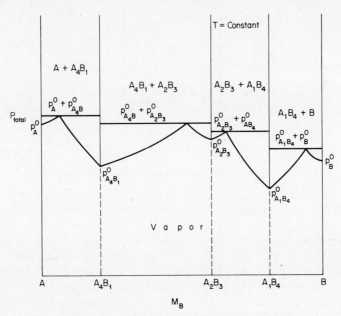

FIG. 2. The solid–vapor diagram for the system A–B in which the congruently vaporizing compounds A_4B_1, A_2B_3, and A_1B_4 are formed.

number of separate subdiagrams of the type depicted in Fig. 1a. Each of these subdiagrams may be treated independently of any other subsystem from which the primitive diagram is constructed.

In Fig. 2, it is also to be noted that all two-phase tie lines intersect a vertical line (the pure compound composition) on one side and a nonvertical vaporous curve on the other side. Note also that no sequential trend in vapor pressures of the pure compounds exists, each of these substances being capable of generating pressures greater than or less than that of the end members at the temperature of the isotherm.

The case may, of course arise, where even though each of the generated compounds is congruently vaporizing, one or more of them is essentially involatile. Based on what was said in Section B, a system of this type may be schematized as in Fig. 3. In Fig. 3, it is assumed that the end

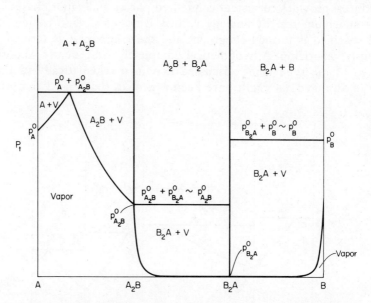

FIG. 3. Isothermal S–V diagram in a system showing involatile compound formation.

members A and B are relatively volatile, exhibiting the pure component vapor pressures p_A^0 and p_B^0 at the temperature of the isotherm. In addition, the compound A_2B, while not developing as high a vapor pressure as either of the end members, is still relatively volatile. As a consequence, the system A–A_2B exhibits well defined S–V fields for both solid A and solid A_2B. The compound B_2A, on the other hand, is assumed to be essentially involatile. Practically, this results in eliminating an A_2B field in the subsystem A_2B–B_2A, and a B field in the subsystem B_2A–B. The maximum pressures developed in S–V equilibria in these latter two subsystems coincide approximately with the vapor pressures of A_2B and B, respectively.

Tie lines in the A_2B–B_2A and B_2A–B Solid–Vapor fields connect on one side with a phase boundary line coinciding with pure solid B_2A. The extension of these tie lines to the vaporous curves intersects the latter at almost pure A_2B in one case and pure B in the other case for the reason previously offered in conjunction with Fig. 1d.

D. Immiscible Solid–Vapor Systems Containing Incongruently Vaporizing Compounds

Frequently, in a binary or pseudobinary system generating immiscible compounds, the vaporization of one or more of these compounds may be incongruent. Thus, when placed in an environment containing vapor available space, a compound of this type will dissociate in the process of developing a vapor pressure. The products of this dissociation will then volatilize preferentially. When equilibrium is established, the vapor composition will be different from the composition of the condensed phase(s). By analogy with the incongruent melting phenomenon in S–L equilibria, such a vaporization is termed incongruent. The compound undergoing this behavior is termed an incongruently vaporizing compound.

As a consequence of the dissociation, no more than two condensed phases may come into equilibrium with a vapor phase in a binary or pseudobinary system at constant temperature. When such a condition prevails, the system is isothermally invariant. Thus, one of the products of the dissociation must volatilize completely, leaving behind the remaining undecomposed starting compound plus one other condensed phase of stoichiometry different from the starting compound. There exists, in the vicinity of the initial starting compound composition, a region of solid solution of varying extent. Within this region, the composition of the starting solid may vary, becoming richer or poorer in one of the end members, without generating a second phase. This region is called a region of variable solid solubility. If we confine our present discussion to systems in which this region of variable solid solubility is vanishingly small, it is seen that attending the formation of any vapor phase at all, a second condensed phase is formed. Thus, for such a case, it is inappropriate to speak of the vapor or dissociation pressure of an incongruently vaporizing compound, but rather, one must refer to the vapor or dissociation pressure of the resulting three-phase system. The latter, made up of two condensed phases and one vapor phase, has a unique pressure at a given temperature. Furthermore, this unique pressure is independent of the overall system composition so long as the same condensed phases are in equilibrium with vapor. This has a direct analogy in eutectic or other two-phase isobaric binary S–L equilibria at constant temperature. Here, if we specify the temperature, both the liquid composition in equilibrium with solid and the solid composition in equilibrium with the liquid remain fixed even though the relative amounts of the two phases vary.

Let us assume that the binary system A–B generates an incongruently vaporizing compound AB. To further simplify matters let us assume that B is essentially involatile. If, to pure B, we add a minute quantity of A vapor such that the solubility of A in B is not exceeded, the vapor that equilibrates with this slightly impure end member will be composed of pure A. A tie line connecting the vapor and solid compositions in the system will intersect pure A on one side of the diagram and essentially pure B on the other side as in Fig. 4a, tie line 1. In the lower left hand

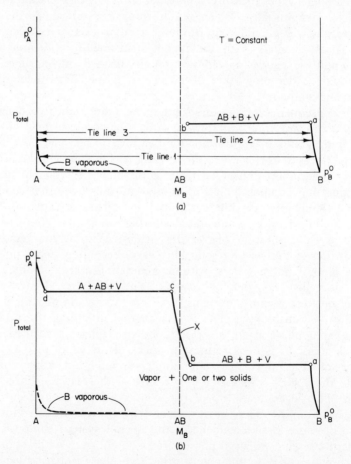

FIG. 4. (a) A tie line in the region in which the solubility of A in B has not been exceeded. (b) The complete diagram for the system A–B in which B is involatile and AB vaporizes incongruently.

D. INCONGRUENTLY VAPORIZING COMPOUNDS

corner of the diagram, the bending of the B vaporous curve is depicted schematically in accordance with the ideas advanced in the preceding section. To all intents and purposes, since B is essentially involatile, the B vaporous coincides with the composition axis and the pure A axis. The line $a-p_B{}^0$ shows the variation in solid solubility of A in B as a function of total pressure: for our purposes it has been exaggerated in terms of total composition extent.

If we continue to add more A vapor to the system (Fig. 4a), allowing for equilibrium to be established, the solubility of A in B will increase slightly, following the curve $p_B{}^0-a$. The total pressure will move along the vaporous $p_B{}^0-p_A{}^0$, tie lines 1, 2, and 3. When the composition a is reached, the solubility limit of A in solid B is exceeded and further addition of A to the system results in the formation of the phase AB. Since there exists some finite solubility of B in AB, there will be a range of variable composition of AB. When this phase first appears, it will contain some excess B. With the appearance of the B-rich AB phase, the system becomes isothermally invariant and so long as the phases B-rich AB, A-rich B, and vapor coexist, the total pressure cannot change. The range of such three-phase coexistence is shown by the line AB–B–V. As more A is added to the three-phase system, more solid AB will form while the total pressure remains constant and the system composition moves from a to b (Fig. 4b). When the point b is reached, all of the B phase will have been used up in forming the AB phase of composition b (B rich), and addition of a little more A will result in a single solid-phase system (in the two-phase system solid–vapor), whose composition with further addition of A varies along the line bc of Fig. 4b. This latter, as with the line $p_A{}^0-a$, shows the effect on pressure of the solubility of A in the condensed phase. Tie lines now intersect the curve $b-c$ on one side and the B vaporous on the other side. Thus AB is now in equilibrium with vapor containing almost pure A, where before, it was B which was in equilibrium with vapor containing almost pure A.

When the point c has been reached, upon continued addition of A to the system, the AB phase moves from a B-rich phase through the stoichiometry point x, to an A-rich AB phase of composition c. Further addition of A results in the formation of an A phase (rich in B) of composition d as the solubility of A in AB is exceeded. If we continue to add A, the total pressure remains constant while the composition of the system as a whole varies along the line $c-d$. When the point d is reached, a single solid phase is once again formed. This time its composition varies along the curve $p_A{}^0-d$ as a function of system composition and its pressure varies in the same interval. The vapor composition is again essentially pure A.

Thus, in a system exhibiting an incongruently vaporizing compound and one involatile end member, we have seen that only a single vaporous curve exists. This curve is the vaporous of the involatile end member and coincides essentially with the diagram boundaries. A tie line between any solid composition curve and a vapor composition curve always intersects this same vaporous, the composition of the vapor being essentially that of the pure volatile end member. Consequently, in moving from one side of the diagram to the other, the pressure must vary monotonically except in three-phase regions. In other words, systems less rich in the volatile end member must exhibit lower pressures than those containing higher compositions of the volatile end member.

Let us now assume that in the system A–B, the compounds A_2B and B_2A are generated, the former vaporizing congruently and the latter vaporizing incongruently. Furthermore, let us assume that B, while not involatile, exhibits a lower vapor pressure than A. This system is depicted in Fig. 5.

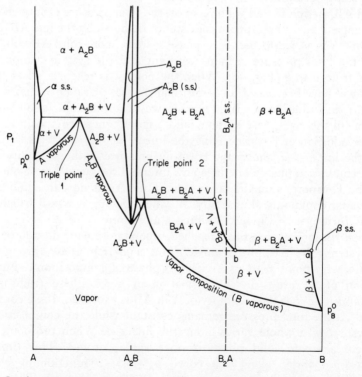

Fig. 5. A system comprising one congruently vaporizing and one incongruently vaporizing compound.

E. HYDRATE SYSTEMS

First it should be noted that the system $A-A_2B$, comprising two congruently vaporizing phases, is similar to such systems previously treated. The system A_2B-B is now to all intents and purposes the same as the system $A-B$ of Fig. 4. The end members (the pseudounary compound A_2B and the unary compound B) are congruently vaporizing and give rise to the incongruently vaporizing B_2A compound. Two distinct vapor-composition curves exist in the system A_2B-B even though another compound, B_2A, is present. The occurrence of two visible vaporous curves is due to the fact that both A_2B and B are volatile.

Furthermore, since both A_2B and B are volatile, and congruently vaporizing while B_2A is incongruently vaporizing, only a single triple point is generated in the A_2B-B system: this is at point 2. The regions of solid solubility for A, A_2B, B_2A, and B are shown as are the all solid regions at higher pressures. Note that above each three-phase line and independent of whether such a three-phase line involves a congruent or incongruent phase, the system is composed entirely of solid phases.

Note also that because of the volatility of both A_2B and B as well as A an all vapor system is possible. (Compare this with Fig. 4 where B is involatile.) Note also that although the invariant pressure of the three-phase system A_2B-B_2A-V is greater than that of B_2A-B-V for the reasons given in conjunction with Fig. 4, no definite relationship exists between A_2B and A. The compound A_2B may exhibit a higher or lower pressure than A even though we have stipulated $p_A^0 > p_B^0$. By the same token $p_{A_2B}^0$ may be less than or greater than p_B^0. All that is required is that tie lines be capable of intersecting the B vaporous in the range in which the pure solid B, incongruent phase plus B, or the incongruent phase alone is in equilibrium with vapor. In Fig. 4 where B is assumed to be essentially involatile, the B vaporous appears to terminate at pure A and p_A^0. However, as we have noted earlier, in the limit, an A vaporous of very small extent must exist in all systems. Consequently, in Fig. 5, the analogous A_2B vaporous is more evident only because both A_2B and B are volatile. The case where $p_{A_2B}^0 < p_B^0$ is given in Fig. 6. Note that tie lines must all intersect either the A_2B or the B vaporous, no B_2A vaporous curve being possible. In order for this situation to be possible, the vapor pressure of the system B_2A-B-V must be less than that of the system A_2B-B_2A-V.

E. Hydrate Systems

Typical of systems exhibiting incongruently vaporizing compounds are those known as hydrate systems. Hydrate systems generate compounds containing removable H_2O. In general, hydrate systems comprise two

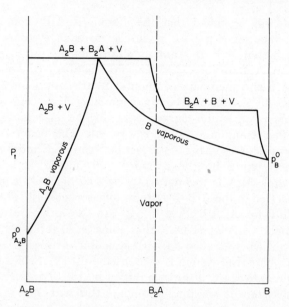

FIG. 6. The system A_2B–B where incongruently vaporizing B_2A is formed and $p^0_{A_2B} < p^0_B$.

pseudounary end members, i.e., $CuSO_4$–H_2O, Na_2CO_3–H_2O that participate in a pseudobinary interaction. One of the end members is essentially involatile, while the vapor phase comprises H_2O alone.

As an example of a hydrate system, we may consider the system $CuSO_4$–H_2O. In it, the hydrates $CuSO_4 \cdot 5H_2O$, $CuSO_4 \cdot 3H_2O$, and $CuSO_4 \cdot H_2O$ are generated. Each of these compounds vaporizes incongruently. From what has been stated in the preceding section, we may expect the vapor composition to be given by the vaporous of the end member $CuSO_4$. Furthermore, in order that all tie lines terminate at H_2O at one end, the systems

$$CuSO_4\text{–}CuSO_4 \cdot H_2O,$$
$$CuSO_4 \cdot H_2O\text{–}CuSO_4 \cdot 3H_2O, \quad \text{and} \quad CuSO_4 \cdot 3H_2O\text{–}CuSO_4 \cdot 5H_2O$$

must exhibit increasingly greater vapor pressures. In other words, starting at pure $CuSO_4$, the vapor pressures of each successively water-rich system must increase in a step-wise fashion in accordance with our previously arrived at conclusions.

To better understand the nature of incongruently vaporizing systems as typified by a hydrate system, and the relationship between the L–V, S–L, and S–V portions of the diagram, we will describe schematically the interaction $CuSO_4$–H_2O.

E. HYDRATE SYSTEMS

If we begin with pure water at room temperature, for example, and add $CuSO_4$ to it, the total pressure will be due solely to H_2O since the $CuSO_4$ is essentially involatile. To a first crude approximation, the partial pressure of H_2O above the solution will be approximated by

$$p_{H_2O} = M_{H_2O} p^\dagger_{H_2O}. \tag{10}$$

Upon continued addition of $CuSO_4$ to the solution, the latter will become more saturated until at some point a solid phase will settle out. Upon analysis, this solid phase is found to have the composition $CuSO_4 \cdot 5H_2O$. This experiment may be visualized by referring to Fig. 7 in which the

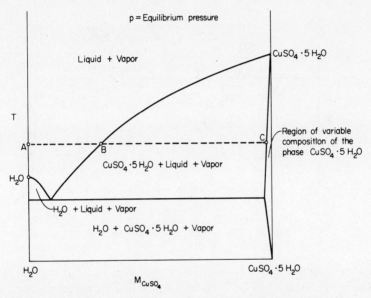

FIG. 7. Schematic representation of the system $CuSO_4$–H_2O.

S–L diagram of the subsystem H_2O–$CuSO_4 \cdot 5H_2O$ is depicted schematically as a simple eutectic interaction. The experiment just described begins at point A and continues to point B where the liquidus for $CuSO_4 \cdot 5H_2O$ is encountered. Upon continued addition of $CuSO_4$, the concentration of the system moves from B to C. In the region A–B, the two-phase, two-component system is isothermally univariant. Thus, p_{H_2O} will decrease according to Eq. (10). At point B, a third phase appears and remains until point C. In this composition interval then, the system is isothermally invariant while the relative amount of solid and liquid change.

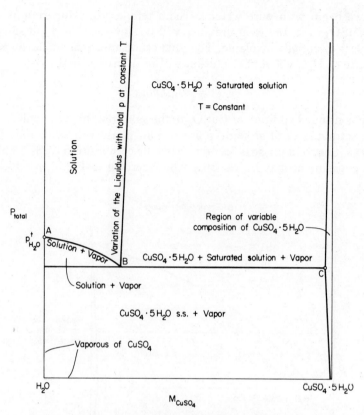

FIG. 8. The p–M_{CuSO_4} diagram for Fig. 5.

Figure 8 shows the variation of total pressure as a function of M_{CuSO_4} for the S–L diagram of Fig. 7. In Fig. 8, the $CuSO_4$ vaporous curve is not evident since it superimposes on the composition and pure H_2O axes as was described for systems containing an involatile component. As before, a small, but finite, region of variable composition for the compound is depicted. The points A, B, and C of Fig. 8 coincide with those of Fig. 7. Upon further addition of $CuSO_4$, the liquid phase vanishes and a water-rich solid having the crystallographic structure of $CuSO_4 \cdot 5H_2O$ results. This solid will have a small range of variable composition. When this univariant S–V composition range is exceeded, a new phase having the stoichiometry $CuSO_4 \cdot 3H_2O$ is generated and the resulting three-phase mixture, $CuSO_4 \cdot 5H_2O$–$CuSO_4 \cdot 3H_2O$–vapor, is isothermally invariant. Sequentially, the composition pure $CuSO_4 \cdot 3H_2O$ is achieved, then a mixture composed of $CuSO_4 \cdot 3H_2O$ and $CuSO_4 \cdot H_2O$,

E. HYDRATE SYSTEMS

pure $CuSO_4 \cdot H_2O$, $CuSO_4 \cdot H_2O$ and $CuSO_4$, and finally, in the limit, pure $CuSO_4$.

The T–M_{CuSO_4} phase diagram for the entire system is given schematically in Fig. 9. Each of the compounds is assumed to be incongruently melting, although this is not a necessary attribute of an

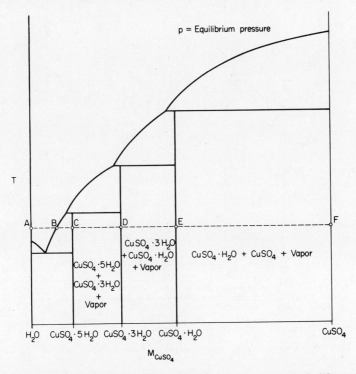

FIG. 9. Schematic T–M_X representation of the system H_2O–$CuSO_4$.

incongruently vaporizing system. The author and others have, however, demonstrated that at least a few of the hydrates melt incongruently.[1] In Fig. 9, the regions of variable composition around each pure phase line have been left out for simplicity. Figure 10 shows the isothermal p_t–M_{CuSO_4} counterpart for the S–L diagram schematized in Fig. 9. Again, for simplicity, the regions of variable composition at each compound location are not constructed. To maintain clarity, however, an exaggerated representation of the $CuSO_4$ vaporous has been included. This vaporous encompasses all of the univariant and invariant regions.

[1] References to this work are given in the Appendix under Chapter 32.

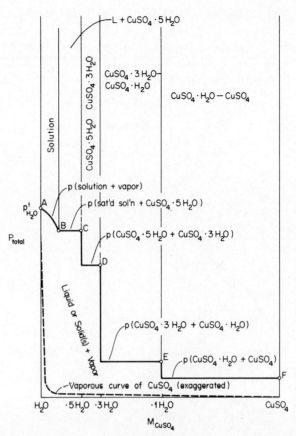

Fig. 10. The p_t–M_{CuSO_4} representation of the isotherm A–F of Fig. 9.

Because of the involatility of the $CuSO_4$, its vaporous terminates essentially at pure H_2O in accordance with previously drawn conclusions.

The stepwise variation of system partial pressures, it is to be emphasized, is a consequence of the vapor composition requirements of incongruently vaporizing systems, not of hydrate systems *per se*.

F. Further Remarks on the Regions of Variable Composition

As has already been noted, all pure compounds may, to a greater or lesser extent, incorporate impurities into their lattices. In a system showing compound formation, it is not unreasonable to expect any one of these compounds to exhibit solubility of one or the other end-member

F. ADDITIONAL COMMENTS

in its lattice, depending upon the total system composition. While the mole fraction extent of a region of variable composition may be trivial, the fluctuation of pressure within the narrow compositional confines of this region may be enormous. The result of this enormous pressure change coincident with an all but nonexistent composition change has undoubtedly been the root of discrepancies in reported vapor pressures for some presumably unary compounds. To indicate better the origins of such pressure fluctuation let us consider a hypothetical system composed of a volatile and an essentially involatile end member. Assume that the system generates but a single incongruently vaporizing compound and assume further that at the temperature of the vapor pressure measurement, the vapor pressure of the volatile end member, e.g., the elements As or I_2, is quite large.

As a case in point consider the S–L diagram of the system A–B

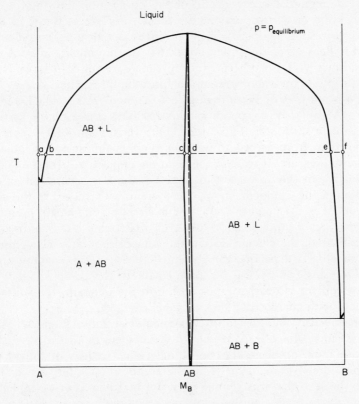

FIG. 11. Schematic representation for a system generating the incongruently vaporizing compound AB, which melts at considerably higher temperatures than the end members.

(e.g., As–Ga) where B is involatile. If in the system A–B, an intermediate, incongruently vaporizing compound that melts at temperatures considerably higher than those of the end members is formed, the two AB liquidi will in general be displaced toward the end member axes as in Fig. 11. Furthermore, the liquidi for the end members will be virtually nonexistent in extent. Typical examples of such systems are As–Ga, Se–Cd, and other pairs from the Groups III–V and II–VI of the periodic table.

Let us now proceed to analyze the probable course of events in vapor-pressure variation as we move from point a to point f in Fig. 11. In this route, we will encounter the three phase lines S–L–V at points b and c that represent the compositions of liquid and solid, respectively, in equilibrium with vapor on the A-rich side of the stoichiometry AB, and a similar set of lines describing S and L compositions on the B-rich side of this stoichiometry. In between, i.e., c–d, there will exist a miniscule region of variable solid-phase composition showing the extent to which AB may incorporate either excess A or excess B into its lattice.

The corresponding p–M_B diagram for the isotherm a–f is presented in Fig. 12. Because of the proximity of the AB liquidus to pure A at the temperature of the isotherm, the univariant region a–b is very small. Thus, if $p_A^\dagger$ is the vapor pressure of pure liquid A, the value p_A at the point b is not very different than $p_A^\dagger$. When point b is reached, and from there all the way to point c, the resulting three-phase system is isothermally invariant. Thus, for slightly less than a 50% change in composition (from point b to point c), there is almost no accompanying change in pressure. If we now focus on the opposite side of the diagram, we see that in the univariant interval e–f, there is, again, only a trivial variation in the composition of the system with regard to the volatile component A. At point e where M_A is already quite small, p_A must be proportionately quite small. Thus even if Raoult's law is only crudely approximated, the change Δp_A in the composition interval e–f must be quite small. Furthermore, the absolute value p_A, at any composition in the interval e–f must also be quite small. In the composition interval d–e, of course, as on the opposite side of the diagram, the value p_A is invariant with composition.

Consequently, in order for p_A whose value at pure A is given by $p_A^\dagger$ to decrease to a value of zero at point f, it is evident that the overwhelming decrease must occur in the vanishingly small region of the phase of variable composition, namely, in the composition interval c–d. It can readily be seen that this change need not be trivial. For example, if we assume that pure A, at the temperature in question, exhibits a vapor pressure $p_A^\dagger = 1$ atm (i.e., As at $\sim$600°C), then as the point f is reached this pressure may have a value of 10^{-8}–10^{-10} atm. Since very little

G. DELIQUESCENCE AND EFFLORESCENCE

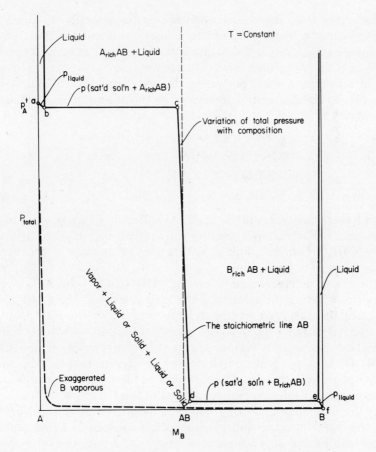

FIG. 12. The p–M_B diagram of the system shown in Fig. 11.

change in pressure occurs in the univariant intervals a–b and e–f the change of 8–10 orders of magnitude of pressure must occur in the even smaller univariant composition interval c–d.

No wonder then, depending upon which point in the interval c–d is being considered, that the determined value of the vapor pressure of a compound may show apparently enormous fluctuations in pressure. An analytical rationale for the above is presented in the Appendix (see additional comments on Chapter 32).

G. Deliquescence and Efflorescence of Hydrate Systems

In general, regions of variable phase composition in hydrate systems are quite small. Consequently, upon exposure to the ambient atmosphere, we anticipate that, depending upon the partial pressure of water, the composition of a hydrate (assuming it began as a stoichiometric phase) will change sufficiently in a short time period to form an invariant system of the next higher or next lower hydrate system.

If in a hydrate system generating the compounds $X \cdot 5H_2O, \cdot 4H_2O$, and $\cdot 1H_2O$, the dissociation pressures of the three phase systems are represented by

1. p (saturated solution $- \cdot 5H_2O) = a$,
2. $p(\cdot 5H_2O - \cdot 4H_2O) \quad = b$,
3. $p(\cdot 4H_2O - \cdot 1H_2O) \quad = c$,
4. $p(\cdot 1H_2O - X) \quad = d$,

each of these systems will be stable at different times relative to the environment. For example, starting with a mixture composed primarily of the $\cdot 4H_2O$ hydrate admixed with a minute amount of the $\cdot 1H_2O$ phase, placed in air whose $p_{H_2O} > c < b$, the system will absorb moisture, i.e., deliquesce, until only a $\cdot 4H_2O$ single-phase of variable composition having the value p_{H_2O} of the air is formed. If $b < p_{H_2O} < a$, water will be absorbed by the system until all of the $\cdot 4H_2O$ phase is converted to the $\cdot 5H_2O$ phase, and the composition will equilibrate somewhere within the region of variable composition of the phase $X \cdot 5H_2O$. If p_{H_2O} in the environment is greater than a, the initial $\cdot 4H_2O$ phase will absorb sufficient water to place the final composition in the univariant solution region adjacent to the H_2O axis.

On the other hand, starting with a solution composition coincident with this latter pressure and placing such a solution in an air ambient atmosphere having an aqueous tension $p_{H_2O} < b > c$, the solution will evolve water, i.e., effloresce, until a $\cdot 4H_2O$ phase of variable composition is formed.

H. Efflorescence and Deliquescence in Aqueous Systems Not Generating Hydrates

In order not to convey the impression that processes of deliquescence and efflorescence can occur only in systems generating hydrates, we note that such processes may occur in any system containing water as one of the end members. A water-containing system not generating a compound would have the p–M_B appearance shown in Fig. 13 for an isotherm intersecting the liquidus of the second end member. It is seen that the diagram has an identical appearance to that of the appropriate region in a system generating hydrates. If placed in an ambient atmosphere at

I. THE VARIATION OF DISSOCIATION PRESSURE WITH T 387

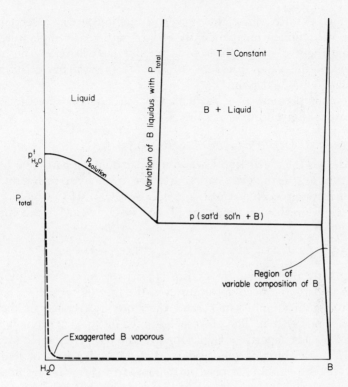

FIG. 13. An isotherm in the system H_2O–B, which cuts through the B liquidus.

partial pressure of H_2O designated by p_{H_2O}, a starting composition in the system H_2O–B will either deliquesce or effloresce exactly as a hydrate system does. The final composition of a nonhydrate-forming aqueous system will be such that the H_2O pressure of the equilibrium system will be precisely that of H_2O in the ambient atmosphere with which it is in contact. The behavior of common table salt in the environment with changing seasons, and the attendant average relative humidities, is a good case in point. Were it not for the fact that NaCl is not too soluble in H_2O, syrupy solutions of NaCl in H_2O would be the rule in salt shakers.

I. THE VARIATION OF p WITH T OF ISOTHERMALLY INVARIANT INCONGRUENTLY VAPORIZING SYSTEMS

Each of the three phase-regions generated in systems of incongruent compounds is in actuality univariant. Thus, given a saturated solution of a hydrate plus the solid hydrate, its vapor pressure will be a function of the temperature. Since the step-wise relationship in a sequence of systems was defined independently of any particular temperature, it is clear that the p–T curves of each of the systems will retain the relative positions of the vapor pressures of the several systems existing in a particular hydrate diagram.

The vapor pressure of a particular hydrate system as a function of temperature is given by

$$\ln p_{\text{diss}} = -(\Delta H_{\text{D}}/RT) + C, \tag{11}$$

where ΔH_{D} is the molar latent heat of dissociation of a particular hydrate system and p_{diss} is the dissociation pressure of the hydrate system. If we consider a system such as $CuSO_4 \cdot 5H_2O$–$CuSO_4 \cdot 3H_2O$ which is being heated up, we realize that only the phase $CuSO_4 \cdot 5H_2O$ is decomposing according to

$$CuSO_4 \cdot 5H_2O(s) \rightleftharpoons CuSO_4 \cdot 3H_2O(s) + H_2O(v). \tag{12}$$

Consequently the term ΔH_{D} in Eq. (11) is the molar enthalpy of the hydrate undergoing decomposition even though the term p_{diss} is referenced to a univariant system, and therefore, referenced to the two-hydrate–vapor system. The process described in Eq. (12) is one in which liberated H_2O is vaporized. Consequently, the term ΔH_{D} includes in it a term ΔH_{v}, the heat of vaporization of H_2O, and a heat of hydration term ΔH_{h}. Equation (12) may be reiterated in step-wise fashion by

$$\tfrac{1}{2}CuSO_4 \cdot 5H_2O(s) \rightleftharpoons \tfrac{1}{2}CuSO_4 \cdot 3H_2O(s) + H_2O(\ell) - \Delta H_{\text{h}} \tag{13}$$

$$2H_2O(\ell) \rightleftharpoons 2H_2O(v) - \Delta H_{\text{v}}. \tag{14}$$

$$\ln K_{5\cdot 3} = -(\Delta H_{\text{h}}/RT) + C, \tag{15}$$

$$\ln K_{\text{H}_2\text{O}} = -(\Delta H_{\text{v}}/RT) + C, \tag{16}$$

$$\ln K_{\text{D}} = \ln K_{5\cdot 3} + \ln K_{\text{H}_2\text{O}} = \ln(K_{5\cdot 3} \cdot K_{\text{H}_2\text{O}})$$
$$= -[(\Delta H_{\text{h}} + \Delta H_{\text{v}})/RT] + C' = \ln p_{\text{diss}}, \tag{17}$$

where p_{diss} is the dissociation pressure.

33

Limited Solid Solubility

A. INTRODUCTION

Except for brief discussions of limited solid-phase solubility, i.e., phases of variable composition, in the the preceding chapter, our focus so far has been confined to one of two boundary states. Thus, we have been concerned with S–L equilibria in which solid phases were either completely immiscible or completely miscible. While numerous examples of each of these boundary states are known, it is obvious that, in the limit, complete immiscibility of solid phases in eutectic systems never occurs. All known materials are impure to a greater or lesser degree. In a binary system, aside from foreign impurities, one might expect contamination of each of the end members by the other.

When one considers the implications of mutual end-member contamination, the following is important; Assuming that the crystal structures of the end members A and B are different, then if B contaminates A, the atoms of B in A conform to the crystal structure of A. Conversely if A contaminates B, the atoms of A in B conform to the crystal structure of B.[1] If the degree of mutual end-member solubility is vanishingly

---

[1] We ignore, for the purposes of this argument, solid solubility in which the solubility is achieved by interstitial location of the solute, and consider only the case where a substitutional event occurs.

small, the perturbations in the disturbed lattice are small. If the degree of mutual solubility is large, the system assumes characteristics of both the eutectic and complete solid solution cases. If, for example, in an interaction of the eutectic type, B is soluble in A to the extent of 0.1 mole fraction at the eutectic temperature, the S–L portions of the A region acquire the appearance of an ascending solid solution of A and B. In the solid solution region, since only a single phase is present, it is a necessary condition that the crystal structures of A and B be the same. Consequently, it would appear that when A and B exhibit significant mutual solubility, they individually appear to be participating in a solid solution interaction with a metastable form of the other, this metastable form exhibiting the same crystal structure as the primary phase.

B. Limited Solubility Systems Exhibiting a Eutectic

Figure 1 depicts a binary eutectic T–M_B equilibrium in which the end members exhibit considerable solid solubility in each other. In the same manner that the curve T_A–T_e indicates the variation of liquid-phase composition with temperature, the curve T_A–1 indicates the variation

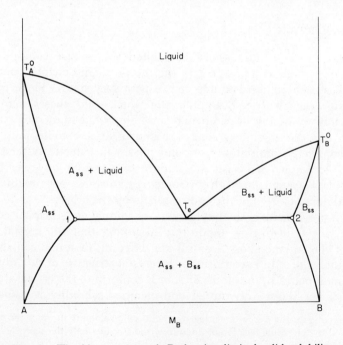

Fig. 1. The binary system A–B showing limited solid solubility.

B. EUTECTIC SYSTEMS WITH SOLID SOLUBILITY

of solid-phase composition of the A phase with temperature. Tie lines from the liquid- to the solid-composition line do not, as in the simple eutectic case, intersect a line of constant composition. In the spaces bounded by $A-1-T_A$ and $B-2-T_B$, single-phase solid solutions exist, the first having the crystal structure of A and the second having the crystal structure of B. The two-phase region T_A-T_e-1 is that in which a solid solution having the A structure is in equilibrium with liquid. The two-phase region T_B-T_e-2 has similar significance for the B phase. Note that each of these regions has the appearance of an ascending solid solution and that the diagram as a whole has the appearance of two intersecting ascending diagrams (Fig. 2).

As a matter of fact, if we develop a model for an ideal system of the type depicted in Fig. 1, each of the associated pairs of solidus and liquidus curves would be given precisely by an equation of the type developed for an ascending solid solution series. The latter, it is to be recalled, is

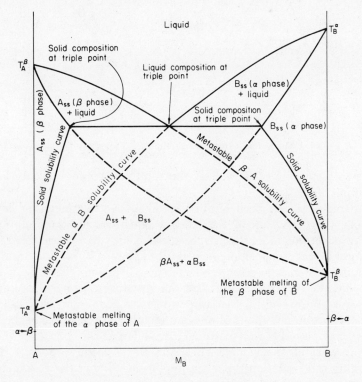

FIG. 2. Metastable extensions of the liquidus and solidus curves of a system exhibiting limited solubility.

derived in the same manner as that for a simple eutectic interaction, but the terms $N_X^{(1)} \neq 1$, and both ΔH_A^0 and ΔH_B^0 terms participate. That such a result is not unexpected can be seen from Fig. 3, where the

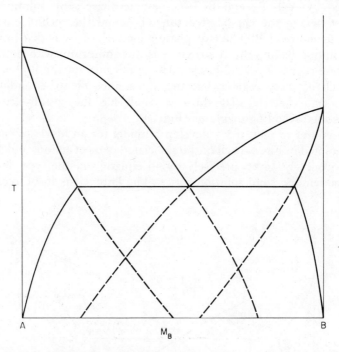

FIG. 3. Metastable extensions of the A and B solubility curves.

metastable extensions of each of the solubility curves must be of the ascending type.

In general, as has been implicit in all of our treatments of eutectic systems, the derivation of the eutectic equations for immiscible solid systems is nothing more than a special case. The general case, where both solid and liquid composition vary as a function of T, is made up of two ascending systems, one for each end member, since each, in general, has a different crystal structure. If the structures are not different, the same situation prevails if the lattice constants are significantly different. Here substitution of atoms of one end member in the lattice of the other cannot take place because of atomic size difference.

Since each of the general curves in the eutectic system is of the ascending solid-solution type, the implication is that the end members participating in each ascending arm are isomorphous. Thus, the terminus of either arm at the opposite end member must be at the metastable

B. EUTECTIC SYSTEMS WITH SOLID SOLUBILITY

melting point of a stable or metastable phase of the opposite end member. Clearly, the metastable melting point must lie above $0°K$. Figure 3 schematically depicts the situation where both end members exhibit a stable phase transformation. The temperatures T_A^α and T_B^β refer to the metastable melting points of the low-temperature A and B phases, respectively.

As the relative temperatures of the stable and metastable melting points change, the eutectic type interaction may acquire what is termed

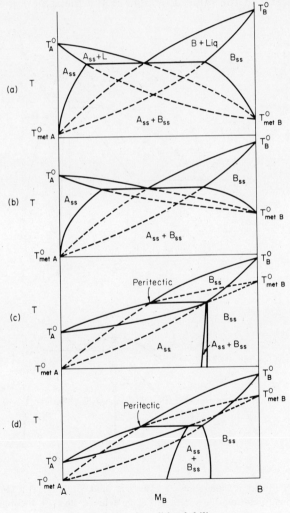

FIG. 4. Variation in solid-solubility curves.

a peritectic configuration. A sequence demonstrating this is shown in Fig. 4. In this figure the symbols T_A^0 and T_B^0, as usual, refer to the stable melting points. The terms $T_{\text{met}A}^0$ and $T_{\text{met}B}^0$ refer to the metastable melting points of the lower temperature stable or metastable phases of A and B, which act as termini for the ascending arms.

Figure 4a and b show the cases where T_A^0 and T_B^0 lie above the metastable melting point of the opposite end member, while Fig. 4c and d show the cases where the metastable melting point of B lies above the stable melting point of A. The net result of this is that both the A and B liquidi ascend in the same direction. If the metastable melting point

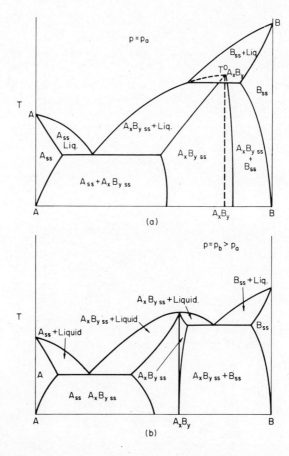

FIG. 5. (a) Peritectic system showing extensive solid solubility. (b) The case where, at sufficiently high pressure, the system of Fig. 5a achieves an A_xB_y melting point below the decomposition temperature of A_xB_y.

B. EUTECTIC SYSTEMS WITH SOLID SOLUBILITY

of B coincides with its stable melting point, then A and B are isomorphous and a single ascending curve results as in the ascending cases treated in earlier chapters. In the solid state, the region of solid solubility, in general, increases as the curves move from the type depicted in Fig. 4a and b to the types shown in Fig. 4c and d, this because of the above.

The isobaric invariant peritectic point is seen, from Fig. 4a–d, to represent the equilibrium between two solid solutions and liquid. It will be recalled that such a phenomenon also appeared in immiscible systems, generating a compound where, under the pressure of the system, the melting point of an incongruently melting solid lay above its decomposition temperature. In all such systems, however, there existed at least one eutectic type triple point in addition to the peritectic point.

As we have indicated, however, a peritectic may be generated in a limited solubility case without the formation of a compound in the system, in which event no eutectic triple point would be generated.

There have been debates as to whether the type of diagram depicted in Fig 4c and d is indicative of compound formation in a system exhibiting extensive solubility. In the opinion of the author, such is never the case. However, systems of the type shown in Fig. 5a are well known, and taking into account all that has been said, such systems are definitely

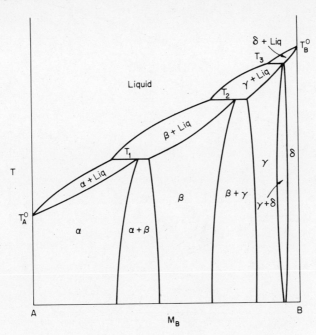

FIG. 6. A complex case of extensive solid solubility.

indicative of compound formation with extensive solid solubility. The argument is not semantic because, presumably, under the right conditions (see Chapter 25), the system of Fig. 5a could exhibit a compound peak (Fig. 5b). The systems of Fig. 4a–d, on the other hand, could never exhibit a compound peak.

The fact, then, that discriminates between the two possibilities in peritectic systems is whether or not a eutectic triple point is also present.

Before leaving this topic, it is of interest to consider systems in which several intermediate phases of variable composition are formed. Metal–metal interactions frequently exhibit this type of behavior. A hypothetical example of such a system is shown in Fig. 6. In metal–metal interactions, phases of variable composition, whether or not the system is of the simple ascending or minimum type, or of the complex type shown in Fig. 6, are termed alloys. Further, these phases are generally designated by greek symbols, α, β, γ, etc., rather than by the assumed stoichiometry of a compound. These symbols represent a structural characteristic of the phase rather than a stoichiometric attribute.

In Fig. 6, we note that a series of peritectic reactions occur at the

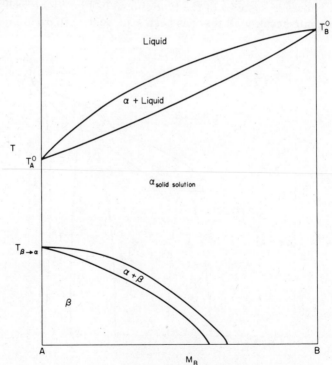

FIG. 7. Limited solubility occasioned by a phase transformation.

C. LIMITED SUBSOLIDUS SOLID SOLUBILITY

temperatures T_1, T_2, and T_3. Furthermore, it is noted that no eutectic triple point is present. What distinguishes the system depicted in Fig. 6 from those presented in Fig. 4c and d is that in the latter, only a single peritectic is observed. As argued previously, this latter instance is not reconcilable with the existence of a compound being formed. In Fig. 6, however, where a number of peritectics are generated, it is the case that at least one intermediate compound is generated. In fact, since the structure of the β and γ phases are different from each of the end members, it is clear that two intermediate compounds are formed. In general, it would be the case that the number of intermediate compounds generated is one less than the number of invariant points generated within the system.

C. THE CASE WHERE LIMITED SOLUBILITY OCCURS BELOW THE SOLIDUS

In systems in which complete miscibility is the rule in the S–L regions, partial miscibility may prevail at lower temperatures. This is particularly the case when one or both of the end members undergoes a phase

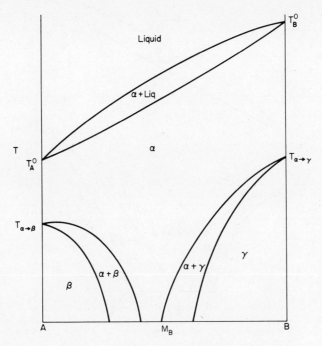

FIG. 8. Limited solubility caused by phase transformation.

transformation. In Fig. 7, the situation is depicted where A and B form an ascending series of solid solutions, but where at lower temperatures, A undergoes a phase transformation. It is assumed that at liquidus temperatures A and B are isomorphous and exhibit the α structure. At lower temperatures, A undergoes an α → β transformation. This gives rise to a two-phase α + β solid-solution mixture, separating the region in which a single β solid-solution phase exists and a region in which a single α phase exists. Presumably, the α + β two-phase region terminates at pure B. In this event there should exist a stable or metastable transformation of the B end member at which it acquires the β structure.

In Fig. 8, a case is depicted where both A and B undergo phase changes, but where the resulting phases are not isomorphic. In this instance, it appears to be the case that the β phase interacts with a metastable or stable β phase of the B end-member while the γ phase interacts with a metastable or stable γ phase of the A end member. These two all-solid two-phase regions give the appearance of a limited solubility eutectic interaction. At lower temperatures they intersect to

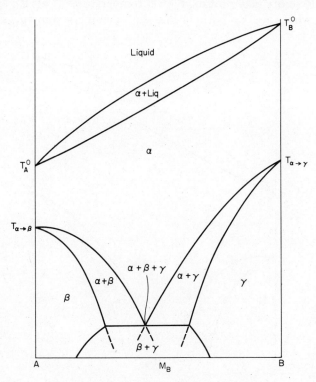

FIG. 9. The eutectoid interaction of Fig. 8.

C. LIMITED SUBSOLIDUS SOLID SOLUBILITY

give an all-solid invariant point as shown in Fig. 9. Such an all-solid interaction is frequently called a "eutectoid" interaction by analogy with its S–L counterpart. In similar fashion, a peritectoid interaction may be generated in the all-solid regions as depicted in Fig. 10. The arguments explaining such a system are quite similar to those advanced for analogous

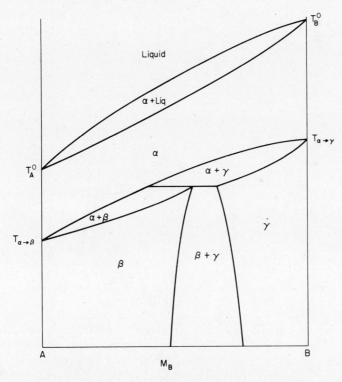

FIG. 10. The case where a peritectoid is formed.

behavior in the S–L regions. It is assumed that the low-temperature phase of the A end-member, the β phase, interacts with a metastable β phase of the B end-member. The γ phase, on the other hand, interacts with a metastable γ phase of the A end-member.

One final case of interest is that in which the phase transformations undergone by the end members lead to isomorphic new phases. These isomorphic new phases may be completely miscible if their lattice constants do not differ appreciably or they may lead to only partially miscible solid systems if lattice constant dimensions differ appreciably. Complete miscibility is depicted in Fig. 11 while only partial miscibility

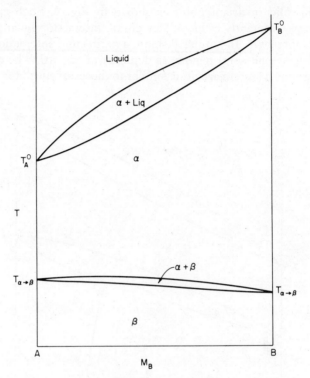

Fig. 11. Isomorphous phases result from phase transformations.

is depicted in Fig. 12. Note that the latter is identical in appearance to the case described in Fig. 10. Now, however, two phases having the same structure equilibrate in a $\beta - \beta'$ miscibility gap. In conjunction with the miscibility gaps generated in partially miscible systems, note that the gap arises because of the limited solubility of the different solids one in another and has no significance other than that.

Sometimes, a S–L system of the ascending type shows MG formation at lower temperatures. This indicates that the solid solution structure cannot accommodate both end member. Alternatively, we may explain such phenomena by assuming that the isomorphic phases leading to continuous solid solubility possess lattices having significantly different thermal coefficients of expansion. When the isomorphic phases enter into only limited solid solution in the liquidus regions, the indication again is that the lattice constants are considerably different. Here the end members behave to a first order approximation as if they were not isomorphous and lead to diagrams of the type shown in Fig. 4a–d.

D. p–T–M_x REPRESENTATIONS

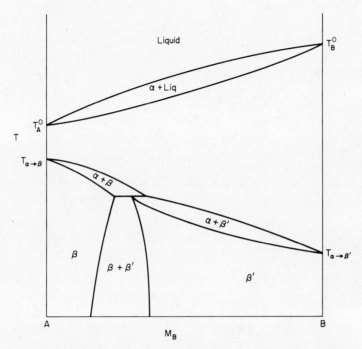

FIG. 12. Isomorphous phases with limited solid solubility are generated.

D. THREE-DIMENSIONAL REPRESENTATIONS OF LIMITED SOLID SOLUBILITY SYSTEMS

Figure 13 shows, in perspective, the three-dimensional space model for the eutectic system A–B. The diagram differs from that of the analogous immiscible solid system in that eutectic and two solid phase tie lines such as the line 1–2 in the isobaric section do not extend to the p–T planes of the end members. Figure 14 shows several isothermal sections through the space model of Fig. 13. Figure 14a is an isothermal section through the L–V portions of the system. Figure 14b is an isothermal section through the L–V portions and the B liquidus field, it being assumed that the triple point of pure B is higher in temperature than that of pure A. It is also assumed that the vapor pressure of liquid A is greater at all temperatures than that of liquid B. In Fig. 14b, it is seen that the three-phase tie line V–L–S does not terminate at pure B, the tie line showing the compositions of vapor–liquid and solid in equilibrium at the liquidus for this temperature. The line vapor–$p_B{}^0$ shows the composition of vapor in equilibrium with B_{ss} of varying composition in

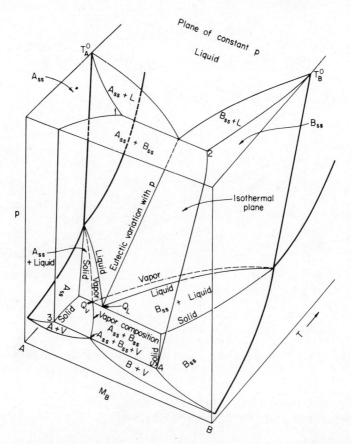

FIG. 13. The three-dimensional space model for a eutectic system exhibiting limited solid solubility.

the composition interval in which the B_{ss} phase may coexist with vapor (see Figs. 1 and 2). Figure 14c shows an isotherm at still lower temperatures, i.e., below the triple points of pure A and pure B, but above the temperature of the quadruple points. The L–V region here is terminated by intersection with the liquidi of both A_{ss} and B_{ss}, and in both instances these phases exhibit composition variation with pressure. In the isothermal section, of course, the two-component three-phase regions S–L–V are invariant, and the triple point compositions of S–L–V are shown for each liquidus. Figure 14d shows an isotherm right through the quadruple points of the system where the invariant compositions of L, V, and two solid solutions are defined by the four points Q_L, Q_V, $Q_{A_{ss}}$, and $Q_{B_{ss}}$.

D. p–T–M_X REPRESENTATIONS

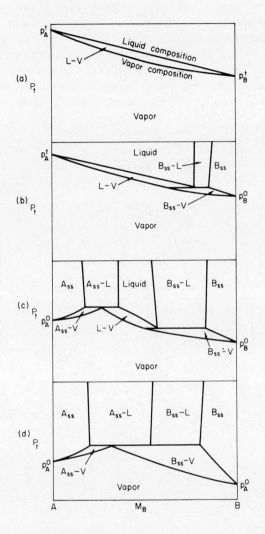

FIG. 14. Isotherms through the p–T–M_B diagram of Fig. 13; (a) above the melting points of A and B; (b) below the melting point of B, but above the melting point of A; (c) below the melting points of A and B, but above the quadruple point; (d) at the quadruple point temperature.

Returning now to Fig. 13, the front isothermal plane in the interior of the space model shows a cut through the S–V regions of the diagram. Unlike the immiscible solids case, the maximum pressure attainable at the triple point A_{ss}–B_{ss}–V is not equal to the sum $p_A{}^0 + p_B{}^0$, where these terms refer to the vapor pressures of pure solid A and pure solid B, respectively, at the temperature of the isotherm. This is because the coexisting solids at this triple point are not the end members, but are solid solutions having the compositions 3 and 4. In the ideal case $P_{\max}$ is less than that for the immiscible solids case. In the simplest instance, i.e., where species and mole numbers in coexisting phases are equal, both $p_{A_{ss}}^{(1)}$ and $p_{B_{ss}}^{(1)}$ are given by the Raoult equation. Thus at the maximum pressure

$$p_{A_{ss}}^{\max} = p_A{}^0 N_A^{(1)}, \tag{1}$$

$$p_{B_{ss}}^{\max} = p_B{}^0 N_B^{(1)}, \tag{2}$$

and

$$P_{\max} = p_{A_{ss}}^{\max} + p_{B_{ss}}^{\max}. \tag{3}$$

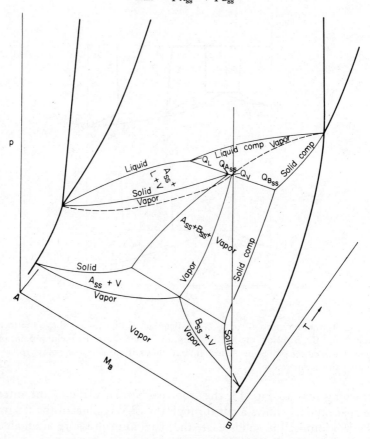

FIG. 15. The p–T–M_B representation of the peritectic system A–B.

D. p–T–M_x REPRESENTATIONS

Having considered the three dimensional model for the eutectic system showing limited solid–solid solubility, we shall consider the peritectic case depicted in Fig. 14c. The p–T–M_B representation is given in Fig. 15

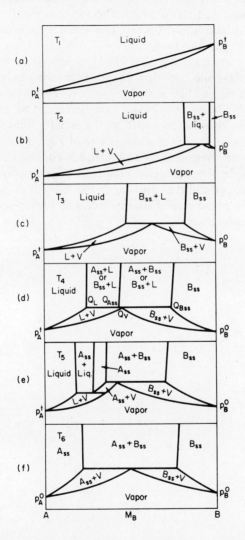

FIG. 16. Isothermal planes through Fig. 15; (a) above the melting point of A and at the triple point of B; (b) above the melting point of A, but below the triple point of B; (c) same as (b), but closer to the quadruple point temperature; (d) at the quadruple point temperature; (e) below the quadruple point temperature, but above the melting point of A; (f) below the melting point of both A and B.

and isothermal sections are shown in Fig. 16. We have assumed that the vapor pressure of liquid B is greater than that of liquid A in constructing Figs. 15 and 16. In view of preceding discussions, neither Fig. 15 nor 16 will be described further. It should be noted that the alternative possibility exists, i.e., that in which the vapor pressure of A is greater than that of B in the liquid phase. The construction of this variant is left as an exercise for the reader.

Because the constructions of $p-T-M_B$ space models for the more involved cases described in this chapter are not easily seen on two-dimensional drawings, they will not be depicted. In general, however, the more complex systems offer no new features and may be built up from sections like those shown in Figs. 13 and 15. The space model for Fig. 5a, for example, is almost a composite of Figs. 13 and 15.

The discussions presented in this chapter conclude the remarks on binary systems. In the next two chapters, a descriptive discussion of condensed-phase equilibria, and a more quantitative discussion of condensed-vapor phase equilibria in three and higher-order component systems will be presented.

34

Three-Component Systems

A. INTRODUCTION

In this and the following chapter we intend to provide an introduction to both the graphical representation and analytical description of selected cases of ternary and higher-order systems. The present chapter will be concerned chiefly with the graphical aspects of ternary diagrams. The next chapter will focus on analytical treatments of univariant two-phase, three-component, solid–vapor equilibria. The extension of such analytical approaches to quaternary and higher-order univariant cases will not be explicitly developed but should be obvious.

A three-component equilibrium may be established in as few as one or as many as five phases. Coincident with the one-phase situation there exist four degrees of freedom, the temperature, the pressure, and two of the three component concentrations. When five phases are present the three-component system possesses no degrees of freedom. Intermediate between the boundaries of tetravariance and univariance, bivariant and trivariant equilibria are, of course, possible.

For systems constrained by a condition of constant pressure, the maximum variance in a three-component interaction is reduced by one to three, the minimum again being zero. A further constraint might be the setting of the mole ratio of two of the three components constant relative to one another. This would reduce the maximum variance to two, this variance coinciding with a one-phase equilibrium. For the doubly constrained system, a two-phase equilibrium is univariant and a three-phase equilibrium is invariant. Having constrained the mole ratio of two of the components relative to one another, the univariant case may be described in terms of the variation of the mole ratio of the third component relative to either of the other two as a function of T. In essence, the double constraint involves starting with a fixed binary mixture and varying the amount of a third component. For each new starting composition relative to the constrained binary system, the ternary univariant extension will be different. More will be said about such cases in Chapter 35 when we discuss solid–vapor equilibria at constant total pressure. The concept is introduced here lest it not be realized that is applicable to condensed phase equilibria as well.

B. Representation of Three-Component Condensed-Phase Equilibria

For systems in which the vapor phase may be ignored due to the involatility of condensed phases and upon which a constant total pressure is imposed, the most convenient method of representation of equilibrium states uses as its basis the equilateral triangle (Fig. 1). Each of the apexes of this triangle represents a pure component, each side represents a binary system, and the interior of the diagram represents the ternary

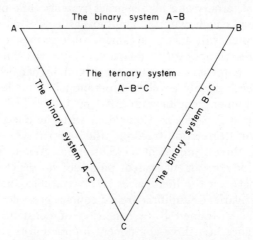

Fig. 1. Representation of compositions in a ternary interaction.

system. The temperature axis is perpendicular to the plane of the paper in Fig. 1. Consequently, the equilateral triangle shown in Fig. 1 is an isotherm in the three-dimensional diagram A–B–C versus T. Alternatively, if all of the intersections of surfaces in the three-dimensional diagram are projected down upon the composition plane in Fig. 1, the triangle may represent a polythermal projection rather than an isotherm. Furthermore, if each of the binary diagrams X–Y versus T that lie perpendicular to the plane of the ternary composition triangle are rotated 90° so that they lie in the same plane as the composition triangle (a collapsed diagram), we may, on a single plane, represent the binary and ternary interactions as a function of T. In Fig. 2, such a

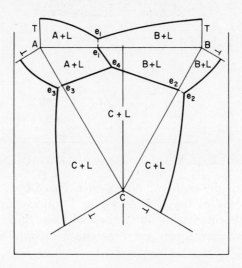

FIG. 2. The diagram of A–B–C–T showing collapse of the binary diagrams A–B–T, B–C–T, and C–A–T onto the plane of the ternary projection composition plane.

collapsed diagram is depicted for the case where each of the binary diagrams A–B, B–C, and C–A versus T are collapsed onto the plane of the ternary composition triangle A–B–C. These binary diagrams give rise to the binary eutectics e_1, e_2, and e_3. The binary systems are complete in that they represent temperature–composition planes as has been the approach previously employed. The projections of these eutectic locations onto the ternary triangle sides A–B–C are shown via the dotted lines extending from the eutectics in the binary T–M_X diagrams to the sides of the ternary triangle. The fate of these binary eutectics when the third component is added is shown by the lines e_1–e_4, e_2–e_4, and e_3–e_4, all of which extend into the interior of the triangle. We shall delay explanation of their significance until later, for the present it being sufficient to provide some feeling for the means available for representing the phase diagrams of ternary systems.

A ternary projection diagram for the type of system shown in Fig. 2 may be represented without explicitly depicting the binary diagrams that form the sides of the triangle. Sometimes this is necessary because the binary diagrams are unknown, only the ternary diagram or parts of it being understood. This situation is shown in Fig. 3a. In other instances,

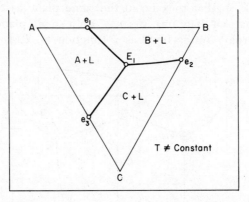

FIG. 3a. The diagram A–B–C–T projected on the A–B–C composition plane.

where only a single isotherm rather than the polythermal projection on the ABC plane of the A–B–C versus T interaction is known, the triangle is representative of an isothermal slice through the surfaces of the ternary system. Again, for a system of the type shown in Fig. 2,

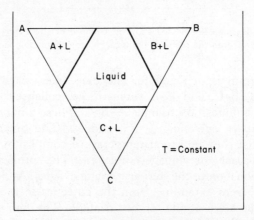

FIG. 3b. An isothermal slice is taken through the three-dimensional diagram of Fig. 3a.

B. REPRESENTATION OF THREE-COMPONENT SYSTEMS

such an isotherm, taken below the melting points of the three components but above the temperature represented by the point E_1, might have the appearance shown in Fig. 3b. We offer the latter without explanation, deferring its examination until later.

An extension of the methods of representation of ternary systems involves perspective drawings of right prisms, in much the same manner as was done in defining the space models of binary systems. Now, however, the diagrams are closed since the basal plane is a compositional plane alone. The vertical direction is still open ended, representing the temperature variable. Such a perspective construction for a system like that depicted in Fig. 2 is shown in Fig. 4.

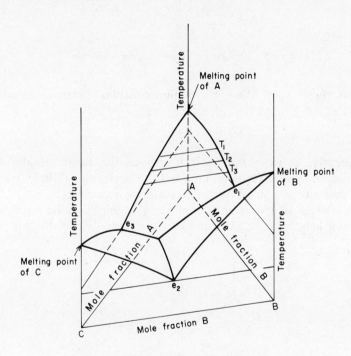

FIG. 4. Perspective view of the system depicted in Fig. 2.

Before continuing, it is worthwhile to consider some of the properties of an equilateral triangle that make it attractive for depicting ternary systems. The line drawn from any apex perpendicular to the side opposite is referred to as the height of the triangle (see Fig. 5). If we specify a point within the triangle, such as the point a of Fig. 6 and construct perpendiculars to the three sides of the equilateral triangle, it is always

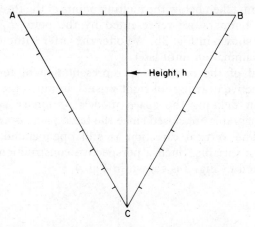

FIG. 5. The height of an equilateral triangle.

true that the sum of these three perpendicular distances equals the height, i.e.,

$$ab + ac + ad = h. \tag{1}$$

Consequently, if we assign unit length to the height, each of the perpendicular distances may be used to designate a mole fraction, a mole percentage, a mass fraction, or a mass percentage.

If, instead of assigning unit length to the height of the triangle, we assign unit length to the sides and construct parallels to the three sides

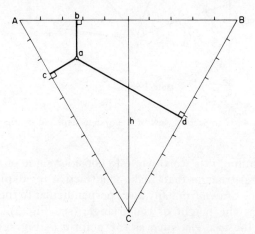

FIG. 6. Perpendiculars from a point in the triangle to the sides.

B. REPRESENTATION OF THREE-COMPONENT SYSTEMS

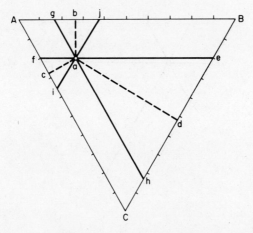

FIG. 7. Parallels to the sides passing through a point.

all passing through the point a, the parallels fe, gh, and ij in Fig. 7, it is the case that

$$ab/h = Be/BC = Be/\text{side}, \qquad (2)$$

$$ac/h = Ag/AB = Ag/\text{side}, \qquad (3)$$

$$ad/h = Ci/CA = Ci/\text{side}. \qquad (4)$$

In other words,

$$Be + Ag + Ci = \text{side}, \qquad (5)$$

and the intersections of parallels through a point with the sides may be used for expressing the mole or mass fractions of this point with respect to the ternary system. This is shown in Fig. 8 where the ternary composition is designated by the point 1. The parallels through this point are designated ab, dc, and ef. Since each side is assigned unit length, we are free to subdivide each side decimally. The parallel opposite an apex, i.e., the parallel ef opposite the apex C intersects the two sides adjacent to this apex at a certain fraction of the side length, the corresponding apex itself representing unit length. We see in Fig. 8 that the parallel ef intersects the sides AC and BC at the second division. Consequently, if we are operating in the mole fraction regime, this stipulates that the mole fraction of the component C in the ternary mixture represented by the point 1 is 0.2. The parallel opposite apex A intersects sides AC and AB 0.6 of the way to pure A. Consequently, the mole fraction of A represented by point 1 is 0.6. By difference then,

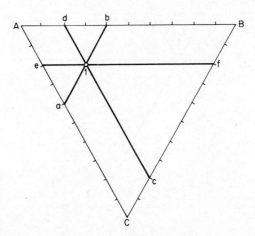

Fig. 8. Determining the mole fraction of the components A, B, and C in a mixture at point 1.

the mole fraction of the component B is 0.2. That this is the case can be seen by inspection of the parallel *dc* which lies opposite the apex B. This parallel intersects the pertinent sides BA and BC 0.2 of the way to the apex B.

One final geometrical consequence is of interest to us, that for a line extending from an apex to an opposite side. This is shown in Fig. 9. Such a line, C*a*, for example, intersects the side AB, representing the composition axis of the system AB at the point *a*. In effect, as we move from point a in the binary system to point 1 and then point 2 in the

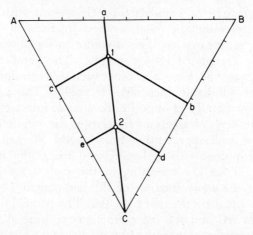

Fig. 9. A line of constant mole ratio of the components A and B in the system A–B–C.

ternary system we are taking a fixed ratio of A to B and adding a quantity of the third component.

In an equilateral triangle, a line from an apex to a side opposite has the property that the ratio of perpendicular distances to the two remaining sides from any point on this line is a constant. Thus the ratio $1b/1c = A/B$ in Fig. 9 is the same as the ratio $2d/2e = A/B$ in this figure even though $1b + 1c > 2d + 2e$. For the case shown, it is evident from the location of point a that the ratio of A/B is $0.6/0.4 = 1.5$. This, then, is also the ratio of A/B in the compositions 1, 2, or any other point on the line aC.

C. Simple Ternary Eutectic Interactions

Consider a ternary system A–B–C, each of whose binary pairs generates a eutectic interaction. Assume also that the addition of the third component to each binary system does not result in the formation of a compound, or in any other way disturb the simple eutectic nature of the binary system. The projected appearance of such a system is that shown in Fig. 3a. The point e_1 on the binary side A–B represents the projected eutectic in this system onto the composition basal plane, similarly for the eutectics e_2 and e_3. Each of the binary eutectics comprises two solid phases and one liquid phase, and each is isobarically invariant. If a third component is added to the binary system, the binary eutectic loses its invariant nature and becomes univariant, that is, the binary eutectic in which two solids are in equilibrium with liquid exhibits a composition that is a function of temperature. In Fig. 3a, therefore, the line e_1–E_1 represents the variation in composition with temperature of pure solids A and B in equilibrium with liquid in the ternary system, similarly for the lines e_2–E_1 and e_3–E_1. Each point on a projected binary eutectic extension line such as the line e_1–E_1 represents a different temperature. Since along this line, pure A and pure B are crystallizing simultaneously with temperature change, analytic description of the variation of liquid composition which is in equilibrium with these two pure solids would involve the heats of fusion for pure A and pure B, i.e., ΔH_A^0 and ΔH_B^0.

It will be recalled that in a binary eutectic interaction, the addition of the second component to the first component causes a lowering of the latter's melting point. In the binary system, the variation in this melting point with composition gives rise to the liquidus curves. In the ideal binary eutectic case, it is the fact that the nature of the component causing the lowering of the melting point is immaterial, only its concentration being important. In the ternary system, the addition of components B and C to pure A again causes a lowering in the melting point of the latter. This time, rather than generating a single liquidus line coincident with a condition of univariance, the addition of two components to the third generates an infinite number of liquidus curves coincident with a condition of bivariance. This infinite number of liquidus curves generates a liquidus surface of crystallization for the component A. Similar surfaces are generated for the cases where pure B is in equilibrium with liquid and where pure C is in equilibrium with liquid. Where pairs of these three surfaces intersect, we have generated the binary eutectic intersections in the form of single univariant curves. In other words, when the A–L surface intersects the B–L surface, each of which was isobarically bivariant, we generate a univariant line for the equilibrium A–B–L. These surfaces and their line intersections can be visualized by referring to Fig. 4. Planes of constant temperature intersecting these surfaces form surface isotherms which are shown, in projection, in Fig. 10, generating a gradient map. In the ideal eutectic

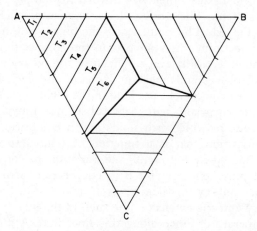

FIG. 10. A gradient map of projected ternary surfaces in an ideal system.

case, where the *quantity* of the second component rather than its *nature* is important, a given temperature drop in the liquidus temperature of one of the components coincides with a given mole fraction of the component crystallizing with respect to the liquid-phase composition at that temperature. In an ideal ternary system where components B and C, for example, are added to component A, again it is the total concentration of B and C that determines the melting point depression of A rather than the nature of B and C. Thus, if the mole fraction of A

C. SIMPLE TERNARY EUTECTIC INTERACTIONS

in the liquid phase of a binary solid–liquid equilibrium has the value $M_A^{(2)}$, its melting point will be lowered by an amount given by the simple van't Hoff equation. In a ternary system where the mole fraction of A is again $M_A^{(2)}$, the melting point of A will be lowered by precisely the same amount as in the binary case, the sum total of the mole fractions of B and C acting as if only a single additional component were present. In real cases, of course, this situation is perturbed.

It will be noted in Fig. 10 that the gradient lines are straight and parallel to one of the sides. The implication then, is that a given lowering of temperature of the component A, for example, in the binary system A–B or A–C or in the ternary system A–B–C, depends solely on the mole fraction of A in the two-phase regions A–L and is independent of the nature of the other component(s). Each of these lines of constant temperature on a surface is an isothermal isobaric univariant situation. In other words, in order to define the state of the system completely we must specify one composition. For example, consider the polythermal projection (Fig. 11) in which only the isotherm T on the A liquidus

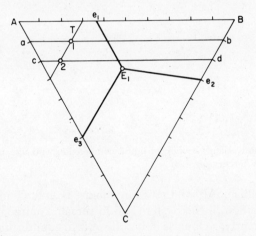

FIG. 11. Univariance of a surface isotherm.

surface is shown explicitly. Let us specify the composition with respect to the component C via the parallel line ab. This intersects the isotherm T at point 1 and defines the composition of the entire system for the case where solid A is in equilibrium with liquid having a composition of the component C specified by the line ab. The parallel line cd intersects this same isotherm at the point 2, etc., for an infinite number of other parallel lines, the sum total of their intersections with the A–L surface generating the isotherm T. An isotherm such as that shown in Fig. 11

gives the compositions of liquids in equilibrium with pure solid A or B or C. Only one point on this isotherm need be specified in order to define either and therefore both of the remaining compositions. In Fig. 3b we see the intersection of an isothermal plane with the solid–liquid surfaces of each of the components. The interior of the diagram represents the isothermal-isobaric bivariant liquid region in which two compositions must be specified in order to define the third composition since only a single condensed phase is present.

In the region above the surface A–L, the system is entirely liquid. Suppose we start with a composition in this region specified by the point 1 of Fig. 12. Upon reaching the A–L surface, pure solid A would

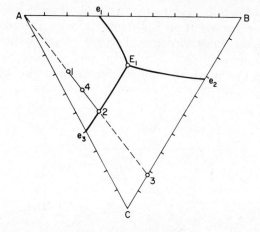

FIG. 12. The univariant variation of liquid composition with temperature in a two-phase region of the ternary diagram.

begin to crystallize. All of the components B and C in the original starting composition of this system would remain confined to the liquid phase. Consequently, so long as only solid A is in equilibrium with liquid, the ratio of B/C in the liquid would have to remain constant. The composition of the liquid then would of necessity have to remain confined to a line extending from the apex A to the side BC and passing through the point 1 (see Fig. 9 and the accompanying explanation). In fact, the liquid composition would vary between points 1 and 2 of Fig. 12, the composition of the liquid having the value at point 1 when the two-phase equilibrium is first established and the composition of the liquid having the composition 2 when a second solid phase, pure solid C, begins to crystallize at point 2. In other words, in the temperature interval represented by the line 1–2, the two-phase equilibrium A–L is

C. SIMPLE TERNARY EUTECTIC INTERACTIONS

univariant since the liquid-phase mole ratio of B/C is constant and has the value C3/3B, i.e., a composition of B relative to C given by the point 3. An infinite number of such univariant lines whose locus is the A apex exist on the A surface. The set of such lines, in fact, defines the A surface, similarly for the B and C surfaces.

The extension of the point 1 in Fig. 12, to the apex A shows that the solid in equilibrium with liquid having a composition along 1–2 is pure solid A. If we cool our liquid starting composition (point 1), upon encountering the A–L surface, it forms a small quantity of solid A. If the temperature is cooled further to point 4, the liquid has the composition of this point while the solid still has the composition A. Obviously the liquid still has the original B/C ratio, but is becoming less rich in A. During cooling to point 4 the fraction of solid has increased from a value approaching zero at point 1 to a value 1–4/A–4 at the temperature coinciding with point 4. The solid-to-liquid ratio has increased from essentially zero at point 1 to a value 1–4/A–1 at the temperature coinciding with point 4. A portion of the plane perpendicular to the triangle in which line A–3 of Fig. 12 is shown is represented in Fig. 13. The tie arms just referred to are depicted as are the several

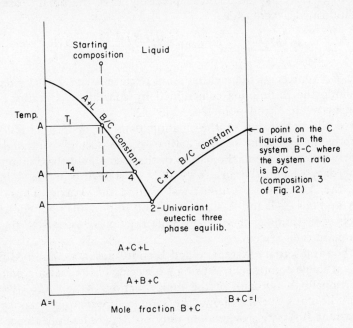

FIG. 13. A slice through the liquid surfaces of Fig. 12 having a constant ratio of B/C.

regions cut by this plane. Figure 13 has much the same appearance as a simple binary eutectic, but with certain notable exceptions. First, the composition axis does not extend from 0–1 mole fraction of C since the line AB terminates at a composition in the system B–C where $M_C < 1$. The maximum melting point on the C liquidus does not coincide with the melting point of C. Instead, it coincides with the liquidus temperature for the equilibrium C–L in the binary system B–C when the mole ratio of B to C in this binary system is B/C, i.e., the point 3. Furthermore, at temperatures below point 2 the system is not solidified, but consists of primary crystals of solid A, and a eutectic mixture of solid A and solid C and liquid. With reference to Fig. 12, it is not until a temperature coincident with E_1 is reached that all liquid vanishes and a three-phase solid equilibrium is generated, this solid consisting of primary crystals of A that have deposited during the liquidus crystallization, eutectic crystals of pure A and pure C, and ternary eutectic crystals of pure A, pure B, and pure C. This will be expanded upon subsequently.

For univariant liquidus curves of the type shown in Fig. 13, where the univariance is caused by the isobaric system and isoplethal liquid conditions, the variation of A in the liquid, $N_A^{(2)}$ or $M_A^{(2)}$, with temperature in an ideal case may be deduced precisely as in the binary case as has been indicated, leading to

$$\ln (M_A^{(2)})_{B/C\,\mathrm{const}} = \exp\left[\frac{-\Delta H_A^0}{R}\left(\frac{1}{T} - \frac{1}{T_A^0}\right)\right], \tag{6}$$

where T_A^0 represents the melting point of pure A and T is any temperature along the A liquidus constrained by constant B/C ratio, i.e., the region between A and point 2 of Fig. 12.

Since the depression of the melting point of A in an ideal binary system is independent of the nature of the second component and similarly independent of the nature of the second and third components in a ternary system (because only A is crystallizing in the A liquidus field), the A surface is symmetrical around the A axis. Thus, in the binary systems A–B and A–C, the drop-off in liquidus temperature of A is solely a function of the system mole fraction of A and is, therefore, the same in each system.

In the ideal ternary system, added B and C act in a combined fashion as only a single component, causing a lowering of the melting point of A proportionate to the sum of the mole fractions $M_B + M_C$. Thus the ideal A liquidus surface shows the same temperature dependence for all constant B/C ratios depending only on M_A. The appearance of the A surface on each side of the diagram differs since the binary eutectic

C. SIMPLE TERNARY EUTECTIC INTERACTIONS

compositions and temperature are determined by both participating components. In Fig. 10, therefore, we see that the A liquidus surface extends to lower temperatures on the AC side of the diagram than on the AB side.

Just as the intersections of any two of the bivariant isobarically constrained liquidus surfaces give rise to univariant curves representing extensions of binary eutectics into the interior of the triangle, intersection of the three liquidus surfaces occurs at a point. This point is termed a ternary eutectic, the point E_1 in Fig. 12. It occurs at the terminus of the crystallization paths of three different solids and the four-phase equilibrium A–B–C–L is isobarically invariant. This is represented in Fig. 13 by the lowest temperature region marked A–B–C.

When the starting mixture depicted in Fig. 12 has been cooled from a temperature coinciding with point 1 to a temperature coinciding with point 2, the line of intersection of the A and C liquidus surfaces is reached. At the point 2 solid C starts to crystallize along with solid A and the three-phase mixture solid A, solid C, liquid is generated. Upon further cooling, solid A and solid C continue to crystallize. For the first time, the ratio of B/C in the liquid phase changes since the component C is no longer confined to the liquid phase. The B/C ratio in the liquid increases as a consequence. The liquid composition, therefore, must move away from the apex C toward B and does so by moving along the line 2–E_1 of Fig. 12. The total solid composition is no longer pure A, becoming a mixture of solid A and solid C. The extension of point 1 can no longer be to the apex A but instead must intersect the side AC at compositions richer in C. This is shown in Fig. 14. The dotted line represents the crystallization path during the A liquidus crystallization process. When point 2 is reached solid C starts to crystallize along with solid A and the liquid becomes depleted in C moving to point 3 toward E_1. The total solid concentration no longer being pure, A moves along AC to point 4. As the temperature is lowered still further, the liquid composition moves to point 5, the solid composition moving to point 6. During this process, the tie lines pivot around the starting composition point 1. The length of the line from point 1 to the line e_3–E_1 increases while the length of the line from point 1 to the side AC decreases. The ratio of these two linelengths gives the solid-to-liquid ratio, or the ratio 1–5/5–6 gives the ratio of solid to the total quantity of solid and liquid (the total quantity of moles or mass depending upon the scale used for describing the system). The ratio of solid A to solid C may be given at any point, e.g., at point 6 it is given by the lengths C–6/6–A. Finally, when the temperature is lowered to point E_1, the solid has a composition defined by point 7

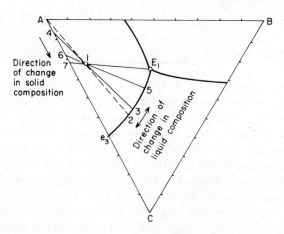

Fig. 14. The change in liquid and solid compositions as the binary eutectic line e_3–E_1 is followed.

while the liquid has the composition E_1. At this point solid B crystallizes. The composition of the solid must now reflect a composition increasing in B from its zero value up to this point to the value in the starting mixture at point 1. The solid composition thus moves from point 7 to point 1 during this isothermal crystallization. Upon reaching the point 1, the solid-to-liquid ratio is infinite and a three-phase solid system is present. This three-phase system exhibits three metallographically distinguishable crystalline agglomerations: primary crystals of A, secondary crystals of A and C, and a ternary eutectic crystallization of A, B, and C.

It will be recalled that the solution for the location of the eutectic in a binary system cannot be obtained analytically; It requires a graphical solution because of the nature of the solubility curve equations. For the same reason, the variation of M_A and M_B with temperature along the extension of the binary eutectic into the ternary system also requires a graphical approach in order to estimate M_A and M_B as a function of T.

The same arguments advanced descriptively for cooling behavior in the A liquidus region of the ternary diagram are applicable to the B and C liquidus regions.

It should be noted that the second material to crystallize simultaneously with solid A, following the primary crystallization in the A field, need not be pure solid C. For example, with reference to Fig. 15, which is a duplicate of Fig. 14, it is seen that if the initial composition of the system is given by point a, the crystallization path will lie along A–b. When the liquid reaches a composition having the value at point b,

C. SIMPLE TERNARY EUTECTIC INTERACTIONS

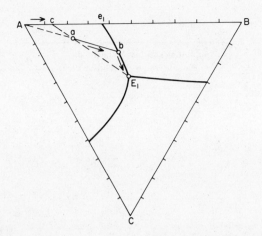

FIG. 15. Cooling behavior in which A and B crystallize along the binary eutectic extension line e_1–E_1.

solid B will crystallize out and the tie-line terminus showing the solid phase composition will move along the binary side AB until the point c is reached. The liquid composition in equilibrium with solid having the overall composition given by the point c will again be E_1, the composition of liquid at the ternary eutectic.

To this point, our discussion of ternary eutectic interactions has focused on projection diagrams. It is of interest now to examine isothermal sections through this type of diagram, since such sections frequently represent the experimental approach used in resolving questions of ternary-system behavior. One such isotherm was offered in Fig. 3b without explanation. In essence, if an isothermal plane is inserted across the diagram of Fig. 4, it will intersect different surfaces depending upon the temperature of the isotherm and the temperatures of melting of the end members A, B, and C. For illustrative purposes let us assume that $T_A^0 > T_B^0 > T_C^0$. An isotherm taken at a temperature greater than T_A^0 will intersect none of the liquidus surfaces and will terminate at the sides of the prism. An isotherm taken below T_A^0 but above T_B^0 and T_C^0 will intersect the A liquidus field as shown in Fig. 16. All tie lines within the region designated A–L terminate at pure A on one side and with the line a–b on the other side. This line a–b is the intersection of the isothermal plane with the A liquidus and, for reasons previously given, is symmetrical around the A apex in ideal interactions.

Figure 17 presents an isotherm at a temperature below the melting points of both A and B but above the melting point of C. As can be

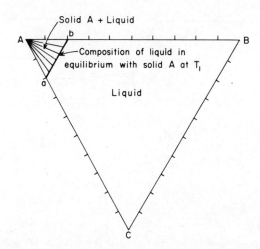

Fig. 16. An isotherm in the system A–B–C just below the melting point of A but above the melting points of B and C.

visualized with the aid of Fig. 4, the extent of the one-phase liquid region becomes smaller in isotherms at successively lower temperatures.

In Fig. 18 we again show an isothermal section for a temperature below T_A^0 and T_B^0 but above T_C^0. The difference between Fig. 18 and Fig. 17 is, that we shall assume that the isotherm of Fig. 18 is precisely at the temperature of the binary eutectic e_1 in the system A–B. We shall also assume that e_1 lies at a temperature greater than the temperature

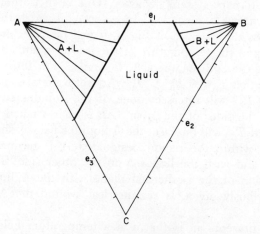

Fig. 17. An isotherm below T_A^0 and T_B^0, but above T_C^0.

C. SIMPLE TERNARY EUTECTIC INTERACTIONS

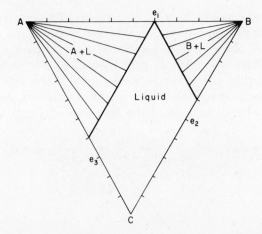

FIG. 18. An isotherm at the temperature of the eutectic e_1.

of the binary eutectic e_2 for the binary system B–C and that $T_{e_2} > T_{e_3}$, where T_{e_3} is the binary eutectic temperature in the system A–C.

In Fig. 19, an isotherm is shown for a temperature below $T_A{}^0$, $T_B{}^0$, and T_{e_1}, but above the other pertinent binary and unary invariant temperatures. This isotherm intersects the extension of the binary eutectic e_1 into the ternary diagram. The intersection of the isothermal plane with this extension is at a single point, the point a, as might be expected since when the phases solid A, solid B, and liquid coexist isothermally and isobarically, the three-component, three-phase equilibrium is invariant. At all starting compositions of the component C less than that bounded by the tie lines Aa and Ba, the system temperature is below that at which a second phase has begun to crystallize. Consequently, any starting composition within the triangle ABa when cooled to the temperature of the isotherm would have deposited either pure solid A or pure solid B initially, followed by subsequent simultaneous precipitation of a mixture of pure solid A and pure solid B. For all starting compositions within this triangle then, the liquid composition achieves a value given by the point a when the temperature of the isotherm is reached, while the solid composition lies along the side AB. All tie lines terminate at point a on one side, passing through the point of starting composition and terminating at its opposite end on the A–B binary composition axis. For example, assuming a starting composition such as that at point 3 in Fig. 19, at the temperature of the isotherm the liquid composition is a and the solid composition is 4. The primary crystallization for the starting composition would have been pure B. Note further that if the starting compositions had lain along either

Aa or Ba, i.e., the points 1 or 2, at precisely the temperature of the isotherm, the liquid composition would have achieved the composition a and the second phase A or B, respectively, would have just begun to crystallize simultaneously with the primary solid phase. Since all starting compositions within the triangle AaB would have achieved liquid compositions lying along the extension of the binary eutectic e_1 at temperatures higher than that of the isotherm of Fig. 19, this entire

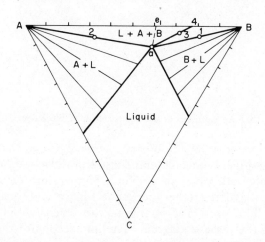

FIG. 19. An isotherm cutting through the extension of the binary eutectic e_1 in the ternary system A–B–C.

region must represent the three-phase region L–A–B at the temperature of the isotherm.

In Fig. 20, an isotherm is depicted that intersects the right prism below $T_A{}^0$, $T_B{}^0$, and $T_C{}^0$, and that also intersects each of the binary eutectic extensions e_1, e_2, and e_3. The description of each of the regions is the same as already given. The region marked L includes the liquid composition of the ternary eutectic, this composition lying at the point intersection of the three eutectic extension lines.

In Fig. 13, we showed a plane intersecting the ternary T–M_A–M_B–M_C prism perpendicular to the basal plane. Figure 13 was constructed primarily to demonstrate the logic behind Eq. (6). As such it was not a complete description of the slice in question since the region above the temperature of the ternary eutectic temperature and below the A and C liquidus curves was not defined. In this region, such a slice would intersect regions in which the three-phase equilibria A–C–L or B–C–L existed. Furthermore, the slice chosen for Fig. 13 is special in that one of its two intersections with the boundaries of the basal triangle is with an

C. SIMPLE TERNARY EUTECTIC INTERACTIONS

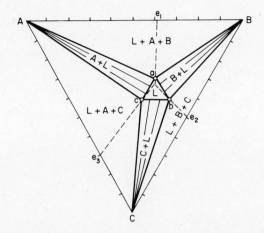

FIG. 20. The three liquidus surfaces and the three binary extension space curves are intersected.

apex while the other is with a side. Such a special case enables us to arrive at the conclusion given in Eq. (6) since within the A liquidus field a tie-line extending from a point on a projection graph to the A apex, for example, lies in the plane of the slice. Thus, in Fig. 12, when the temperature has dropped to the value indicated by point 4, tie lines designed to show solid-to-liquid ratios would extend between point 4 and apex A and lie in the plane of the slice A–3, which is shown in Fig. 13. The composition range A–2 of Fig. 12, therefore, is precisely the same as the distance A–2 of Fig. 13 and in the A liquidus region of Fig. 13, tie lines may be used in a normal manner. In the region 2–3 of Fig. 12 such a treatment is invalid since tie lines in the region 2–3 do not lie in the plane of the slice but extend rather from the apex C to either the line e_3–E_1 or e_2–E_1 of Fig. 12. Each tie line, therefore, rather than lying in the plane of Fig. 13, intersects it at a single point. This can be seen in Fig. 21 where the starting compositions a, b, c, and d, all lying within the C liquidus field, but still on the slice A–3 of Fig. 12, are designated. Whereas tie lines for all compositions along A–2 lie along A–2 it is seen that none of the tie lines for the compositions a, b, c, and d lie along 2–3. Thus, in Fig. 13, we cannot obtain tie-line data for any region except the one in which the tie lines lie in the plane of the slice. We can, however, use a diagram such as that partially shown in Fig. 13 to define the temperature range of given polyphase regions of existence and the composition intervals of such existence, within limits. For example, from a slice, such as that defined in Figs. 12, 13, and 21, which passes through the apex A, we will always be able to tell the mole

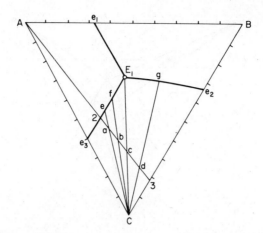

Fig. 21. The lines for the compositions a, b, c, and d along the slice A–3 of Figs. 12 and 13.

fraction of A present in the liquid, but not the mole fraction of B or C individually. We will be able to define the sum B + C if the diagram is constructed correctly.

With reference to Fig. 21, we see that starting with different compositions along the interval A–2, the first crystallization will always be pure A and that for all such starting compositions a liquid will finally be obtained whose composition is that given by the point 2. At the temperature coincident with that of point 2, the ternary eutectic extension of the binary eutectic e_3 will be reached, at which time pure solid C will crystallize simultaneously and univariantly with pure solid A, both in equilibrium with liquid, along the line 2–E_1. This univariant curve of simultaneous crystallization will be the same in the composition interval A–2, independent of the original composition with which we started. The crystallization paths for both pure A and pure A and C in the two temperature-intervals just considered are shown in the T–M_x slice of Fig. 22, which reiterates part of what was already shown in Fig. 13.

Along the line 2–E_1 of Fig. 21, where the discussed univariant three-phase condition prevails, a temperature will be reached, that coincides with point E_1, below which the system is completely solid, and at which all of the remaining A and C in solution crystallizes invariantly and simultaneously with all of component B. The latter, to this point, had been confined to the liquid. The binary eutectic temperature for the section in question, as well as the ternary eutectic temperature for all sections, is shown in Fig. 22. For all compositions in the interval A–2 of Fig. 21 considered, these are seen to delineate the regions of two and

C. SIMPLE TERNARY EUTECTIC INTERACTIONS

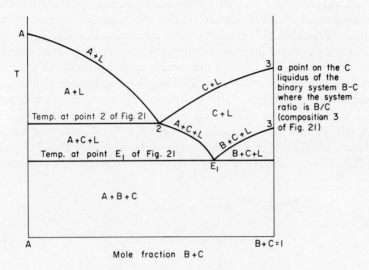

FIG. 22. The slice A–3 of Figs. 12 and 21.

three-phase coexistence. The region A–L is univariant because the B/C ratio in the liquid has been set constant by the choice of the slice A–3, and the region A–C–L is univariant because three phases coexist. For starting compositions in the interval 2–3 of Fig. 21 two different situations may prevail. For all compositions lying along 2–3 that are encompassed by the triangle e_3–E_1–C (Fig. 21), there is, following a primary crystallization of pure solid C, a univariant simultaneous crystallization of pure solids A and C in equilibrium with liquid. Starting with compositions along 2–C, the univariant A–C–L crystallization path must be the same as that attained by starting compositions along A–2, since the same three phases are in equilibrium. This univariant situation, starting at the temperature of point 2 in Figs. 21 and 22 is shown by the curve 2–E_1 in Fig. 22. The second situation that is possible in the interval 2–3 is, that following a primary crystallization of C, a simultaneous crystallization of C and B occurs. The lowest temperature for such a coexistence is again the temperature of the ternary eutectic E_1. Furthermore, this curve must intersect with the curve 2–E_1. The composition interval in which this second situation applies is contained by the triangle C–e_2–E_1 of Fig. 21. Thus, for all compositions between c and 3 on the slice A–3, a univariant three-phase equilibrium B–C–L is generated that terminates at the temperature and composition of E_1. The primary C liquidus crystallization curve as well as the B–C–L equilibrium just discussed is depicted in Fig. 22 by the curves 2–E_1

and E_1–3, respectively. For all compositions richer in A than point 2, the primary crystallization is pure A. For all compositions poorer in A, the primary crystallization is pure C. For all compositions lying between point 2 and the point of intersection of the tie line E_1–C with the line A–3, i.e., the point a of Fig. 21, the univariant equilibrium A–C–L follows the primary crystallization of A or C. For all compositions poorer in A, the univariant equilibrium B–C–L follows the primary crystallization of solid C. In an ideal system all T–M_X slices that cut through the apex A will exhibit an A liquidus curve, on a T–M_X diagram such as that shown in Fig. 22, that exactly superimposes over the A liquidus of Fig. 22. Thus, with reference to Fig. 23, the original slice considered,

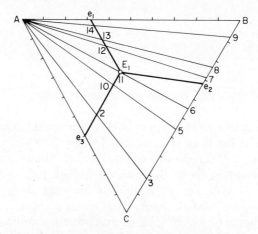

FIG. 23. Different slices intersecting the apex A.

i.e., slice A–3, will exhibit the same M_A–T graph as those for the slices A–5, A–6, A–7, etc., in the region in which pure A coexists with liquid. Three-phase coexistence will begin at lower temperatures for each of these slices and the nature of the three-phase coexistences will be of two types. Thus either A–C–L or A–B–L equilibria may be generated depending upon in which of the triangles A–e_3–E_1 or A–e_1–E_1 of Fig. 23 the slice is located. When the slice lies in the triangle A–e_3–E_1, the three-phase curve will superimpose on the curve 2–E_1 of Fig. 22, except that the latter curve will be extended to temperatures approaching that of the binary eutectic e_3 for slices lying in the triangle A–e_3–2. For a slice close to the slice A–e_3, the composition and temperature intervals of the curve A–C–L are large and the A–L liquidus extends over a composition and temperature range almost identical with its range in the binary system A–C. Figure 24 provides T–M_X graphs of

C. SIMPLE TERNARY EUTECTIC INTERACTIONS   431

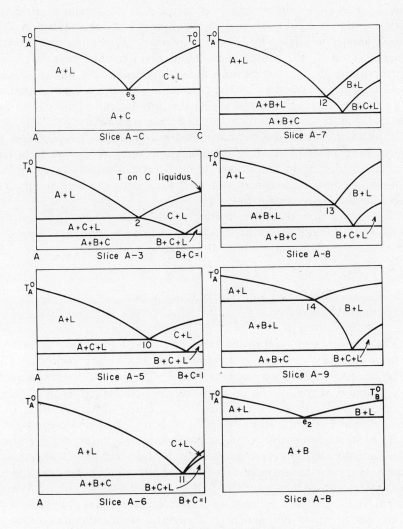

FIG. 24. T–M_X representations of the slices in Fig. 23.

the slices shown in Fig. 23. Corresponding slices may be deduced by the reader for slices through the B and C apexes.

In Fig. 23, the only slice requiring further explanation is A–6 that cuts through the ternary eutectic E_1. Following a primary crystallization of pure A in the interval A–E_1, no secondary three-phase crystallization involving two solids and a liquid occurs. Instead, the primary crystallization is followed by a simultaneous invariant crystallization of solids A, B, and C. For compositions in the interval E_1–6, following a primary crystallization of solid C, there occurs a simultaneous crystallization of B and C along the line e_2–E_1. For a point along E_1–6 lying close to the binary line B–C, the temperature along E_1–e_2 is close to that of e_2. Furthermore, the temperature on the C liquidus in the binary system B–C, i.e., that represented by the point 6, is only slightly above that of e_2 so that in the slice A–6 of Fig. 24, the liquid composition curves for the C–L and B–C–L equilibria are close to one another.

In all of the slices, the terminus, at one end of the A liquidus, is at the melting point of A, $T_A{}^0$. Excepting for slices A–C and A–B, the termini in the region of primary C or B crystallization are along the C liquidus curve in the binary system B–C or along the B crystallization curve in the binary system B–C.

Slices other than those terminating at an apex are left for the reader to deduce. It should be realized that for obvious reasons, no regions in which tie-line constructions are valid exist, in such slices.

D. Ternary Diagrams for Systems in Which Binary Congruently Melting Compounds Are Formed

The extension of treatments developed in the preceding section to cases where one or more congruently melting binary compounds is generated within the binary system A–B, A–C, or B–C, is straightforward. For example, if the system A–B gives rise to the congruently melting compound A_xB_y, it is the case that the primitive system A–B may be divided into two subsystems A–A_xB_y and B–A_xB_y. Since compound A_xB_y melts congruently, it functions as a pseudounary end-member. Consequently, when a new component C, for example, is added, we expect that two independent ternary-systems will be generated, i.e., the systems A–A_xB_y–C and B–A_xB_y–C. The possible pseudobinary system A_xB_y–C that may arise as a consequence of the addition of C to the pseudounary component A_xB_y creates a situation analogous to that in the binary system, in that two ternary subsystems are generated, each of which is separately definable, and each of which is describable in the same way as was the primitive ternary system discussed in Section C. Figure 25 depicts the projection diagram of such a ternary interaction. We see that in the pseudobinary system A_xB_y–C a new binary eutectic e_5 is generated, and that the entire ternary equilateral triangle encompasses two ternary eutectics, one within the subtriangle A–A_xB_y–C, the other within the subtriangle B–A_xB_y–C. For purposes of

D. BINARY CONGRUENTLY MELTING COMPOUNDS

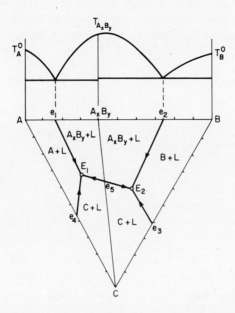

FIG. 25. A ternary interaction where in the binary system A–B, the congruently melting compound A_xB_y is generated.

clarity, the collapsed binary diagram A–B is shown along the triangle side A–B. In all ternary diagrams involving subdiagrams as in Fig. 25 (or incomplete subdiagrams when incongruent phenomena are involved as in Fig. 29 et seq.), the direction of falling temperature along three-phase lines is readily ascertained. It is always the case, as we shall see, that the temperature rises along such three-phase lines toward the line that connects the compositions of the two coexisting solids. Thus, in Fig. 25 along e_5–E_2 where A_xB_y and C coexist with liquid, the temperature rises in the direction toward the line connecting A_xB_y and C, i.e., one side of the subsystem A_xB_y–C–B.

Within the subtriangle A–A_xB_y–C, the ternary eutectic E_1 represents the lowest temperature at which a liquid may be present and similarly for the other ternary eutectic E_2. Arrows are placed on each of the univariant projected binary eutectic extensions to indicate the paths followed during cooling. It should be realized that beginning at T_C^0 for each of the binary or pseudobinary systems involving C, the C liquidus is identical in each case in an ideal interaction. The extent of stability for the C liquidi varies in each of the three binary interactions involving C since the three eutectics involved, e_3, e_4, and e_5 occur at different compositions. A more complicated version of the case in question is that where several congruently melting binary or pseudobinary compounds are generated, e.g., where along A_xB_y–C a compound such as $A_xB_yC_z$ is formed. Figure 26 shows the case where each of

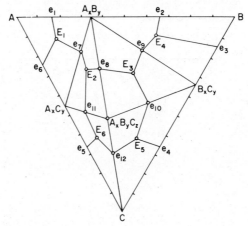

FIG. 26. The case where congruently melting binary and ternary compounds are formed.

the binary and pseudobinary interactions generates a compound. This causes the evolution of six possible ternary subsystems, each having a ternary eutectic E. Again, each of the ternary subsystems

$$A-A_xB_y-A_xC_y, \quad A_xB_y-A_xC_y-A_xB_yC_z, \quad A_xB_y-A_xB_yC_z-B_xC_y,$$

$$A_xB_y-B-B_xC_y, \quad A_xB_yC_z-B_xC_y-C, \quad \text{and} \quad A_xC_y-A_xB_yC_z-C$$

may be described in the same way as the simple system discussed in Section C.

A simpler case than that depicted in Fig. 26 is one in which the binary compounds A_xB_y, B_xC_y, and A_xC_y are not formed, but in which a ternary compound $A_xB_yC_y$ alone is generated. This compound may form a pseudobinary interaction with each of the unary end members A, B, and C and generate three subsystems as shown in Fig. 27.

A complication arises when there exists the possibility of mutually exclusive binary interactions among pseudounary and unary components. Here, experimental resolution is necessary in order to determine what situation exists since a qualitative prediction is impossible. Consider, for example, the case shown in Fig. 28 where each of the binary diagrams gives rise to a congruently melting compound. It is seen that a number of possible pseudobinary interactions among these pseudounary

D. BINARY CONGRUENTLY MELTING COMPOUNDS

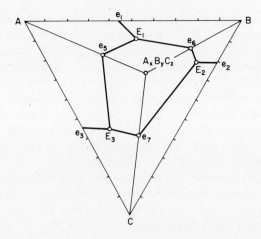

FIG. 27. A ternary congruently melting compound is formed.

components is possible with each other, and with the unary end members. These possibilities are shown by the dotted and solid lines in Fig. 28. For clarity, projections of binary eutectic extensions as well as the intersections of these at ternary eutectics have been omitted. If, for example, the ternary subsystem A_xB_y–A_xC_y–B_xC_y exists, this precludes the existence of the ternary subsystems A–A_xC_y–B_xC_y, A_xC_y–A_xB_y–C, A_xB_y–C–B_xC_y, B–A_xC_y–B_xC_y, etc. We see by inspection of Fig. 27 that delineation of which method of triangle division is applicable may

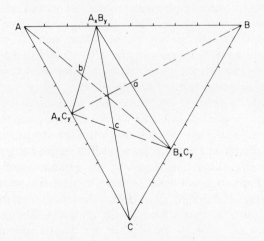

FIG. 28. A case where qualitative prediction of the nature of the ternary system cannot be made without further information.

be made experimentally, relatively simply. Note that existence of the pseudobinary system A_xB_y–C precludes the existence of the pseudobinary systems A_xC_y–B_xC_y, A_xC_y–B, and B_xC_y–A. If a composition of point C is examined crystallographically, it should consist of crystals of A_xB_y and C, if the system A_xB_y–C exists. If these two phases are not found, then at this composition one should detect a pseudobinary eutectic-type solid mixture of A_xC_y and B_xC_y or some other mixture. Similar considerations are applicable to the points a and b of Fig. 27. We assume, of course, that no complications other than the one being considered, exist for the system in question.

The number of possible diagrams about which we may speculate and in which only simple eutectic interactions occur is infinite and need not be entertained further. In all cases these involve a trivial extension of what has been described previously.

E. Ternary Diagrams for Systems in Which Binary Congruently Melting Compounds Occur, One of Whose Eutectic Extensions Generates a Ternary Peritectic

In the preceding section, all of the cases described had in common the fact that the binary compounds which formed generated pseudobinary interactions with the third component, or with another pseudobinary component. The situation may arise, however, where such a binary congruently melting compound does not participate in a pseudobinary interaction with one of the original ternary components or a compound derived from among them. That such a situation could arise might have been anticipated from what was stated in conjunction with Fig. 27 where certain pseudobinary interactions were mutually exclusive of other pseudobinary interactions.

For the cases treated in the preceding section, the intersection of binary eutectic extensions always lay within a ternary subsystem triangle. If, however, such extensions do not lie within the appropriate subsystem triangle, then clearly one ternary eutectic that might have been expected cannot be present. When this happens, the generation of two potential ternary subsystems is precluded. Concomitantly, the potential pseudobinary interaction that might have been expected is also precluded.

Figure 29 shows the case where, in the system A–B–C, a congruently melting compound A_xB_y is formed in the binary system A–B. The binary eutectic extensions e_1–P_1 and e_4–P_1 intersect outside of the triangle A–A_xB_y–C at the point P_1. Consequently, the subtriangle

E. TERNARY PERITECTIC FORMATION

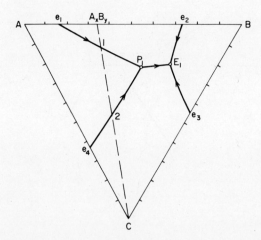

FIG. 29. The eutectic extension on one side of a binary compound partakes of a peritectic interaction on the ternary system.

$A-A_xB_y-C$ cannot be treated as a separate ternary subsystem, nor can the join A_xB_y-C be treated as a pseudobinary system since within the composition range 1–2, a phase other than A_xB_y or C (namely solid A) makes an appearance. The binary eutectic extensions of the binary eutectics e_2 and e_3 cannot terminate at P_1 since the total number of phases, excluding the vapor phase, implied at this point would number 5, a violation of the phase rule. Therefore, the extensions of e_2 and e_3 must be at another invariant point, specifically at the ternary eutectic E_1.

The point P_1 is a ternary peritectic at which three solid phases and a liquid may coexist. In order for the temperature to drop below that uniquely defined under an isobaric constraint at this point, one of the phases must vanish. Once this happens, the system becomes univariant again. By analogy with the binary peritectic phenomenon, one of two occurrences is possible. Either all of the liquid will react with some or all of a previously crystallized phase to form a new solid phase or mixture of solid phases, or it will partially react with all of the previously crystallized phase to form the new phase. In the former case the liquid will disappear. In the latter case, the previously crystallized phase will disappear. Whichever of the two phenomena ensues at the ternary peritectic depends on the starting composition of the ternary system.

The sequences of crystallization paths for the above as well as other starting compositions is described in the following:

Figure 30 is concerned with crystallizations that occur with the primary deposition of solid A. Within the projected surface $A-e_1-P_1-e_4$, and in a fashion similar to that considered for uncomplicated ternary

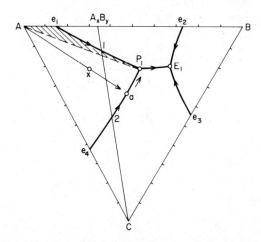

FIG. 30. The region of primary crystallization of A.

eutectic crystallizations, the first crystals to be deposited when a liquid is cooled comprise pure solid A. For all starting compositions within the shaded region A–e_1–P_1, solid A will continue to crystallize until the liquid becomes saturated with the compound A_xB_y at a composition along the line e_1–P_1. When any such composition is reached, A_xB_y will coprecipitate with A and the liquid composition will move toward P_1 with decreasing temperature. As previously stated, such a three-phase coexistence in an isobarically constrained ternary system is univariant.

For any composition within the region A–P_1–e_4, the primary crystallization will again be pure solid A. Now, however, the direction of crystallization is such that the liquid composition moves toward the line e_4–P_1 rather than toward e_1–P_1. When the line e_4–P_1 is contacted, the liquid has become saturated with respect to solid C and further cooling is attended by the coprecipitation of pure solids A and C. During either of these univariant coprecipitation processes, the liquid composition moves toward P_1. Upon encountering P_1, a peritectic reaction occurs in which either all the liquid is consumed and three solids are left or in which some of the liquid is consumed and all of the solid A that has previously crystallized reacts with the liquid to form A_xB_y and C. For all starting compositions within the region A–1–2–e_4, it is the liquid phase that disappears at P_1. For all compositions within the region 1–2–P_1 it is the previously crystallized phase A that disappears.

To demonstrate these two cases, consider first a composition x within the region A–1–2–e_4 of Fig. 30. Upon encountering the A

E. TERNARY PERITECTIC FORMATION

crystallization surface, solid A will begin to crystallize and the liquid composition will move toward the line e_4–P_1. The tie line showing solid and liquid overall composition will, as before, always pass through point x, the starting composition. In Fig. 30, the terminus of this tie line is the apex A at one end and is the point a at the other end. Upon further cooling, the liquid composition achieves the value of point a on the binary eutectic extension e_4–P_1 and the liquid becomes saturated with respect to solid C. Simultaneously then, solids A and C crystallize while the liquid composition moves from point a to P_1 along the line e_4–P_1. Since the solid phase is no longer made up of pure A, the terminus of the tie line at its other end cannot be the apex A but must be a point along the binary line A–C. As this tie line moves counterclockwise it does so by pivoting on the point x as shown in Fig. 31. The area swept

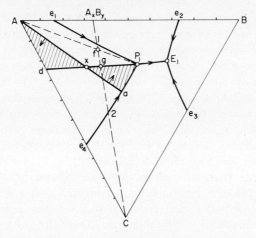

FIG. 31. The region of primary crystallization of A following intersection of the line e_4–P_1 by the line A–a.

out during the entire crystallization process that has been going on up to the time the liquid achieves a composition P_1 is shown by the shaded area in Fig. 31.

At this time, the total solid composition is given by the point d on the binary join A–B. The ratio of solid A to solid C is Cd/Ad and the ratio of solid-to-liquid is xP_1/xd (assuming concentrations are expressed in mass fractions).

A liquid of composition P_1 in equilibrium with crystalline A in the proportions A–f/P_1–f has a net overall composition f in Fig. 31. This hypothetical mixture could have been synthesized from A_xB_y and C in the ratio $A_xB_y/C = (C-f)/(A_xB_y-f)$. In other words, had we started with an initial system composition at point f we could express this composition in terms of either solid A–liquid P_1 or in terms of the stoichiometry A_xB_y–C. In our case, where the quantity of A and liquid are not, in general, in the correct proportions to yield a net composition coinciding with point f, only sufficient A and liquid will participate to be consistent with the tie-arm lengths required. At this point P_1, therefore, we have the invariant intersection of three surfaces, i.e., those for primary crystallization of A, C, and A_xB_y. Up to this point only A and C have crystallized. At point C, A_xB_y must begin to crystallize. It does so via the reaction of all of the remaining liquid with enough of the previously crystallized A so that a ratio of A_xB_y to C results which coincides with the point f. In order for the process to proceed, in which A reacts with liquid to form A_xB_y and additional C other than the amount of C previously crystallized, the quantity of liquid phase must obviously decrease. The liquid-to-solid ratio is dx/xP_1 at P_1, the overall solid composition being given by the terminus of the line xd at point d. During the peritectic reaction, to generate an A_xB_y/C ratio specified by the point f, the total solids composition moves from d toward x, and as the arm d–x becomes smaller, the liquid-to-solid ratio decreases. Finally, when the total solid composition achieves the value at point x, the liquid-to-solid ratio is zero. Since the final composition does not lie on A_xB_y–C, namely, the point g, it is clear that not all of the solid A has been reabsorbed in the peritectic reaction and we are left with a mixture of A, A_xB_y, and C.

To summarize then, for compositions lying within the region A, 1–2, e_4 of Fig. 29, the final result of a crystallization sequence is the disappearance of liquid at the ternary peritectic point with reabsorption of some of the previously crystallized A. The solid phases that coexist following the peritectic reaction are A, A_xB_y, and C independent of which of the extensions e_1–P or e_4–P is intersected following the primary crystallization of A.

The second general possibility for crystallization within the primary field of crystallization of solid A is for compositions lying within the region bounded by 1–P_1–2 in Fig. 29. As in the case just discussed, following the primary crystallization of A, one or the other of the binary eutectic extensions, e_1–P_1 or e_4–P_1, will be intersected after the temperature is lowered sufficiently, depending upon whether the starting composition lies within the shaded or unshaded portions of the A field, as shown in Fig. 31. As before, the only difference encountered is that, in one instance solid A_xB_y coprecipitates with A, and in the other instances solid C coprecipitates with A. In either event, the liquid composition moves toward the point P_1 while the solid composition

E. TERNARY PERITECTIC FORMATION

moves along A–B or A–C. As a case in point, let us examine a crystallization sequence in which the three-phase univariance lies along e_4–P_1 as it would be starting with the composition y in Fig. 32. Upon encoun-

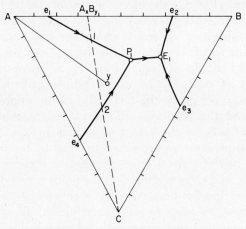

FIG. 32. Crystallization paths for a starting composition lying within the region 1–P_1–2 at the temperature of contact with the A surface.

tering the A surface as the temperature is cooled, a liquid of composition y comes into equilibrium with a solid of composition A. Upon continued cooling the liquid composition moves from y toward its intersection with e_4–P_1 at the point a, Fig. 33. When this latter point is attained, the system exhibits a solid-to-liquid ratio $(y-a)/(y-A)$ and becomes

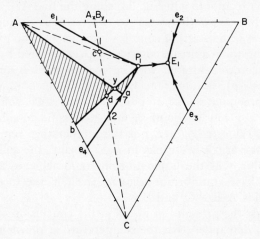

FIG. 33. The crystallization path for the starting case depicted in Fig. 31.

saturated with solid C. Further cooling results in the univariant coprecipitation of A and C as the liquid composition moves toward P_1, and the total solid composition moves toward b on the binary join A–C. The shaded area (Fig. 33) shows the sweep of the tie lines around the pivot point y during this process. When the point P_1 is reached, we have a liquid of composition P_1 and a two-phase solid containing pure A and pure C. The A solid present in the mixture can react in appropriate proportions with liquid of composition P_1 as previously described to give a mixture of A_xB_y and C in the ratio specified by the point c. As this reaction takes place, the total solid composition moves along bd from b to d and the liquid-to-solid ratio, as represented initially by $(b-d)/dP_1$, decreases as the arm $b-d$ shrinks in size. When the total solid concentration achieves the value specified by the point d, this total solid content being derived from previously crystallized C and the mixture of A_xB_y and C coinciding with the point c, we note the following: The liquid phase has not been completely consumed, the liquid-to-solid ratio being given by $(d-y)/(y-P_1)$. In addition, all of the previously crystallized A phase has been consumed, the solids containing only A_xB_y and C. The four-phase isobarically invariant condition that prevailed once the liquid attained the composition represented by point P_1, has been replaced with the disappearance of solid A by a three-phase univariant equilibrium. The temperature is now free to drop again, and as the temperature begins to drop, the liquid composition moves from P_1 toward E_1 along the line P_1–E_1. During this process, solid A_xB_y coprecipitates with solid C and the solid composition which at the conclusion of the peritectic reaction had the composition d of Fig. 33, moves from d to e in Fig. 34, along the line A_xB_y–C. In Fig. 34, the shaded areas show the sweep of the tie lines during this process, where, as always, the starting composition y acts as the pivot point.

When the liquid acquires a composition represented by the point E_1, the liquid becomes saturated with respect to solid B and, as a consequence, a four-phase invariant condition is once more established. As solid B precipitates out at point E_1, the total solid composition moves along $e-y$ toward point y, and the liquid-to-solid ratio which exhibited the value $(e-y)/(y-E_1)$ at the temperature represented by E_1 decreases as the arm $e-y$ shrinks in size. Finally, the solid composition achieves a value y as the liquid-to-solid ratio achieves a value of zero and a three-phase equilibrium reasserts itself in the form of the three coexisting solids A_xB_y, C, and B.

We note that the presence of liquid at the conclusion of the peritectic reaction at P_1 demanded that the temperature at the point E_1 be lower than the temperature at P_1, a fact not obvious from treating a starting

E. TERNARY PERITECTIC FORMATION

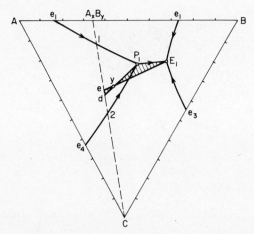

FIG. 34. The crystallization path for the composition of Figs. 31 and 32 following the peritectic reaction point P_1.

composition within the region A–e_1–1–2–e_4. It should have been previously evident that the temperatures at e_1 and e_4 must be higher than the temperature at P_1 because of the presence of a liquid phase at least up to contact with the point P_1. Similarly, e_2 and e_3 must lie at higher temperatures than E_1 in order that the liquid phase in the regions with which they are concerned be able to disappear. e_2 and e_3 are not constrained to obey such a relationship with P_1 since no starting compositions that yield liquids whose compositions lie along e_2–E_1 or e_3–E_1 become involved with the invariance at point P_1.

Starting compositions lying within the primary crystallization fields for either A_xB_y or C proceed by one of three generally analogous paths so that consideration of one of these primary fields will suffice to define by analogy what occurs in the other. We will, for our example, choose the A_xB_y field as a point of focus. In Fig. 35, this field is divided into three regions, a shaded unhatched region A_xB_y–E_1–P_1. An unshaded region A_xB_y–P_1–E_1 and a shaded cross-hatched region A_xB_y–E_1–e_2. The latter two regions are without complexity in that all starting compositions lying within these fields give rise to a single invariant condition upon cooling, that at the ternary eutectic E_1. The difference between compositions in these two regions is that in one case, following primary crystallization of A_xB_y, a coprecipitation of C occurs (for those compositions within A_xB_y–P_1–E_1) and in the other case coprecipitation of B occurs (for those compositions within A_xB_y–E_1–e_2). In either event crystallization paths never become involved in peritectic phenomena and the system appears as a simple ternary eutectic case. For composi-

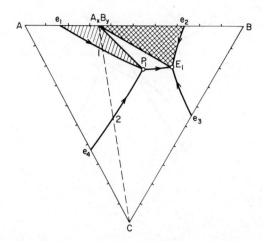

FIG. 35. The three composition intervals in the A_xB_y primary field of crystallization.

tions within the region A_xB_y–e_1–P_1, the peritectic reaction at P_1 is always encountered since liquid compositions along e_1–P_1 are always generated.

Furthermore, within the region A_xB_y–e_1–P_1 two types of peritectic behavior are observable, as might have been anticipated from our previous discussion. Thus, in the region e_1–1–A_xB_y all existing liquid is consumed at P_1 and some of the coprecipitated A along e_1–P_1 is reabsorbed. In the region A_xB_y–1–P_1 all of the coprecipitated A is reabsorbed and some liquid remains following the peritectic reaction. For compositions within this latter region then, the system undergoes two invariant phenomena during crystallization. In Fig. 36, a cooling cycle is shown for a composition z lying within the region e_1–A_xB_y–1. Cooling a melt of composition z results in liquid compositions that initially move along z–a toward a, and a solid of composition A_xB_y. Upon encountering e_1–P_1, the liquid becomes saturated with respect to solid A, and once this occurs, A_xB_y and A coprecipitate simultaneously. The liquid composition moves along a–P_1 toward P_1, and the solid composition moves toward b from A_xB_y along the line A_xB_y–b. The region swept out by the tie lines is shown as a shaded area. When the liquid achieves a composition P_1, the total solid composition has the value b. The liquid becomes saturated with solid C at this point and the liquid of composition P_1 is able to react with solid A in the proportions $(4\text{–}A)/(P_1\text{–}4)$ to form a mixture of A_xB_y and C having the ratio consistent with point 4. The liquid-to-solid ratio z–b/z–P_1 at P_1 decreases during the peritectic reaction, the arm z–b decreasing in size during

E. TERNARY PERITECTIC FORMATION

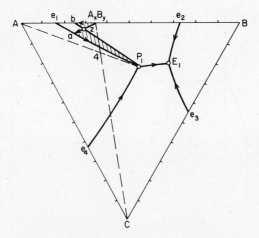

Fig. 36. Crystallization of a composition z.

the process, and the solid composition moves from b toward z along b–z. When the solid acquires the composition at point z the liquid-to-solid ratio equals zero, and the resulting all-solid system containing solids A, A_xB_y, and C is univariant.

One point in the above discussion may be emphasized. It has been pointed out in the treatments of the peritectic cases that a liquid of the peritectic composition P_1 reacts with a solid A in certain fixed proportions to give a mixture of solids A_xB_y and C defined by the intersection of a tie line constructed between P_1 and A with the line A_xB_y–C. Since, in all of these cases other solid in addition to A, has crystallized, the final outcome of the peritectic reaction is never a solid represented by this intersection point. Consequently, except for special cases, either excess A or excess liquid will remain. On the other hand, during the peritectic process just discussed, where all the liquid is consumed, it is clear that the point of origin is not one particular phase of solid, but rather the solid composition as a whole. Thus, liquid of composition P_1 reacts with the solid A contained in a solid whose total composition is b (Fig. 36) to give a solid of total composition z (the starting composition). This relationship involves all of the solid so far as its composition is concerned, and all of the liquid as opposed to just a portion of one or the other.

Compositions lying within A_xB_y–1–P_1 in Fig. 35, are again not unlike a previously treated case as to crystallization sequence and the processes in this region are left as an exercise for the reader to deduce.

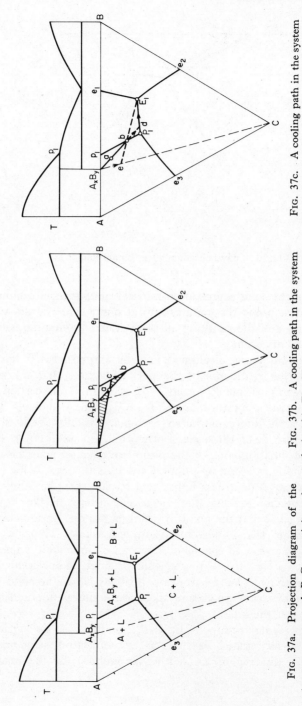

Fig. 37a. Projection diagram of the ternary system A–B–C containing the incongruently melting binary compound A_xB_y.

Fig. 37b. A cooling path in the system depicted in Fig. 37a.

Fig. 37c. A cooling path in the system depicted in Fig 37a, to a lower temperature than shown in Fig. 37b.

F. Ternary Systems in Which Incongruently Melting Binary Compounds Are Present

In Section E, the formation of a ternary peritectic at the intersection of two binary eutectic extensions was considered. In this section, a similar case will be examined briefly, that in which a peritectic is already present in a binary system that forms one side of the ternary triangle.

Figure 37a shows a collapsed diagram of the system A–B in which the compound A_xB_y melts incongruently at p_1. The ternary projected diagram showing the presence of the ternary peritectic P_1 is also depicted. We immediately see that the ternary diagram is almost identical to that considered in the preceding section, and it is a simple task to demonstrate that the nature of each of the several crystallization paths for different starting compositions is quite similar to those treated in Section E with the exception of starting compositions lying within the triangle A_xB_y–B–C where solid A is the primary solid of crystallization, i.e., point a of Fig. 37b. If we start with a liquid of composition a in Fig. 37b, the first solid that crystallizes is pure A. The liquid composition moves from point a to point c upon further cooling. When point c is reached, the binary compound A_xB_y starts to crystallize at the expense of previously crystallized A. While the liquid composition follows c–b during this peritectic reaction, the overall solid composition moves along A–A_xB_y toward A_xB_y. When the liquid achieves a composition b we note that all of the solid A originally crystallized has been reabsorbed, and only a single solid having the composition A_xB_y coexists with the liquid of composition b. At this point, the univariant nature of the system gives way to a bivariant state since only two phases are present. The cooling path, as a consequence, is no longer constrained to the three-phase line p_1–P_1, and the liquid composition moves from the point b toward the point d along the straight line A_xB_y–d, in Fig. 37c. When the liquid composition achieves the value at point d, the liquid is saturated with respect to pure C, and the three-phase equilibrium A_xB_y–C–L obtains. The liquid composition now moves from d to E_1, while the overall solid composition moves from A_xB_y toward C along the line A_xB_y–C. Finally, when the liquid achieves the composition E_1 and the solid achieves the composition e, the system becomes invariant due to the appearance of solid B. The remaining liquid solidifies at this temperature resulting ultimately in a mixture of A_xB_y, B, and C.

G. TERNARY COMPOUND FORMATION—CONGRUENTLY MELTING TERNARY COMPOUNDS

Figure 38 shows the projected ternary diagram A–B–C in which a congruently melting ternary compound $A_xB_yC_z$ is generated. This results in the formation of three ternary subsystems, i.e., $A-A_xB_yC_z-C$,

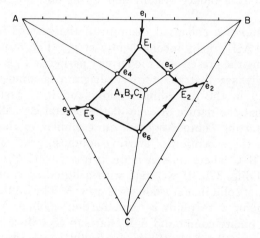

FIG. 38. The ternary compound $A_xB_yC_z$ in the system A–B–C.

$B-A_xB_yC_z-C$, and $A-A_xB_yC_z-B$, each of which may be treated separately. Three ternary eutectics are generated as are three additional binary eutectics. The direction of falling temperature is shown by the arrows on each of the univariant three-phase lines. Further comment is unnecessary.

H. TERNARY COMPOUND FORMATION—AN INCONGRUENTLY MELTING TERNARY COMPOUND IS PRESENT

If an incongruently melting ternary compound is generated, two general phenomena may be observed depending upon the location of the ternary invariant points. One of these, an extreme case, shown in Fig. 39a, exhibits an invariant point in each of the subtriangles, not unlike the case represented in Fig. 38. Unlike the case where the ternary compound melts congruently, the diagram (Fig. 39a) cannot be treated as three separate subsystems since there exists a composition region within the triangle $A_xB_yC_z-B-C$ where pure A is the primary crystallization. It should be noted in comparing Figs. 38 and 39a,

H. INCONGRUENTLY MELTING TERNARY COMPOUNDS

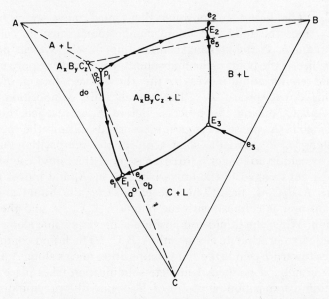

FIG. 39a. The incongruently melting ternary compound $A_xB_yC_z$.

that one of the interior pseudobinary eutectic points present in Fig. 38 is absent in Fig. 39a, giving way to the peritectic point p_1. As stated in Section D, the direction of rising temperature along three-phase lines is easily ascertained by determining in which direction we must move along a three-phase line to reach the tie line (a triangle edge) connecting the compositions of the solid phases coexisting along the three-phase line. That this is true can be deduced from the following: Consider the starting composition point a in the field of primary crystallization of C (Fig. 39a). Upon encountering the C liquidus surface, and after further cooling, a simultaneous crystallization of $A_xB_yC_z$ and C must take place when the liquid acquires a composition along E_1–e_4. The solid now is no longer pure C as it was during the initial stages of primary crystallization. In fact, it becomes a mixture of $A_xB_yC_z$ and C, the liquid composition line rotating along e_1–E_1 counterclockwise toward E_1, while the solid composition moves along C–$A_xB_yC_z$ toward $A_xB_yC_z$. Note that if we had assumed that the liquid composition moved away from E_1 toward e_4, the solid composition would move along A–C. Since it is $A_xB_yC_z$ that must coprecipitate with C along E_1–E_3, the direction of falling temperature must be toward E_1. On the other hand, consider a starting composition b in Fig. 39a. Upon encountering the curve e_4–E_3, the solid $A_xB_yC_z$ begins to coprecipitate. The solid composition must again move away from point C and must move along

$C-A_xB_yC_z$. The liquid composition then must move away from e_4 toward E_3 since the solid and liquid compositions must lie colinearly with the point b. If the starting composition for the system depicted in Fig. 39a lies at the point c, the first solid to precipitate is pure solid A. The liquid composition moves to the three-phase line p_1-E_1, and follows this down toward E_1. The overall solid composition moves along $A-A_xB_yC_z$ during the process. Before E_1 is reached, however, all of the solid-A phase becomes reabsorbed during the peritectic process along p_1-E_1, and the solid acquires the composition $A_xB_yC_z$. Since the system is no longer a three-phase one, the liquid cooling path is no longer constrained to the univariant line p_1-E_1 and moves instead across the $A_xB_yC_z-L$ field. The line connecting the liquid and solid composition now is colinear with the points $A_xB_yC_z$, c, and the liquid composition. When the liquid composition achieves a value along e_4-E_3, solid C begins to precipitate along with $A_xB_yC_z$. The liquid composition then moves toward E_3 while the solid composition moves along $A_xB_yC_z-C$ toward C. Finally, at point E_3 complete solidification takes place.

A starting composition at point d also initially precipitates pure solid A. When the temperature has dropped sufficiently, the liquid achieves a composition along p_1-E_1, and $A_xB_yC_z$ begins to precipitate. The overall solid composition moves along $A-A_xB_yC_z$, while the liquid composition moves toward E_1. Before all of the A phase can be

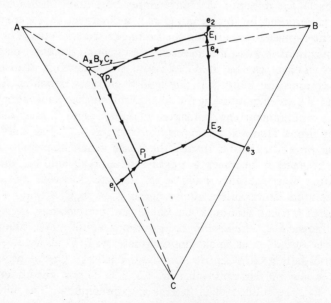

FIG. 39b. The incongruently melting ternary compound $A_xB_yC_z$.

I. ISOTHERMAL SECTIONS FOR TERNARY SYSTEMS

reabsorbed, the liquid composition achieves the value E_1 at which complete solidification takes place.

Thus, we can see that for starting compositions within the triangle $A_xB_yC_z$–C–B, whether A or $A_xB_yC_z$ is the primary crystallization, the invariant point E_3 is the terminus of liquid compositions. For all compositions within the triangle A–$A_xB_yC_z$–C or A–$A_xB_yC_z$–B, the invariant points E_1 or E_2 are the termini of liquid compositions.

The second general case for systems exhibiting an incongruently melting ternary compound is presented in Fig. 39b. It is seen that one of the invariant points present in Fig. 39a lies within the triangle $A_xB_yC_z$–B–C. An extension of this case is that where all three invariant points lie within this triangle. For the case shown, P_1 does not represent a ternary eutectic but rather a ternary peritectic. It is always the case that, in a single triangle only one ternary eutectic is possible. For all starting compositions lying within the triangle A–$A_xB_yC_z$–C, P_1 represents a final solidification point as may be deduced by the reader, e.g., consider a cooling path for the point a in Fig. 39b. For all compositions within $A_xB_yC_z$–B–C the point P_1, if it is encountered, represents the place where crystallized A is completely reabsorbed. Such compositions do not solidify completely until the liquid achieves a composition at E_2. Since no new phenomena arise in association with cooling paths in Fig. 39b, we leave to the reader the exercise of deducing them.

I. Isothermal Sections for Selected Ternary Systems

While the polythermal projections utilized in Sections D–H are of considerable value in presenting an overall picture, they represent an optimistic approach relative to the knowledge available to us. Frequently, all that are available to the experimenter are the data for one or more isotherms. It is of interest, then, to examine briefly some selected sections through the right prism as was done for the simple case treated in Section C.

1. *Isothermal Sections for Systems Forming Binary Compounds*

Since ternary systems containing one or more binary compounds may be divided into subsystems, each of which may be treated separately, isothermal sections represent a combining of the sections of two or more subsystems. Figure 40, for example, shows a combination of two isothermal sections through Fig. 25 at a temperature just above both ternary eutectic temperatures. The liquid field in the triangle A–A_xB_y–C is depicted as smaller than that remaining in the triangle

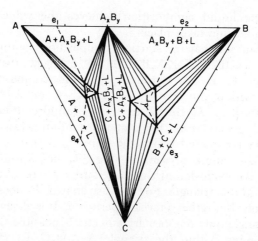

FIG. 40. An isotherm through the diagram of Fig. 25.

A_xB_y–B–C. This is indicative of a sequence of compound melting temperatures $T_A^0 > T_{A_xB_y}^0 > T_B^0 > T_C^0$.

2. *Isothermal Sections for Systems Forming Binary Compounds and Ternary Peritectics*

Figures 41a–e show a sequence of isothermal sections for a system like that discussed in Section E and shown in Fig. 29. For simplicity it has been assumed that $T_A^0 > T_{A_xB_y}^0 > T_B^0 > T_C^0$. Figure 41a shows the intersection of the isothermal plane with only the A liquidus surface. Figure 41b shows the intersection with the A, A_xB_y, B, and C surfaces below the A–A_xB_y–L univariant eutectic extension, but above the others. Figure 41c shows the intersection at a temperature below all of these eutectic extensions, but above both the peritectic and ternary eutectic points. Because the diagram is not divisable into subsystems, Fig. 41c shows a liquid field bounded tetragonally rather than triangularly as in Fig. 40. Figure 41d shows a section at a slightly lower temperature than that in Fig. 41c, just above the ternary peritectic point P_1. Finally, in Fig. 41e, a section just above the ternary eutectic temperature is depicted. It is seen that only three two-phase fields are now present.

3. *Isothermal Sections through a Ternary System in Which an Incongruently Melting Binary Compound Is Present*

The system in question is that described in Section F and depicted in Fig. 37. Of interest is the fact that for all starting compositions

I. ISOTHERMAL SECTIONS FOR TERNARY SYSTEMS

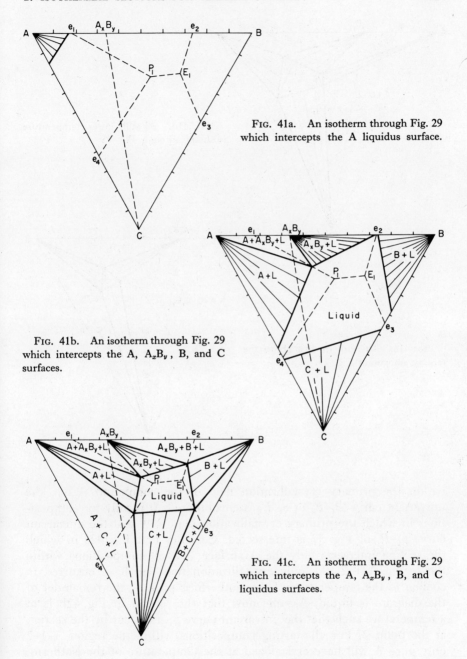

Fig. 41a. An isotherm through Fig. 29 which intercepts the A liquidus surface.

Fig. 41b. An isotherm through Fig. 29 which intercepts the A, A_xB_y, B, and C surfaces.

Fig. 41c. An isotherm through Fig. 29 which intercepts the A, A_xB_y, B, and C liquidus surfaces.

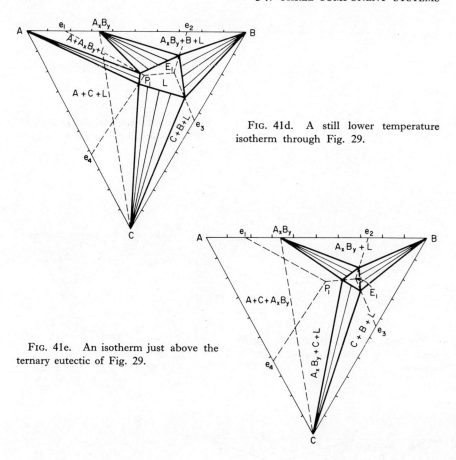

Fig. 41d. A still lower temperature isotherm through Fig. 29.

Fig. 41e. An isotherm just above the ternary eutectic of Fig. 29.

within the primary crystallization field of the compound A_xB_y, the univariant curve P_1-E_1 or e_1-E_1 is encountered. It is only for compositions in which the primary crystallization is pure A that the univariant curve p_1-P_1 of Fig. 37 is intersected. Consider first Fig. 42a in which the section intersects only the A surface. For all compositions within the region A–1–2, the only crystallization that will have occurred in cooling to the temperature of the isotherm is pure A; the remainder of the diagram is liquid. Assume now that the section in Fig. 42b is at a temperature such that the univariant curve p_1-P_1 is cut by the section at the point 2. For all starting compositions within the region A–1–2 only pure A will have crystallized at the temperature of the isotherm. For all compositions within the region p_1–2–3 only A_xB_y will have crystallized. Note the dotted-line extensions in this region to the A_xB_y

I. ISOTHERMAL SECTIONS FOR TERNARY SYSTEMS

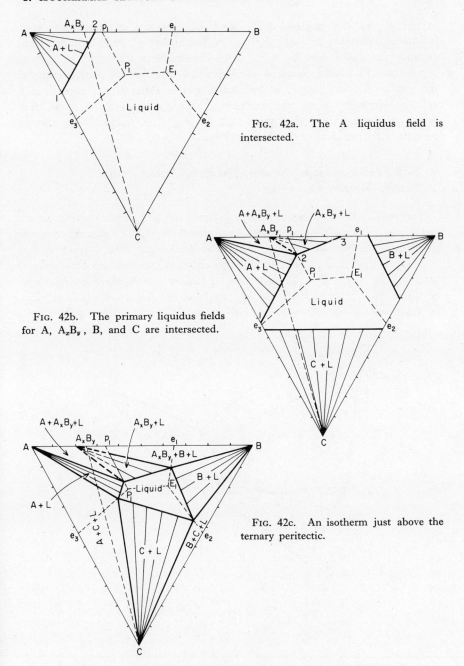

FIG. 42a. The A liquidus field is intersected.

FIG. 42b. The primary liquidus fields for A, A_xB_y, B, and C are intersected.

FIG. 42c. An isotherm just above the ternary peritectic.

FIG. 42. Sections through a system exhibiting a binary peritectic.

terminus. In the region A_xB_y–p_1–2 all starting compositions will, following a primary crystallization of A, become involved in a three-phase equilibrium involving A, A_xB_y, and liquid. Similarly, the A–A_xB_y–2 region would also exhibit a three-phase equilibrium. Consequently, the entire region A–p_1–2 is one involving a three-phase equilibrium at the temperature of the isotherm. Figure 42c shows a somewhat lower temperature section than that of Fig. 42b. Lower temperature sections become similar to those in Fig. 41.

4. *Isothermal Sections for Systems Exhibiting Incongruently Melting Ternary Compounds*

Figure 43 shows a perspective drawing of a system exhibiting an incongruently melting ternary compound $A_xB_yC_z$. We have seen previously in Section H, Fig. 39a, that if the nonbinary-system axes $A_xB_yC_z$–C and $A_xB_yC_z$–B cut through either of the ternary three-phase lines E_1–E_3 or E_2–E_3, the direction of falling temperature on one side of these axes is toward E_1 or E_2, and on the other side toward E_3. A case not to be considered by us is that where the subtriangle

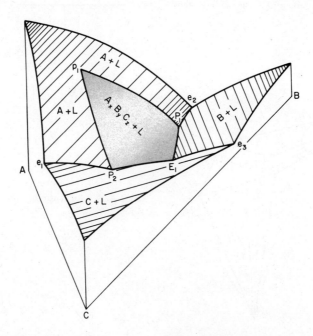

FIG. 43. Perspective view of a ternary system containing an incongruently melting compound and a single ternary eutectic.

I. ISOTHERMAL SECTIONS FOR TERNARY SYSTEMS

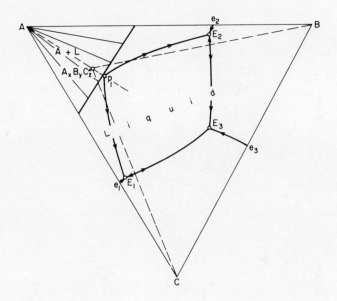

FIG. 44a. Section through the system A–B–C in which the incongruently melting compound $A_xB_yC_z$ is formed.

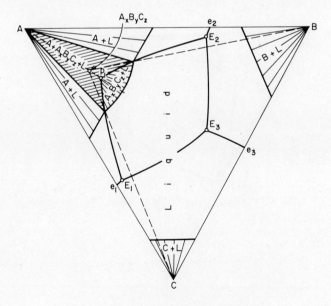

FIG. 44b. A lower temperature section.

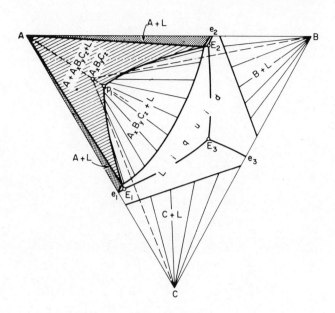

Fig. 44c. A still lower temperature section.

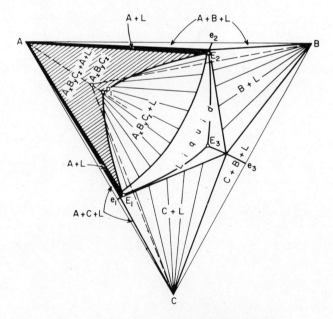

Fig. 44d. A still lower temperature section.

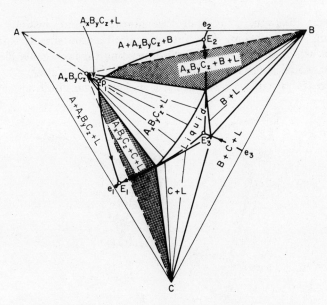

Fig. 44e. A section just above the ternary eutectic E_1.

$A_xB_yC_z$–B–C encompasses two or three of the quadruple points as contrasted to the case treated where E_1 and E_2 lie outside this subtriangle.

The latter type of system, as we have seen in Section H, always shows decreasing temperatures from E_1 or E_2 toward point E_3. Consequently, E_1 or E_2 must lie at higher temperatures than E_3. In the system we have treated in Section H, Fig. 39a, the maximum temperatures along E_1–E_3 and E_2–E_3 lie at the intersection of these three-phase lines with the axes $A_xB_yC_z$–C and $A_xB_yC_z$–B, respectively.

Figures 44a–e show a series of isotherms through the incongruent system in question. The figures are self-explanatory in view of previous discussions (Section H) and need not be described individually. It is to be noted that the tie lines within the field of primary crystallization of $A_xB_yC_z$ terminate at the three-phase lines E_1–p_1–E_2, since the A field is not encountered. The direction of falling temperature along three-phase lines is shown only in Fig. 44a.

J. Systems Exhibiting Solid Solubility

Let us consider the case where, of the three components, two are completely miscible in all proportions in the liquid and solid state, while the third is immiscible with these two in the solid state. Thus, the binary diagrams consist of two simple eutectic systems and one ascending solid solution. The limiting state for the eutectic interactions is that in which there is no solid solubility, but the more general condition is one in which a certain degree of solid solubility occurs.

Figure 45 provides a perspective view of the limiting state. The

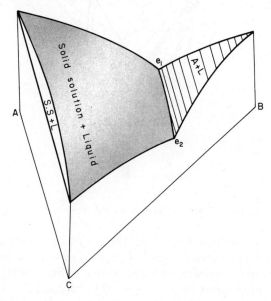

FIG. 45. The ternary system A–B–C where A–C forms a continuous series of solid solutions and the binary systems A–B and B–C are of the simple eutectic variety.

liquidus surface extending from the solid-solution binary diagram A–C terminates at the three-phase line e_1–e_2. The line e_1–e_2 extends from the binary eutectic in the system A–B to the binary eutectic in the system C–B. Since B is insoluble in the solid solution A–C, the solidus line in the binary system A–C has no extension into the ternary triangle. Consequently, all tie lines between the ternary liquid phase and the solid phase coexisting with it in the region A–e_1–e_2–C have as their termini the binary side A–C and the liquidus surface in the interior of the ternary triangle. In the region e_1–B–e_2, the ternary liquid is in equilibrium with pure solid B so that tie lines in this region all extend from the apex B to the B liquidus surface.

Notably then, two features serve to distinguish the ternary system under discussion from those that we have previously described. Since only two solids, one a solid solution and the other a pure phase, are ever present, no invariant four-phase equilibrium may occur. Secondly,

J. SYSTEMS EXHIBITING SOLID SOLUBILITY

neither the A nor C end-members have regions of primary crystallization. Of practical significance is the fact that the termini of tie lines on the binary side A–C cannot be predicted in advance by simple extension of tie lines to an apex, such extensions being dependent on the exact shapes of the solidus and liquidus curves in the binary A–C diagram.

Figure 46 shows a polythermal projection of the system shown in

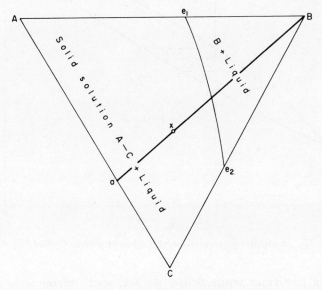

FIG. 46. The polythermal projection of Fig. 45.

Fig. 45. For all discussions, we will assume that $T_A^0 > T_C^0$ and $T_{e_1} > T_{e_2}$. Consider a starting composition x lying within the solid solution region A–e_1–e_2–C. After having undergone an equilibrium solidification process, a tie line showing the compositions of coexisting phases must include the point x. Alternatively, if we begin with pure B and a solid solution of composition a the composition x may be synthesized by mixing B and a, respectively, in the proportions of $|a - x|/|x - B|$, this ratio representing the proportions of B-to-solid solution. In fact, all compositions on this tie line may be synthesized by combining proper proportions of B and a solid solution of composition a.

In Fig. 47, the binary diagram A–C is depicted schematically. It will be recalled that if a melt, such as composition a of Fig. 46, is cooled, a solid richer in the higher melting component will be deposited when the liquidus temperature is encountered. Remembering also that the solidus describes the composition of the total solid that coexists with

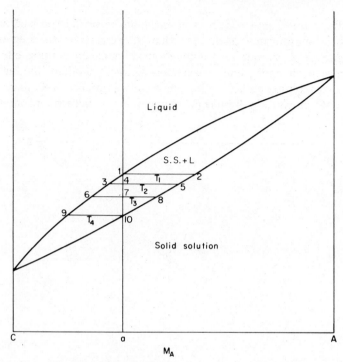

Fig. 47. Schematic representation of the binary system C–A of Fig. 45.

liquid at a particular temperature, we see, with reference to Fig. 47, that the initial precipitation provides a solid having the highest composition of A. With continued cooling, and assuming that equilibrium obtains at all temperatures, the total solid composition moves from this starting point to values poorer in A while the liquid moves in composition to values richer in C. As this cooling process continues, the solid-to-liquid ratio increases until finally at the temperature T_4, the solid composition achieves the value a, the liquid composition achieves the value 9, and the solid-to-liquid ratio becomes infinite. Neither pure A nor pure C is ever obtained, and the freezing process terminates after the cooling has progressed through the temperature interval $T_1 \rightarrow T_4$.

This same behavior occurs in the interior of the ternary system. Consider, for example, an isotherm at the temperature T_3 (Fig. 47) cutting through the ternary diagram (Fig. 48). This isotherm will cut through the binary diagram at points 6 and 8, these compositions representing the coexisting binary liquid and solid compositions, respectively. The ternary liquidus surface at the temperature T_3 termi-

J. SYSTEMS EXHIBITING SOLID SOLUBILITY

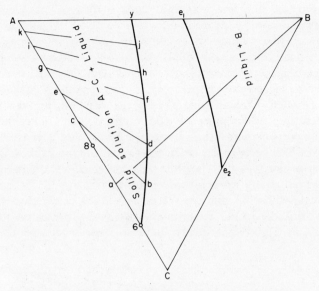

FIG. 48. Isotherm at the temperature T_3 of Fig. 47.

nates therefore at point 6 of Fig. 48. The tie line showing coexisting solid and liquid composition at this terminal location must lie on the binary line A–C and this tie line is the one extending from point 6 to point 8. With reference to Fig. 47, it is seen that all starting compositions of A relative to C lying within the interval 6–8 relate to this same tie line at the temperature T_3. On the binary side A–B, the point y of Fig. 48 represent the composition of liquid in equilibrium with pure A at T_3. It is, in fact, the liquidus for the A component. The tie line in question for this binary side is given by the line A–y. All binary compositions in the system A–B having an A/B ratio within this tie-line extent will have pure A in equilibrium with liquid of composition y.

For ternary compositions, however, the compositions of solids and liquids in equilibrium with each other cannot be represented by tie lines lying on a binary side. Thus, as the system composition moves along the line a–B toward B, for example, the binary solid in equilibrium with liquid at the temperature T_3 will be different than that where no B component is present. The liquid will become poorer in C (richer in A) and so will its coexisting solid. The tie lines b–c, d–e, etc., each show schematically the composition of ternary liquid in equilibrium with binary solids lying along the binary side A–C.

Before examining other isotherms in the ternary diagram, it is instructive to examine the state of affairs that exists along the line e_1–e_2.

Each point on this line coincides with a different temperature, the temperature dropping monotonically from e_1 to e_2. At each temperature along e_1–e_2, there exists an equilibrium mixture of a solid solution, a liquid whose composition follows the line e_1–e_2, and a pure-B phase. The three-phase condition along e_1–e_2 is isobarically univarient. In other words, if we specify the temperature, the compositions of the coexisting phases are fixed. One of these phases, the B phase, is pure, so that at all temperatures encompassed by the range T_{e_1}–T_{e_2}, a tie line between the liquid composition (lying along e_1–e_2) and the B phase terminates at the apex B. The opposite tie line between e_1–e_2 and the solid-solution phase move along A–C. The infinite number of such three-phase ties is represented at any temperature in the interval T_{e_1}–T_{e_2} by a concentration triangle whose termini are the compositions of the three coexisting solid and liquid phases. Three such triangles, pivoting on the apex B are shown in Fig. 49. At the temperature represented by the point 2,

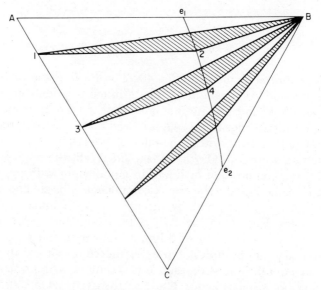

Fig. 49. Composition triangles in the three-phase temperature interval T_{e_1}–T_{e_2}.

for example, the significance of the triangle is as follows. At this temperature, the liquid has the composition 2 and is in equilibrium with two solids, one having the composition 1 on the binary side A–C, the other having the composition, pure B. As the temperature is lowered to that represented by point 4, a liquid of this composition is in equilibrium with two solids, one a pure phase having the composition B,

J. SYSTEMS EXHIBITING SOLID SOLUBILITY

the other a solid solution having the composition denoted by the point 3 on the binary side A–C. For each of the three triangles depicted, the temperature in question is below that shown in Fig. 48 since the latter only depicts a two-phase equilibrium. As we shall see shortly, within a triangle such as that defined by the points 1–B–2, all starting compositions will yield a three-phase equilibrium of solids 1 and B coexisting with liquid, once the temperature coinciding with point 2 obtains. What will be different about such compositions are the relative amounts of the coexisting phases. This is the ternary analog to the binary case described for Fig. 47 where all starting compositions lying within the interval 6–8 achieve an equilibrium between a solid of composition 8 and a liquid of composition 6 at the temperature T_3, the relative amounts of the coexisting phases again depending upon the starting composition.

With the above in mind let us proceed to examine a series of isotherms at temperatures lower than that of Fig. 48 (the temperature T_3 of Fig. 47). Figure 50 shows an isotherm at a temperature below the

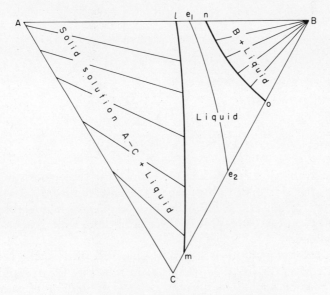

FIG. 50. An isotherm below T_A^0, T_B^0, and T_C^0.

melting points of A, B, and C. Note that since the isotherm lies below T_C^0, there is no possibility of a tie line lying on the binary side A–C as in Fig. 48, since, in the binary system A–C, complete solidification would have occurred. On the binary side A–B, however, a tie line A–l

does exist. It defines the composition of binary liquid in equilibrium with the pure A phase. All tie lines in the region A–*l*–*m*–C extend between the liquidus intercepts *l*–*m* and the binary side A–C. In the region *n*–B–*o*, however, all the tie lines extend between the liquidus intercept *n*–*o* and the apex B.

An isothermal plane cutting through the ternary surfaces at T_{e_1} is depicted in Fig. 51. The three-phase triangle for this temperature is

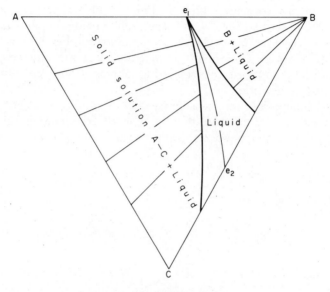

FIG. 51. An isotherm at T_{e_1}.

compressed to a line lying along A–B since the three coexisting phases all lie on the side A–B. Thus, when the temperature T_{e_1}, is reached, all starting compositions lying on the side A–B yield a three-phase mixture of pure A and pure B in equilibrium with liquid, i.e., the binary eutectic.

When the temperature is lowered somewhat further, the isothermal plane intersects the line e_1–e_2 at the point *n* in Fig. 52. A liquid having the composition *n* is in equilibrium with a solid solution of composition *m* and a pure phase of composition B. All starting compositions within the interval A–B–*m*, along the line *c*–B, for example, will have completely solidified by the time the temperature has been depressed to the value of the isotherm. All starting compositions lying within the region *m*–B–*n* will be made up of liquid *n* and solids B and *m*. All starting compositions within C–*m*–*n*–*x* will consist of liquid whose composition

J. SYSTEMS EXHIBITING SOLID SOLUBILITY

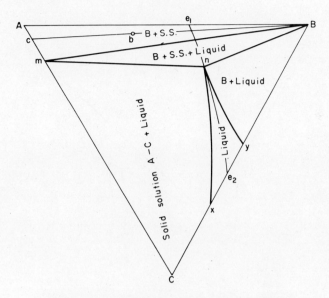

FIG. 52. An isotherm below T_{e_1} and above T_{e_2}.

lies along n–x and a binary solid solution whose composition lies along C–m. All starting compositions within n–B–y will consist of liquid whose composition lies along n–y and pure B. In the interval x–n–y all starting compositions will consist of a single liquid phase.

In the solidified region A–B–m, it will be noted that the tie lines extend between the side A–C and the apex B. This is in line with the argument presented in conjunction with Fig. 46. A composition b, for example, on the line c–B of Fig. 52, may be synthesized from a mixture of pure B and solid solution c and will have the proportions B/c = | c–b |/| b–B |.

In Fig. 53, an isotherm at a temperature just above T_{e_2} is shown. The all-liquid region is now small, confined to the area a–n–b. Concomitantly, the all-solid region is large and is bounded by the sides of the triangle A–B–m.

If a liquid of composition x, as shown in Figs. 46 and 54, is cooled until it encounters the liquidus, let us assume it gives rise to a solid having the composition 2 of Fig. 47. The liquid at this first contact with the liquidus surface essentially has the composition x. The tie line 2–x of Fig. 54 is such that the liquid-to-solid ratio at T_1, the temperature of first contact with the liquidus, is given by | 2–x |/| point x | $\to \infty$. As the temperature is lowered, the composition of the solid will move along the line 2–a, since the solid solution is binary in nature. At

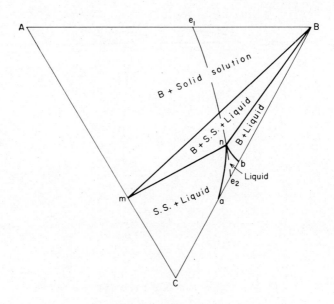

Fig. 53. An isotherm just above T_{e_2}.

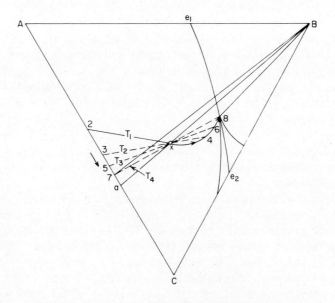

Fig. 54. Cooling path of the composition x of Fig. 46.

J. SYSTEMS EXHIBITING SOLID SOLUBILITY

any temperature below T_1, tie lines showing liquid and solid solution coexisting compositions must include the point x. For this to be the case, the cooling path that the liquid composition follows cannot be straight, unlike the case for a pure-phase in equilibrium with a liquid. That this is so can be seen from the following: Starting with a maximum concentration of A in the solid-solution phase, given by the point 2, the total solid-solution phase must change its composition in the direction toward point a of Fig. 54. When the composition of the solid solution acquires this value, as it must, since it comprises only A and C, and, in fact, contains all the available A and C, the system will consist of a mixture of solid solution of composition a and pure B in the proportions B/solid solution $= |\,a-x\,|/|\,x-B\,|$. Consequently, for all starting compositions in the region A–e_1–e_2–C, solidification cannot be completed until the line e_1–e_2 is encountered and B coprecipitates. If the projection of the cooling path from point x to the line e_1–e_2 were straight, the composition of the solid could not continuously move from point 2 to point a, since the point x must lie in all tie lines connecting liquid and solid-phase compositions. Similarly, if the trace of the cooling path were convex with respect to the side C–B, the composition of the solid solution could not move toward point a, again all tie lines being straight and having to include point x. Consequently, the cooling path must be concave relative to the side C–B as shown in Fig. 54. Thus, when the temperature achieves the value T_2, the liquid of composition 4 is in equilibrium with solid of composition 3, the points 3, x, and 4 being colinear. At a still lower temperature, solid solution 5 is in equilibrium with liquid 6. The liquid-to-solid ratio has decreased from a value approaching infinity at T_1 to a value $|\,x-5\,|/|\,x-6\,|$ at T_3. At the temperature T_4, the temperature of the three-phase equilibrium, solid solution–B–L is reached. At just this temperature, the liquid has a composition 8 in equilibrium with a solid solution of composition 7 and a minute quantity of pure B. The solid-to-liquid ratio effectively is $|\,x-8\,|/|\,x-7\,|$, since x lies along the side 7–8 of the composition triangle. Once the line e_1–e_2 is encountered, a single tie line can no longer connect solid-liquid compositions since two solids are present.

At a temperature slightly lower than T_4, the composition x falls within the solid solution–B–L coexistence triangle and the following obtains (Fig. 55). A solid solution of composition 9 is in equilibrium with a liquid of composition 10 and a solid of composition B. The ratio, solid solution-to-liquid–B can be determined as follows with reference to Fig. 55. A line drawn through points 9 and x, and extended to the three-phase triangle side 10–B (the latter connects the composition of liquid in equilibrium with both solid solution and B) intersects

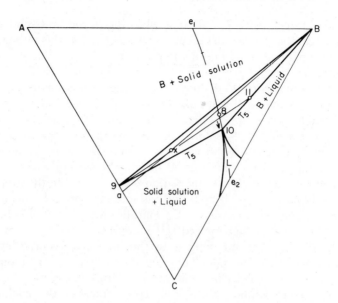

FIG. 55. Tie lines at a temperature T_5 where solid solution, B, and liquid coexist.

this side at point 11. A mixture of liquid and B having the overall composition at point 11 when mixed with a solid solution having the composition 9 in the proportions solid solution/L + B = | x–11 |/| 9–x | will give a composition x.

Since the system contains only solid solution, pure B, and liquid, the tie line 9–11 enables us to compute the solid solution-to-liquid + B ratio. The fraction of the B component that has precipitated at the temperature T_5 can be deduced as follows: As stated, a mixture of B + L having the total composition of point 11 may be admixed with solid solution of composition 9 to give a system composition of point x in Fig. 55. The ratio | 9–x |/| 9–11 | then, is the sum B + L divided by the total moles or mass, depending on the dimensions chosen, necessary to give a synthetic mixture of composition x. Thus,

$$\frac{B+L}{B+L+\text{solid solution}} = \frac{|9-x|}{|9-11|}.$$

We must now deduce a means for separating B from the sum B + L. The line 10–B in Fig. 55 shows the composition of the B phase in equilibrium with liquid of composition 10. A mixture of B and liquid of composition 10 in the proportion B/L = | 10–11 |/| 11–B | would give a system of composition 11. Specifically, the fraction of solid B

J. SYSTEMS EXHIBITING SOLID SOLUBILITY

in such a mixture would be, B relative to liquid alone is $|\,10\text{–}11\,|/|\,10\text{–}B\,|$. The fraction of B desired, however, is relative to the total quantity of solid and liquid not just relative to liquid. The total quantity of solid + liquid, as we have seen, may be synthesized from the tie line 9–11 which includes the system composition x. Therefore, B relative to the entire system is

$$\frac{B+L}{B+L+\text{solid solution}} \cdot \frac{B}{B+L} = \frac{|\,9\text{–}x\,|}{|\,9\text{–}11\,|} \cdot \frac{|\,10\text{–}11\,|}{|\,10\text{–}B\,|}.$$

With the above in mind, it is now a simple matter to evaluate such fractions at the temperature T_4 where the three-phase line e_1–e_2 was first encountered. We see from inspection of Fig. 54 that the ratio $B/L \sim 0$ at the temperature of first encounter.

Upon cooling the system further to some temperature T_6, the solid solution acquires a value equal to point a and the liquid acquires a composition y, as shown in Fig. 56. At this point the ratio A/C is precisely

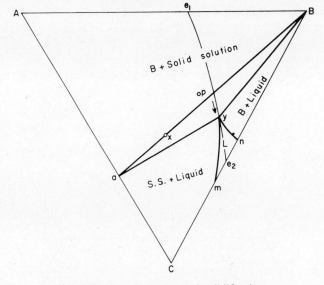

FIG. 56. The final stage of solidification.

that in the starting mixture and the fraction of B in the total solids is that of the starting mixture. Solidification, therefore, must be complete. This is shown in Fig. 56. The tie line showing the composition of the two coexisting solid-phases includes the system composition x. Consequently, the ratio $B/L \to \infty$, its value being approximately $|y\text{–}B|/|\text{point B}|$.

Therefore, the ratio of

(B + L)/(solid solution + B + L) $\sim$ B/(solid solution + B) $\sim$ | a–x |/| a–B |, etc.

Thus, we see that although the last liquid to solidify has the composition y, the quantity of this liquid is vanishingly small.

For any starting composition within a–A–B, i.e., the point p of Fig. 56, this point would have become colinear with a side of the composition triangle at a higher temperature than that represented by point y. Consequently, for all starting compositions within a–A–B, it is clear that solidification would have been completed once T_y had been achieved. For all compositions along a–B, the last trace of liquid would be present at T_y. For all starting compositions within a–B–y, however, the system would be three-phase and solidification would not be completed until a temperature lower than T_y were achieved.

Finally, we may examine extension of the special case of two simple eutectic systems plus one solid solution to the more general case where the eutectic interactions exhibit a degree of solid solubility. A schematic representation of such binary systems is offered in Fig. 57 and is assumed representative of the systems A–B and B–C. Contrasted with the special case just described, the ternary diagram will exhibit two solid-solution regions, one emanating from the A–C binary side, the other from the B–C binary side. Furthermore, whereas the solid solution in the special

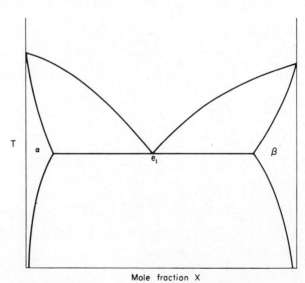

FIG. 57. A schematic representation of the systems A–B or B–C.

J. SYSTEMS EXHIBITING SOLID SOLUBILITY

case was binary in nature, the two solid solutions, in the general case, are ternary in nature. As a consequence of this fact, the solidus line which, in the system A–C, was confined to the binary side A–C must now generate a surface which extends into the ternary diagram. This extension does not continue to the three-phase line e_1–e_2 since the solubility of the component B is assumed finite (as was inferred from Fig. 57). In perspective, the ternary system has the appearance shown in Fig. 58. By and large, most of the discussion relating to the special

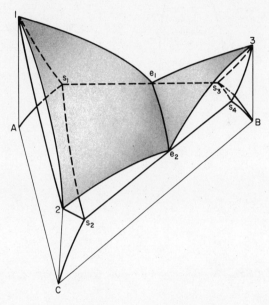

FIG. 58. A ternary system in which the eutectic interactions exhibit limited solid solubility.

case where the eutectic systems exhibit miniscule solid solubility is applicable to the present system. A notable difference in the appearance of the isotherms is that, now, two ternary surfaces are cut in each region, one being the liquidus, which has an exact counterpart in the special case, and the other being the extension of the solidus, which has no counterpart in the special case. Thus, referring again to Fig. 58, it is seen that extending from the ends of the ascending solid solution A–C, a solidus surface 1–S_1–S_2–2 propagates into the ternary triangle, terminating at S_1–S_2. The temperatures at S_1 and S_2 are identical to those at e_1 and e_2, respectively. The line joining S_1 to S_2 will not, in general, exhibit the same curvature as e_1–e_2. On the opposite side of the diagram, a solidus surface begins at T_B^0 providing the surface 3–S_3–S_4, which terminates in the line S_3–S_4. Again S_3 and S_4 occur at the same temperatures at e_1 and e_2, but the line joining them does not have the general shape of the binary eutectic three-phase line.

Because of the regions of ternary solid solution, tie lines joining liquid and coexisting solid-phase compositions terminate within the ternary triangle rather than at an apex or a side. In Fig. 59, an isotherm

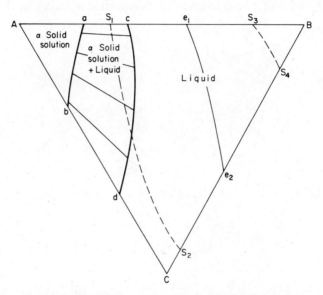

Fig. 59. An isotherm below the melting point of A.

below T_A^0, but above T_B^0, T_C^0, and e_1 and e_2 is shown. The projections of S_1–S_2, e_1–e_2, and S_3–S_4 are also shown. At the temperature of the isotherm, all starting compositions within the single phase α solid-solution region will be completely solid, those within the region a–b–c–d will consist of liquid in equilibrium with a ternary solid solution whose composition lies along a–b, and the remainder of the compositions will be liquid.

Figure 60 depicts an isotherm below the melting points of A, B, and C.
It should be noted that in the B-side of the diagram where the β solid-solution is formed, tie lines do not converge at the apex B, but rather along the line r–p. Extension of the discussion to lower temperature isotherms is unnecessary since our previous discussion is applicable to the two- and three-phase regions.

J. SYSTEMS EXHIBITING SOLID SOLUBILITY

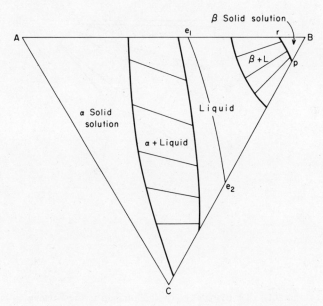

FIG. 60. An isotherm below the melting points of both A and B.

In tracing the course of freezing on a polythermal projection, note that, unlike the special case, not all starting compositions result in a three-phase α–β–L equilibrium prior to complete solidification. For temperatures above those for which liquid may participate in equilibrium with solids, the solubility limits B–S_3–S_4 and A–C–S_1–S_2 of Fig. 59 define the composition boundaries within which solidification is completed prior to coprecipitation of a second solid phase. The remaining two regions, following a L–solid equilibrium, encounter the line e_1–e_2, at which time solidification continues with simultaneous precipitation of two solid phases. Three-phase triangles denoting solid compositions in equilibrium with liquid terminate at the lines S_1–S_2 and S_3–S_4, as shown in Fig. 61. It is seen that these triangles do not pivot around the B apex as in Fig. 49.

Below the temperature at which liquid can be an equilibrium participant, the lines S_1–S_2 and S_3–S_4 move toward the side A–C and the apex B, respectively (Fig. 58). Thus, compositions that froze as a single solid-solution phase break up into conjugate α and β solid solutions.

Finally, in Fig. 61, freezing of a composition denoted by the point x is shown. Upon encountering the liquidus, a solid of composition 2 that is rich in the highest-melting-point component comes into equilibrium with liquid, having essentially the starting composition x. Cooling to a

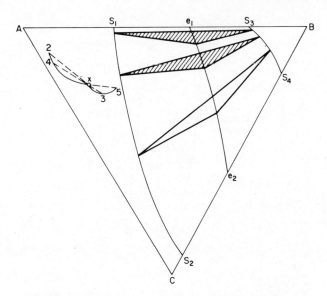

FIG. 61. Tie triangles and a cooling path in the 7 region.

lower temperature, the solid acquires a composition 4 while the liquid moves in composition to the point 3. The tie line 3–4, contains the point x and the solid-to-liquid ratio acquires a value $|\,3{-}x\,|/|\,4{-}x\,|$. Upon cooling further, the total solid composition moves along the curve 2–x/toward x, ultimately achieving a value x. The liquid moves along the curve x–5 during this time. Just prior to complete solidification, the solid-to-liquid ratio approaches infinity, i.e., $|\,5{-}x|/|\text{point }x|$. As before, the projection of the liquid-composition path is curved. Unlike the preceding consideration of cooling (Fig. 54), the solid composition is not confined to the side A–C since A, B, and C are in solid solution. The solid-composition path becomes curved with decreasing temperature. The direction of curvature of both the liquid and solid is such that tie lines connecting the compositions of the coexisting phases include the point x.

35

Three-Component Solid–Vapor Equilibria— Chemical Vapor Transport Reactions

A. INTRODUCTION

The advent of electronic technology with its requirements for materials of heretofore unheard of purity and thin-film single crystals has provided great impetus to the study of metals and special types of solid–vapor equilibria. Many semiconductors of practical importance are relatively involatile but give rise to halogen-containing compounds having considerable vapor pressures. Certain of these halogen compounds exhibit the properties such that, when species derived from them at one temperature are transported to a lower or a higher temperature, they undergo internal oxidation–reductions, disproportionating in the process. In desirable cases, the fragments resulting from such disproportionations include the pure element. Consequently, one is able to extract the element from an impure mixture at a "source" and redeposit it at some other location in the system, the "seed site," by applying an appropriate temperature gradient. Such techniques have been very useful in extracting semiconductors from starting ores as a first step of purification prior to zone refining. A second, equally important use of chemical vapor transport processes involves the passage of a halogen or halogen acid over a source of pure or doped semiconductor, at which point, a reaction occurs and semiconductor halogen species are brought into the vapor. Again, by applying a suitable temperature gradient, these species are made to disproportionate and the semiconductor is precipitated from the vapor. At the point of precipitation one locates a nucleating single crystal "seed," in the form of a flat wafer slice of the semiconductor. Under proper conditions of flow rate, gas phase concentration, etc., the precipitated semiconductor grows epitaxially upon the wafer seed, i.e., it deposits as a single crystal conforming in periodicity to the underlying wafer or substrate periodicity. Frequently, disproportionation processes may be effected under quasi-equilibrium conditions and such transport processes take place at lower temperatures than the other types of vapor transport reactions to be discussed below.

A second vapor transport process having great practical utility is that referred to as a hydrogen-reduction process. This technique is most extensively used for epitaxial single crystal growth and purification of single-element materials while the disproportionation scheme has had its greatest application in transporting binary compounds. In hydrogen-reduction processes a room-temperature stream of hydrogen carrying a tetrahalide or trihalide is carried over a heated pedestal where reduction of the vapor species occurs. The desired element precipitates out, and under proper conditions, epitaxial growth may be effected upon a seed wafer. The prevailing conditions do not approximate quasi-equilibrium to the same extent that disproportionation processes do and require temperatures that are, in general, higher. Growth conditions are, however, mass transport rather than kinetically limited by surface phenomena.

A third method for achieving purification and epitaxial growth is the so-called hydride pyrolytic technique. The method has been used primarily for Ge and Si reactions and involves the cracking of GeH_4 (germane) or SiH_4 (silane) in an apparatus similar to that employed in hydrogen reduction processes. Pyrolytic techniques are generally applicable at temperatures intermediate between the other two and represent the poorest approximation to equilibrium conditions.

Typical reactions for the three types of vapor transport processes, respectively, are

$$2GeI_2(v) \rightleftarrows Ge(s) + GeI_4(v) \quad \text{disproportionation,} \tag{1}$$

$$GeCl_4(v) + 2H_2(v) \rightleftarrows 4HCl(v) + Ge(s) \quad \text{hydrogen reduction,} \tag{2}$$

$$GeH_4(v) \stackrel{\Delta}{\rightleftarrows} Ge(s) + 2H_2(v) \quad \text{pyrolysis.} \tag{3}$$

A. INTRODUCTION

The disproportionation reaction [Eq. (1)] may be conducted in a closed system, i.e., a sealed tube, that is subjected to a temperature gradient. For example, if, at one end of this tube, we place crushed Ge and add iodine to the tube, and if this Ge is placed in the hotter portions of a system having a gradient, e.g., 600°C → 350°C, the Ge will react with the iodine to form a mixture of GeI_2 and GeI_4. The GeI_2 will disproportionate in the cooler regions according to Eq. (1). The liberated GeI_4 will move to the hotter portions, react with Ge to form more GeI_2 and the cycle will be repeated. Eventually, all of the Ge will have been transported to the cool portions of the tube (see Fig. 1). Conditions are

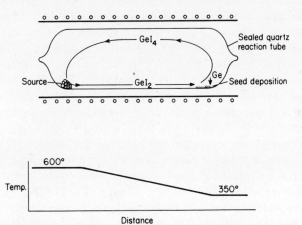

FIG. 1. A closed-tube iodide disproportion process.

so chosen that the dew points of the GeI_2 and GeI_4 are never exceeded. Consequently, the halogen is confined to the vapor phase at all times, the solid phase consisting of pure Ge.

Most usually, the three classes of vapor-transport processes are utilized in an open tube apparatus at constant pressure (atmospheric). The reactive species are transported in either an inert or reactive carrier gas. In the following discussion we will restrict ourselves to a consideration of open tube disproportionation reactions. It should be noted that the hydrogen-reduction process may also become involved in a disporportionation process if equilibrium is permitted to prevail at the heated pedestal mentioned. For example, from Eq. (2) we see that a product of the reduction reaction is HCl. If the HCl is not carried away rapidly enough, it will, in effect, use the deposition site as a source region and etch the material already deposited and form primarily $GeCl_2$. This $GeCl_2$ will then disproportionate downstream depositing

Ge in the process. Obviously, if the pedestal is the desired seed site, the secondary process described is unwanted.

B. Disproportionation Processes

Associated with Eq. (1) is an enthalpy of process ΔH_p. Since the equilibrium constant for Eq. (1) as a function of T is

$$\ln K_p = -(\Delta H^0/RT) + C, \quad (4)$$

it is seen that if ΔH_p^0 is positive for the reaction as written (endothermic), the reaction will be driven further to the right with increasing temperature. On the other hand, if ΔH_p^0 is negative for the reaction as written, the reaction will be driven to the left. The significance of these two possibilities is as follows. With ΔH_p^0 positive for the disproportionation reaction given in Eq. (1), it is necessary that the seed site be at a higher temperature than the source site. Such processes are termed cold-to-hot vapor transport processes. With ΔH_p^0 negative, it is necessary that the seed site be at a lower temperature than the source in order for transport to occur. These processes are termed hot to cold. Equation (1), as written, has a negative ΔH_p^0 and is consequently a hot-to-cold process.

An interesting practical application of cold-to-hot processes is in the manufacture of long-life tungsten-filament light bulbs. The filaments tend to evaporate during use, thinning down nonuniformly along their length. These thin regions, having higher resistance are hotter and it is at these "hot spots" that failure is likely to occur. It was found that if small quantities of Cl_2 are added to the light bulbs, the Cl_2 will react with the evaporated tungsten which has condensed out at the cooler walls of the bulb, i.e.,

$$W + 3Cl_2 \rightleftarrows WCl_6. \quad (5)$$

The hexachloride, then, transports to the hottest portion of the system (the hot spots on the filament) and deposits tungsten, thus renewing the filament. For reaction 5, as written, ΔH_p^0 is negative. Consequently, the reverse process which is what is desired at the hot filament has a positive ΔH_p^0 and K for the reverse process increases in magnitude with increasing temperature, giving rise to a cold-to-hot transport process.

An extension of the reaction depicted in Eq. (1) involves the sequence of steps shown in Fig. 2. Hydrogen is transpired through a bed of iodine. The effluent mixture of H_2 and I_2 is carried through a platinum catalyst tower where the iodine is converted to HI. The gaseous mixture is

B. DISPROPORTIONATION PROCESSES

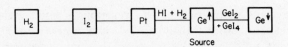

FIG. 2. A block diagram of the vapor transport of germanium by iodine in a hydrogen carrier.

then carried through a packed bed of Ge at some elevated temperature where it reacts to form GeI_2 primarily, plus an amount of GeI_4 required by equilibrium. Conditions are chosen such that, at this packed bed and at a subsequent seed site, both the GeI_2 and GeI_4 are completely volatile. The effluent from this packed Ge bed is then transported to a cooler region of the system where the reaction given in Eq. (1) prevails and Ge is deposited. The pertinent reactions and equilibria involved are given as follows:

$$I_2 \rightleftarrows 2I \quad \Delta H_p^0(+), \tag{6}$$

$$I_2 + H_2 \rightleftarrows 2HI \quad \Delta H_p^0(-), \tag{7}$$

$$I_2 + Ge \rightleftarrows GeI_2 \quad \Delta H_p^0(-), \tag{8}$$

$$2I_2 + Ge \rightleftarrows GeI_4 \quad \Delta H_p^0(-), \tag{9}$$

$$Ge + GeI_4 \rightleftarrows GeI_2 \quad \Delta H_p^0(+). \tag{10}$$

In all of these equations the only condensed phase present is the element Ge. Furthermore, it is to be noted that of the set comprising Eqs. (8)–(10), only two of the three are independent of one another; the remaining equation is deducible from the other two. Similarly, reactions depicting a HI reaction with Ge to form either GeI_2 or GeI_4 are deducible via Hess' law from Eqs. (7) and (8) or Eqs. (7) and (9), respectively.

As indicated, whether a disproportionation process is hot-to-cold or cold-to-hot is determined by whether or not ΔH_p^0 for the overall disproportionation process written from left to right is positive or negative, respectively. For example, if only Eq. (10) governed the process in question, then clearly the transport, i.e., the reverse of Eq. (10), is hot-to-cold. On the other hand, note that Eq. (7), involving the formation of HI from the elements has an enthalpic change of opposite sign. The qualitative effects of simultaneous equilibria can be seen from the following.

Suppose the primary reaction desired is depicted by

$$aA(s) + bB(v) \underset{}{\overset{\Delta H_p^0(+)}{\rightleftarrows}} xAB(v). \tag{11}$$

With increasing temperature, the reaction is driven to the right. In order to effect a vapor transport of A, it is necessary that the source temperature of solid A be greater than the sink temperature for this material.

Suppose now that accompanying the desired reaction given by Eq. (11), there is a simultaneous competing vapor phase equilibrium represented by

$$d\mathrm{D(v)} + b\mathrm{B(v)} \underset{}{\overset{\Delta H_p^{0\prime}(-)}{\rightleftarrows}} y\mathrm{DB(v)}, \tag{12}$$

for which ΔH_p is negative. For simplicity, let us assume that all of the coefficients in Eqs. (11) and (12) are unity so that the ΔH_p^0 represent molar heats of formation of the species AB and DB.

Suppose now, that into a container in which 1 mole of A and 1 mole of B have been placed, we introduce a small fraction of a mole of D and a molar quantity of B equal to it.

The overall reaction that occurs may be represented by

$$a\mathrm{A(s)} + (b + d')\,\mathrm{B(v)} + d'\mathrm{D(v)} \underset{}{\overset{\Delta H_p^{0*}}{\rightleftarrows}} x\mathrm{AB} + y'\mathrm{DB(v)}. \tag{13}$$

The process heat for the overall reaction ΔH_p^{0*} will be positive in sign but slightly smaller in magnitude than ΔH_p^0. If subsequent experiments are performed in which the quantity of D is increased, a point will be reached at which the reaction given by Eq. (11) is secondary to that depicted in Eq. (12) and in which ΔH_p^{0*} is now negative. Transport will then proceed from the cold to the hot regions of the system. Since, in a given experiment, the coefficients a and b may have any value, i.e., Eq. (11) may be multiplied by any number, it is evident that, depending on the quantities of A and B and D present and the relative magnitudes of ΔH_p^0 and $\Delta H_p^{0\prime}$, the effect of D on the sign of ΔH_p^{0*} will be different. Thus, for increasing quantities of A and B, the direction of transport implied by Eq. (11) will be less likely to change when a specified quantity of D is added.

Because, as just shown, ΔH_p^{0*} can have an infinite number of values, depending on the set of experimental conditions imposed, a knowledge of its sign rather than its absolute value is of greatest interest in making use of transport properties. In fact, even the former is less useful than direct knowledge of how much of the material being transported is in the vapor at different temperatures. Since all of the halogen is confined to the vapor phase during the process, a convenient measure of different equilibrium states is in the form of semiconductor-to-halogen vapor-phase component ratios. When this figure increases with temperature,

B. DISPROPORTIONATION PROCESSES

it implies that ΔH_{p}^{0*} is positive and transport will be from hot-to-cold. When this figure decreases, the reverse is the case.

Clearly, another measure of ΔH_{p}^{0*} could be obtained by evaluating the mole fraction of semiconductor component in the vapor relative to semiconductor and halogen contents. Since open-tube systems comprise an inert or reactive carrier gas plus the semiconductor (or metal) and halogen, the systems are at least ternary in nature. Sometimes there is an advantage to employing a mixed carrier gas in which one of the constituents is reactive. In such instances, the system is quaternary. If dopants are also present, the component order is higher still.

In any event, a plot of the ratio $P_{\text{semicond}}/P_{\text{halogen}}$ or something similar (where the values P_x represent the component partial pressures) as a function of T defines the composition of vapor in equilibrium with pure solid and is a useful way of plotting the solid–gas phase diagrams in question. By inspection, one can then define the sign of ΔH_{p}^{0*} and quantitatively define the amount of semiconductor present in the vapor at different temperatures and starting concentration conditions.

As an example of how such phase diagrams may be deduced, let us consider the system already defined by Eqs. (6)–(10).

There are three components present, i.e., Ge, I_2, and H_2, in two phases at constant total pressure. From the reduced phase rule we see that two degrees of freedom are present. We can choose as these degrees of freedom the temperature and the component ratio P_{H_2}/P_{I_2}. In the vapor phase we have present species of I, I_2, HI, H_2, GeI_2, and GeI_4. In order to define the phase diagram P_{Ge}/P_{I_2} versus T we need to know the concentration (in terms of partial pressures) of each of the six species present. Consequently, a set of six independent equations must be defined. All that is required of this set is that it contain the unknowns involved. Clearly, since the system is bivariant, each new P_{H_2}/P_{I_2} component starting ratio will yield a new phase diagram for P_{Ge}/P_{I_2} as a function of T. Consequently, there exists an infinite number of phase diagrams for the system, as we have previously implied when discussing ΔH_{p}^{0*}.

A useful set of six independent equations in which species and component concentrations are related via the conservation equation,

$$\frac{P_{H_2}}{P_{I_2}} = \frac{p_{H_2} + \tfrac{1}{2}p_{HI}}{\tfrac{1}{2}p_I + p_{I_2} + \tfrac{1}{2}p_{HI} + p_{GeI_2} + 2p_{GeI_4}}. \tag{14}$$

The total pressure is conserved in terms of species partial pressures,

$$P_t = 760 \text{ torr} = \sum_{1}^{6} p_x. \tag{15}$$

where p_x is the partial pressure of the six species, i.e., p_{HI}, p_I, p_{GeI_2}, p_{GeI_4}, etc. The equilibria in Eqs. (6)–(8) and (10) are expressed in terms of temperature dependent equilibrium constants as

$$\log_{10}[(p_I)^2/p_{I_2}] = \log_{10} K_{I_2} = 8.362 - 7991/T, \qquad (16)$$

$$\log_{10}[(p_{HI})^2/p_{H_2}p_{I_2}] = \log_{10} K_{HI} = 0.834 + 622.4/T, \qquad (17)$$

$$\log_{10}[p_{GeI_2}/p_{I_2}] = \log_{10} K_{GeI_2} = 2.45 + 434/T, \qquad (18)$$

$$\log_{10}[(p_{GeI_2})^2/p_{GeI_4}] = \log_{10} K_{Ge} = 12.56 - 7936/T. \qquad (19)$$

Equations (14)–(19) may be solved simultaneously, given values of p_{H_2}/p_{I_2} and T. At each specified temperature and starting hydrogen/iodine ratio there will be one equilibrium value for each of the species present. These species pressures may then be employed to provide a vapor-phase component concentration value. As indicated, a useful parameter is the component P_{Ge}/P_{I_2} ratio defined by

$$(P_{Ge}/P_{I_2}) = \frac{p_{GeI_2} + p_{GeI_4}}{\frac{1}{2}p_I + p_{I_2} + \frac{1}{2}p_{HI} + p_{GeI_2} + 2p_{GeI_4}}. \qquad (20)$$

This ratio may be determined for each desired temperature using the same input P_{H_2}/P_{I_2} ratio, thus providing a curve of P_{Ge}/P_{I_2} versus T. The process may be repeated for different values of the input value P_{H_2}/P_{I_2}, thereby providing a family of curves each of which describes the variation in component vapor-phase concentration in the Ge–vapor equilibrium as a function of T at a particular starting reactant concentration. Each of these curves then, is univariant showing the variation of P_{Ge}/P_{I_2} with T at constant input of P_{H_2}/P_{I_2}.

Figure 3 shows such a set of phase diagrams deduced for the system in question. It is seen that the Ge/I_2 ratio first decreases with increasing temperature indicating a cold-to-hot transport process at lower temperatures, then increases with increasing temperature indicating hot-to-cold transport for higher temperature sources. This nonmonotonic behavior is due to the fact that the HI reaction is opposite in sign to the desired disproportionation reaction and competes for I_2 at lower temperatures. With increasing temperature, the HI reaction, which has a negative ΔH^0 of formation of HI, becomes less competitive and the reaction given by Eq. (10) prevails.

For comparison purposes, Fig. 4 shows the results obtained in a system where, despite simultaneous equilibria of opposite sign for ΔH^0, one reaction prevails at low temperatures. The three-component two-phase system $Ge-I_2-He$ at constant total pressure possesses two degrees

B. DISPROPORTIONATION PROCESSES

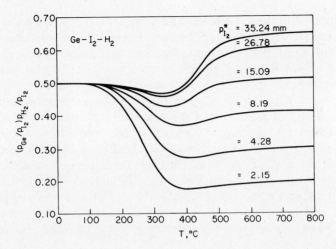

FIG. 3. Vapor phase solubility curves for the system Ge–I_2–H_2 at varying iodine source bed pressures.

of freedom. Consequently, a family of curves is generated here also. As can be seen from Fig. 4, transport is hot-to-cold up to approximately 600°C. Above this temperature, the vapor-phase solubility curves tend to bend downward. At this temperature the dissociation of GeI_2 (note that the Ge/I_2 ratio at 600°C is approximately 1.00 which indicates that the vapor species containing Ge is almost pure GeI_2) starts to become

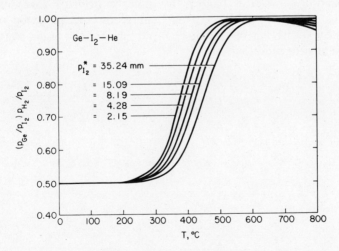

FIG. 4. Vapor phase solubility curves for the system Ge–I_2–He at varying iodine source bed pressures.

significant via Eq. (8). Concomitantly, the prevailing low-temperature equilibrium [Eq. (10)] becomes less important. It is seen that the ΔH_p^0's of Eqs. (8) and (10) are of opposite sign, and actually it is again the existence of simultaneous equilibria that accounts for the bend-over in the vapor-phase solubility curves at higher temperatures.

In setting up and evaluating a vapor-transport process, curves of the types shown in Figs. 3 and 4 are of considerable utility. For example, under conditions where the carrier gas becomes saturated with the halogen and then where equilibrium is permitted to obtain at the source and seed, the rate of transport can be calculated precisely. Knowing the halogen through put/unit time in moles/liter, the pickup of Ge in moles is deducible from Fig. 3 or 4. Furthermore, the temperature intervals and concentration conditions for achieving either cold-to-hot or hot-to-cold transport can be determined by inspection in the case of Fig. 3.

Almost any reaction that results in the volatilization of an essentially involatile substance may be used as the basis for a vapor transport process provided that the desired substance participates in pure form as part of a solid–gas equilibrium. In this respect, for example, transport agents such as O_2, H_2O, and the halogens have been used extensively for purification and single crystal growth of a wide number of elements and binary and ternary inorganic compounds. The method outlined for defining ternary or higher order systems is useful for describing such systems. When the number of components exceeds three, additional concentration constraints have to be imposed in order that univariant situations may prevail. Thus, one can study the behavior of such systems by specifying two or more input component ratios as the total component order increases.

36

Experimental Techniques

A. INTRODUCTION

While the primary purpose of this monograph is to provide an introduction to the principles governing the behavior of heterogeneous systems, this intent will not be satisfied if no mention is made of the methods by which the phase diagrams of real systems are deduced. On the other hand, any attempt to discuss the experimental aspects of the subject in depth would consume an amount of space at least equal to that already expended. In order to provide some rapport with the experimental techniques available, in a relatively brief accounting, only a few methods will be considered in reasonable detail while several other approaches will be mentioned cursorily. The restricted choice of techniques on which we will elaborate is obviously subjective. Nonetheless, it is the case that these particular methods are used extensively for a broad spectrum of investigations.

The description of one-component systems is generally in the form of $p-T$ diagrams of state. The parameters to be measured, then, are the temperature and pressure. The measurement of pressure and temperature is not unique to systems of one component, as can be inferred from the requirements for describing a system of the type discussed in the preceding chapter, or for the description of the behavior of hydrate or other dissociative systems. Obviously also, $p-M_X$ diagrams of state also require measurement of pressure.

Phase diagrams for multicomponent systems are frequently constructed in terms of their temperature–composition relationships. The experimental techniques here must be sensitive to thermal anomalies accompanying crystallization or melting. Again, however, such techniques are not restricted to the study of multicomponent systems, since, if a single component undergoes a phase transformation, reversible or otherwise, a latent heat anomaly accompanies such crystallographic variation also.

Pressure measurements may be grouped into two types—direct and indirect. Within each of these categories different methods are needed for estimating pressures in different pressure intervals. Vapor pressures below approximately 10^{-1} torr (mm of Hg) are most usually determined by the Knudsen effusion method, or some variant of it, or by the so called transpiration method. Neither of these methods is direct, in that assumptions relating to gas ideality are necessary in order to calculate a pressure. In the range from 10^{-1} to 760 torr, the *transpiration* and *Bourdon gauge* methods or some variant of the latter are employed. The Bourdon gauge method, while not strictly speaking a direct approach, may be thought of as such. Above atmospheric pressure, the Bourdon gauge method alone is of considerable utility.

Other pressure measuring techniques that have enjoyed limited usage include, in the submillimeter range, the *effusion torsion* technique and in the range above this, to atmospheric pressure, the *vapor balance* method.

Temperature measuring methods have, for the most part, been confined to resistance thermometry at very low temperatures, and direct or differential thermocouple measurements from, perhaps, liquid nitrogen temperatures to 2200°C, and optical pyrometry from approximately 800°C up.

Accessory techniques of considerable importance involve mass spectrometry associated with effusion techniques, X-ray methods for identification of solid phases, dilatometry for detection of phase changes, electrical conductivity measurements for the same purpose, and density measurements for determining composition ranges of solid-phase existence. A variety of other methods for detecting changes in physical or chemical properties have also been used and, indeed, the number of potential peripheral methods is large.

B. Pressure Measurements

1. *Effusion Methods*

Methods utilizing manometers, barometers, and the many specialized versions of these instruments will not be discussed, primarily because the vapor of the material being measured comes into contact with mercury, oil, or other liquid. A second feature that makes such techniques somewhat limited in scope, is that they cannot be applied readily over wide temperature intervals, for obvious reasons.

Methods involving electronic vacuum gauges will not be discussed since these are of greatest import in measuring pressures of inert gases only, and find greatest use in high vacuum evaporator or sputtering systems. In addition, such techniques are not readily employed over a wide temperature range.

A very useful, albeit indirect, method for evaluating the vapor pressure in a system composed of condensed phases and a vapor is the Knudsen effusion method. The technique is particularly useful for micron-range pressures and is based on the following principle: If a sample exhibiting a small vapor pressure is contained in a heated chamber, then, after sufficient time, condensed vapor-phase equilibrium will be established. If the evaporation rate is sufficiently high (see Chapter 4), and if the chamber contains a small orifice, vapor will effuse through this orifice while the pressure in the chamber retains its equilibrium value. From a knowledge of the weight of the sample w_s, the area of the effusion orifice a_o, the temperature T, an assumed molecular weight of the vapor-phase species $(\text{mol wt})_A$, and the weight loss l, in a time period t, one is able to estimate the vapor pressure of the sample according to

$$p_A = (l/a_o t)(2\pi RT/\text{mol wt}_A)^{1/2}, \qquad (1)$$

where p_A, the pressure, is in dyn/cm^2 when l is in grams, a_o has dimensions of square centimeters, T is in °K, and R, the ideal gas constant, is defined in ergs/mole° (8.31439×10^7).

The method has the disadvantages that it requires assumption of a species molecular weight, requires that no molecules be reflected back from the orifice into the chamber and assumes that the behavior of the vapor is ideal. The method is not easily applied to multispecies systems. Any error in measuring the orifice geometry or failure to apply corrections for the effect of the finite thickness of the orifice on the reflection phenomenon in the orifice leads to further inaccuracies in the method. Nonetheless, when coupled with a mass spectrometer, the effusion technique has enabled study of a host of low vapor-pressure materials over a wide range of temperatures. A sophisticated modification of the Knudsen technique involves suspending a two-orifice Knudsen chamber from a quartz fiber torsion balance. The effusing vapors from both openings tend to rotate the fiber. A knowledge of the degree of twisting, the orifice and fiber geometries, and the temperature, enables one to compute vapor pressures without invoking assumptions concerning the molecular weights of species. The molecular weight may be calculated from the data of the same experiment after the instrument is calibrated with materials of known parameters. Figure 1 depicts schematically such an apparatus.

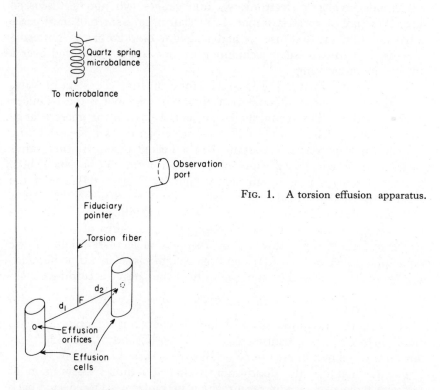

FIG. 1. A torsion effusion apparatus.

The effusion cells have orifices on opposite sides of the arm connecting the cells. During the effusion process, a torque is transmitted to the torsion fiber causing it to twist. The magnitude of the torque is measured using the fiduciary pointer affixed to it and a reference scale.

The vapor pressure may be evaluated by means of Eq. (1) which may be written

$$p_A = (l/t) \, T^{1/2} C_1 / M_A^{1/2}, \tag{2}$$

B. PRESSURE MEASUREMENTS

where the constant C_1 includes the orifice area. The constant C_1 may be determined by the use of a known substance. As seen from the following however, Eq. (2) need not be used for evaluating p_A, but is more useful in estimating M_A. The torque at the fulcrum F due to the recoil forces at the orifices is proportional to the arm lengths d_1 and d_2, the pressure, and the orifice area a_o (see Fig. 1). Any difference in torques due to the fact that the two effusion processes are going on simultaneously is not translated into a torque at point F so that the observed torque is given by twice the smaller of the two torques, i.e.,

$$\text{torque} = 2(\tfrac{1}{2} p a_o d), \tag{3}$$

where the term in the parenthesis is the smaller of the two values. The pressure in this torsion system is given by

$$p = C_2 \phi, \tag{4}$$

where ϕ is the angle of torsion and C_2 is a constant including the torsion constant of the fiber, orifice area, arm lengths, etc.

Using a known substance as a calibrant, C_2 can be determined. Eq. (4) then enables direct measurement of pressure without knowledge of molecular weight. Equations (2) and (4) together may be used to determine the molecular weight (or average molecular weight if several species are present) via

$$M_A = lTC_1^2/tC_2^2\phi^2. \tag{5}$$

The torsion technique for direct pressure measurement and molecular weight determinations is more powerful than the simpler effusion approach. It does, however, depend on the precision of two calibration steps and on orifice reflection phenomena, which are inherent in these calibrations. One must assume that, for example, the values C_1 and C_2 are constants for all substances, which need not be the case.

Experimentally, effusion and effusion torsion techniques have been applied with many modifications involving effusion-cell design, microbalance design, heat shielding, and a host of other features. While potentially powerful, such methods are not trivial experimentally, as evidenced by rather marked discrepancies among data obtained on presumably similar systems.

2. *Transpiration Methods*

A second, indirect method of somewhat greater simplicity that is applicable to much greater pressure ranges than the effusion approach is that known as the *transpiration method*.

In essence, it involves the passage of an inert gas at a known, but not necessarily constant, rate over the material whose vapor pressure is to be determined. The rate of flow is adjusted in such a way that equilibrium is established between the carrier gas and the material, and the former becomes saturated. If we invoke ideal gas assumptions and an assumed molecular weight for the species in the vapor phase, the pressure may be calculated. As with the Knudsen approach, it is most usable when only a single vapor-phase species predominates and where gas ideality is a realistic approximation.

The transpiration technique, because of the use of a flowing gas is effected most usually at a total system pressure of 760 torr (1 atm). A description of its operation in conjunction with the determination of the vapor pressure of a simple substance such as iodine will point up the salient experimental features of the approach. For simplicity, it will be assumed that the environmental pressure is constant during the course of the experiment and is precisely 760 mm. The apparatus is shown schematically in Fig. 2. A reactive carrier gas H_2, is introduced

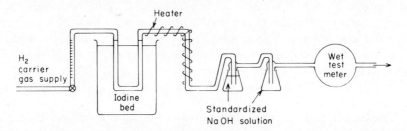

FIG. 2. A transpiration apparatus for the determination of the vapor pressure of iodine.

into a controlled temperature iodine saturating bed via a calibrated flow meter, which is used only to obtain approximate flow rate settings. For the case in question, the transpired iodine emanating from the iodine saturating bed is carried through a platinum wool catalyst tower at 300–400°C where it is converted to HI. In the region between the iodine bed and the catalyst tower, the tubing must be heated to prevent condensation of the iodine.

The HI effluent in the excess H_2 carrier gas is carried through a series of standardized NaOH solutions where the HI reacts to form H_2O and NaI. The carrier gas, stripped of HI is then passed through a gas measuring meter (i.e., a wet test meter) where its volume is determined accurately. The quantity of iodine transpired from the iodine bed in a given time may be determined by back titration of the

B. PRESSURE MEASUREMENTS

NaOH solutions using standard analytical techniques. A sufficient number of NaOH traps must be placed in series so that the last of these functions as a blank.

In any transpiration experiment a means must be devised for determining the quantity of material transpired. In the case under discussion, a convenient method for evaluating transpired iodine is that already described. Alternative methods could have been employed with essentially equal effectiveness.

One test that must be invoked in a transpiration experiment is that which demonstrates that the carrier gas is saturated with the evaporating material. The method for demonstrating this is to decrease the flow rate through the source bed and determine whether the quantity of transpired material/unit volume is constant. Experimentally, one begins at an arbitrary practical flow rate and then repeats the experiment at successively lower flow rates. The maximum practical flow rate must lie on a plateau of a flow rate versus transpired material/unit volume curve (Fig. 3). Such a procedure has to be followed at each new source temperature to insure that equilibrium between the source bed and the carrier gas has been achieved. In general, experience will enable the experimenter to restrict this experiment to as little as two flow rates at each temperature.

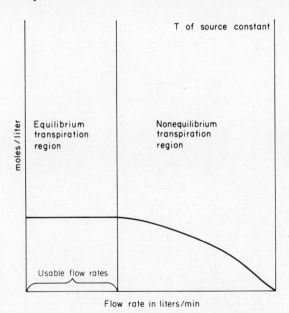

FIG. 3. Determination of the equilibrium region in a transpiration experiment.

Assuming now that the behavior of the system at one atmosphere pressure is ideal and that the species in the vapor phase are known, one may evaluate the vapor pressure of a pure substance at some temperature T_{bed}, knowing the quantity of material m_a carried in a defined volume V.

With reference to Fig. 2, let us assume that the measured volume in the wet test meter at the end of the reaction train is V and the temperature of the meter is T_M (°K). Since the gas being measured becomes saturated with water vapor in the wet test meter, the measured volume V is larger than the quantity of dry gas at that point would be at 1 atm total pressure. The carrier gas after picking up iodine reacts partially with the latter to form HI. When the HI is titrated, the reacted hydrogen is converted to water. Consequently, the measured volume of effluent hydrogen at the output of the entire system due to this effect is smaller than the input amount by exactly the number of moles of iodine transpired. This amount must be corrected for also if a precise answer is sought. If the vapor pressure is relatively small, i.e., the number of moles of carrier gas is very much greater than the number of moles of transpired iodine, the hydrogen used to form HI may be neglected. In determining the vapor pressure of iodine up to its melting point, neglecting this correction will lead to an error of as much as 0.5%.

The moles of dry hydrogen passing through the wet test meter at temperature T_M, where the total measured volume V, is given by the ideal gas law in the form

$$n_{H_2} = (V/RT_M)(1 - p_{H_2O}), \qquad (6)$$

where the value of the partial pressure of water is given in atmospheres, R in liter atm/mole°, T in °K and V in liters.

Since the number of moles of I_2 is independently determined, the quantity of H_2 used to react with the I_2 is also known. The total number of moles of carrier gas $n_{H_2}^t$ is given then by

$$n_{H_2}^t = (V/RT_M)(1 - p_{H_2O}) + n_{I_2}, \qquad (7)$$

where the last term on the right-hand side accounts for H_2 reacted in forming HI.

At the iodine bed, the total volume of I_2 and H_2 that coincides with the sum $n_{H_2}^t + n_{I_2}$,

$$n_t = n_{H_2}^t + n_{I_2} = (V/RT_M)(1 - p_{H_2O}) + 2n_{I_2}, \qquad (8)$$

B. PRESSURE MEASUREMENTS

is given by

$$V_{\text{bed}} = [(n_{H_2}^t + n_{I_2})RT_{\text{bed}}]/P_t = (n_{H_2}^t + n_{I_2})RT_{\text{bed}}. \tag{9}$$

where $P_t = 1$ and T_{bed} is the temperature of the iodine in °K.

Having determined V_{bed} and knowing n_{I_2} and the temperature of the iodine source T_{bed}, the pressure of iodine at this point is

$$p_{I_2} = n_{I_2}RT_{\text{bed}}/V_{\text{bed}}, \tag{10}$$

where the I_2 pressure is in atmospheres. We may now substitute in Eq. (10) the value for V_{bed} given in Eq. (9) to obtain,

$$p_{I_2} = n_{I_2}RT_{\text{bed}}/(n_{H_2} + n_{I_2})RT = n_{I_2}/(n_{H_2}^t + n_{I_2}) = n_{I_2}/n_t. \tag{11}$$

For $n_{H_2}^t$ in Eq. (11) we may substitute the value given in Eq. (7) to obtain

$$(p_{I_2})_{T_{\text{bed}}} = n_{I_2}\{[V(1 - p_{H_2O})/RT_M] + 2n_{I_2}\}^{-1}. \tag{12}$$

If p_{I_2} is desired in torr and p_{H_2O} is also given in this same unit, we may write

$$(p_{I_2})_{T_{\text{bed}}} (\text{torr}) = \{n_{I_2}\{[(V/RT_M)(760 - p_{H_2O}(\text{torr}))/760] + 2n_{I_2}\}^{-1}\} 760. \tag{13}$$

Note that in the solution, the only temperature and volumes required are those at the carrier gas measuring station, those at the iodine bed having canceled out in the derivation.

For the general case where the carrier gas does not react with the transpiring material we may write

$$p_A (\text{torr}) = \{n_A\{[(V/RT_M)(760 - p_{H_2O}(\text{torr}))/760] + n_A\}^{-1}\} 760. \tag{14}$$

For the general case where the dry volume of the carrier gas is determined, we may write

$$p_A (\text{torr}) = \{n_A/[(V/RT_M) + n_A]\} 760. \tag{15}$$

Errors may creep into the experiment if the system pressure differs significantly from 1 atm, or if it varies considerably during the experiments, and if the temperature at the carrier gas measuring point changes significantly during the course of the experiment.

Once having determined p_A as a function of T, a plot of log p_A versus $1/T$ may be used to estimate the heat of sublimation or vaporization, as the case may be, providing ΔH is constant over the temperature interval of interest.

From
$$\log p = -(\Delta H/2.303RT) + C, \tag{16}$$
and the value of ΔH and one pressure value, the constant of integration may be determined to provide an equation describing the straight line graph $\log p_A$ versus $1/T$. This constant of integration is $\Delta S^0/R$, where ΔS^0 is the standard entropy of the process.

3. *Bourdon Gauge Techniques*

The final vapor pressure measuring technique that will be considered is that known as the Bourdon gauge method. In this method, the pressure developed in a chamber is translated to, and results in the distention of a diaphram. The pressure needed to counterbalance this distortion is measured. The technique is powerful in that it does not require any assumptions as to molecular weights, and may be used in certain instances to deduce molecular weights. It is also useful for determining the pressure of multispecies vapors.

To demonstrate the use of the technique both as a pressure measuring tool and also to indicate its utility in studying vapor-phase equilibria, the following example is of interest: A schematic representation of a modification of the Bourdon gauge, known as a kidney or sickle gauge, is shown in Fig. 4. The gauge is constructed entirely of quartz and is thus usable to about 1100°C. Starting at the bottom of Fig. 4, there is a sample holder e which terminates at one end in the distortable kidney d. This kidney is made by blowing a small region of molten quartz. The temperature of the sample holder is monitored via the thermocouple well f, which is centrally located within it. The top portion of the gauge is isolated from the lower portion by a partition g, the kidney extending into the top portion. Attached to the kidney is a pointer c which terminates in a flag B. The flag has a reference mark on it. If a sample is placed in the sample holder and the system is heated, a change in pressure in the sample holder will cause the kidney to distend, thereby causing a motion in the flag. If a counter pressure is exerted through the control port a, the distortion can be counterbalanced until the pointer returns to its original or null point position. The counterpressure may be measured and corresponds to the pressure within the sample holder. Automated designs for such equipment have been reported where either manual or automated back pressure readings may be taken.

The sensitivity of this type of pressure measuring equipment is a function of the thickness of the kidney wall, the accuracy of the ability to return the pointer to the null position, and the ability to read the

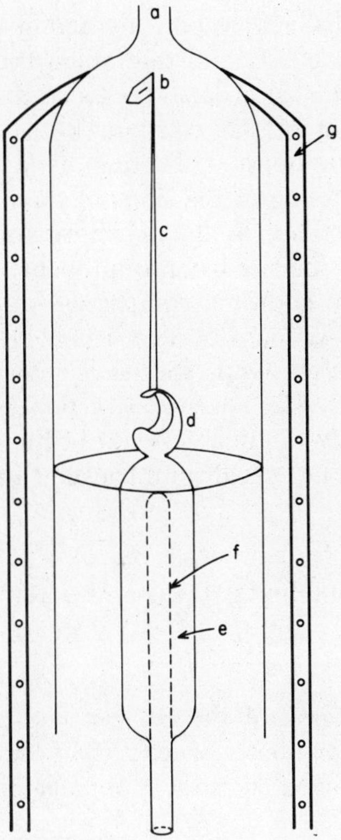

B. PRESSURE MEASUREMENTS

FIG. 4. A sickle gauge. (a) Pressure balance control and pressure measure; (b) flag with reference mark; (c) pointer; (d) sickle; (e) sample chamber; (f) thermocouple well; (g) partition.

back pressure. A usable sensitivity range is of the order of ± 0.1 torr, although room temperature nonvitreous versions of the gauge may be an order of magnitude more sensitive. The direct measurement of the pressure of a pure substance is relatively straightforward and requires no further discussion. An additional use involves obtaining species information in simple chemical equilibrium processes, an example of which is given in the following:

When Ge reacts with $GeCl_4$, a number of possible pathways, each resulting in a two-phase, two-component equilibrium, may be visualized. These are tabulated as

(a) $3Ge(s) + GeCl_4(g) \rightleftarrows 4GeCl(g)$, (d) $3Ge(s) + GeCl_4(g) \rightleftarrows 2Ge_2Cl_2(g)$,
(b) $Ge(s) + GeCl_4(g) \rightleftarrows 2GeCl_2(g)$, (e) $Ge(s) + GeCl_4(g) \rightleftarrows Ge_2Cl_4(g)$,
(c) $\frac{1}{3}Ge(s) + GeCl_4(g) \rightleftarrows \frac{4}{3}GeCl_3(g)$, (f) $\frac{1}{3}Ge + GeCl_4(g) \rightleftarrows \frac{2}{3}Ge_2Cl_6(g)$.

The differentiating feature among several of the equilibria considered is that the number of gas molecules present changes while a reaction is taking place. Experimentally, it was observed that starting with Ge and $GeCl_4$ vapor, a reaction began between the two at approximately 500–600°C and that when the reaction had subsided, the slope of the p versus T data was twice that at the lower temperatures prior to the onset of the reaction. Both the low and high temperature p–T data were linear indicating good adherence to ideal gas behavior. This doubling of the slope of the linear portions of the curves indicates that reactions (b) and (d) above are potential models, the remaining specified possibilities being improbable.

In addition, comparison of the data with observations of the system Ge–I_2 via Bourdon gauge and weight-loss measurement techniques, where a Ge_2I_2 species has not been observed, argued somewhat against a Ge_2Cl_2 species. Reaction (b) was, therefore, considered the most likely candidate. A test of the model was obtained as follows:

The equilibrium constant for reaction (b) is given by

$$K_p = (p_{GeCl_2})^2/p_{GeCl_4}, \qquad (17)$$

while the total pressure in the gauge is given by

$$P_t = p_{GeCl_4} + p_{GeCl_2}, \qquad (18)$$

since all of the halogen had previously been shown to be confined to vapor-phase species. Molecular conservation of the component Cl (designated as the monomer) in terms of the assumed species is given by

$$M_{Cl}RT/V = 4p_{GeCl_4} + 2p_{GeCl_2}. \qquad (19)$$

Simultaneous solution of Eqs. (18) and (19) for the species $GeCl_2$ and $GeCl_4$ yields

$$p_{GeCl_4} = (M_{Cl}RT/2V) - P_t, \qquad (20)$$

$$p_{GeCl_2} = 2P_t - (M_{Cl}RT/2V). \qquad (21)$$

For each sample used, the term $M_{Cl}RT/2V$ is a constant and may be determined from the observed linear portions of the p–T curves. Thus, at low temperatures where only $GeCl_4$ is present in the vapor, Eq. (20) may be used to determine $M_{Cl}RT/2V$. In the high-temperature linear region only $GeCl_2$ is present in the vapor, and Eq. (21) may be used to determine the constant. At all intermediate temperatures where a mixture of the two species is present, the pressure of each is now readily evaluated, as is K_p as a function of T. It was found that K_p was constant with varying pressure and that $\log K_p$ was linear with respect to $1/T$, indicating that reaction (b) is a satisfactory model. If other species such as GeCl, $GeCl_3$, or Ge_2Cl_3 were present in significant quantities

C. MEASUREMENT OF THERMAL ANOMALIES

and not accounted for, K_p would not be expected to be independent of total pressure and neither would linear semilog plots be expected. The Bourdon gauge technique just described has an obvious advantage over the transpiration method in that pressures considerably in excess of atmospheric may be measured. It suffers from being somewhat less sensitive, and from being usable only to approximately 1100°C. Together, however, the transpiration and Bourdon techniques represent a powerful combination for investigating a wide variety of substances over extensive pressure and temperature intervals. When these techniques are used in conjunction with mass spectrometric and ultraviolet or other absorption techniques, they become even more powerful. For example in the Ge–Cl$_2$ study cited, ultraviolet absorption experiments indicated the presence of only one species in addition to GeCl$_4$, thus further substantiating our conclusions.

C. Measurement of Thermal Anomalies

1. *Conventional Thermocouple Thermometry*

Two techniques are most usually employed to measure latent heat anomalies in condensed-phase systems as a function of varying temperature. These are the so-called *thermal analysis method* and a powerful modification of this method called *differential thermal analysis* (DTA). To understand the underlying principles of the latter, it is of value to discuss the first method, although in practice DTA represents the preferable tool.

When two dissimilar metal wires are connected at junctions at either end as shown in Fig. 5, an electromotive force (emf) is generated if the junctions are at different temperatures. The combination of two

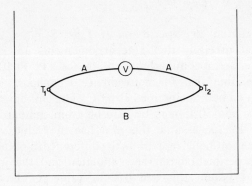

FIG. 5. A thermocouple: (A) one metal, (B) another metal.

dissimilar metals (pure elemental metals or alloys), as shown in the figure, is known as a thermocouple. In all instances in which the thermocouple is employed as a temperature measuring element, one of the junctions is maintained at a reference temperature. This reference temperature is purely an arbitrary choice since an emf develops only when the temperature at the second junction is different from that at the first junction. For convenience in setting up tables for specific thermocouples, e.g., Pt–Pt 10% Rh, the emf's developed by the measuring junction are compared to a reference junction at 0°C or 0°F. When put into use, however, the reference junction may be maintained at any steady convenient temperature. The temperature of the measuring junction may then be evaluated by measuring the emf developed between the two junctions, adding to this value the emf difference between the reference value listed in the table and the reference value used in the experiment and then determining the temperature from the table. Figure 6 shows a schematic representation of a thermocouple as it might

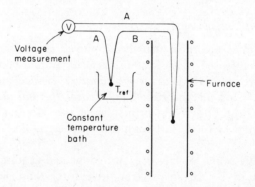

FIG. 6. A schematic representation of a thermocouple experiment.

appear physically in an experiment. If Figs. 5 and 6 are compared, it is noted that precisely the same arrangements are shown. Let us assume that the thermocouple shown in Fig. 6 is a Pt–Pt 10% Rh couple. Let us also assume that for convenience we choose as a reference temperature an oil bath maintained at 50°C. The emf developed between a 0°C reference and 50°C may be read off from standard table for this thermocouple. Let us assume this emf has the value x. The voltage measured on the voltmeter represents the difference in potential generated between the experimental reference junction and the junction in the furnace. Let us assume this emf has the value y. The total emf that would have been measured if the reference junction had been at 0°C is

C. MEASUREMENT OF THERMAL ANOMALIES

$x + y$. This sum may be looked up in the table to provide the temperature value in the furnace.

It is seen, in both Figs. 5 and 6, that only one of the thermocouple wires is interrupted by the voltage measuring device, i.e., a voltmeter or potentiometer. It is highly improbable that either the contacts to this meter or the internal wiring comprise the same metal as the interrupted leg of the thermocouple. In all probability, several different metals are interspersed in the path of the interrupted thermocouple leg as shown schematically in Fig. 7. The thermocouple is made up of

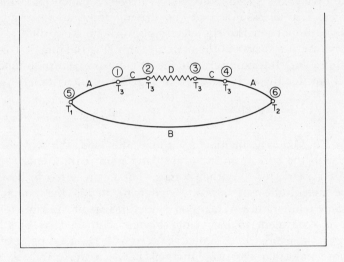

FIG. 7. The presence of other metals in a thermocouple.

the dissimilar metals A and B, the junctions between these metals being at temperatures T_1 and T_2. The measuring meter is assumed to be made up of metals C and D, junctions between these metals and between metal C and A being formed at the positions noted. It is seen that each dissimilar metal forms two junctions with every other dissimilar metal with which it comes into contact. So long as each of these junctions is at the same temperature, the metals C and D cause no emf effect on the measuring thermocouple made up of the metals A and B. Junctions 1 and 4 which represent contact between metals A and C need not be at the same temperature as junctions 2 and 3 which represent the contacting of metals C and D. Junctions 1 and 4 must, however, both be at the same temperature and junctions 2 and 3 must be at the same temperature if they are to have no effect on the emf difference between junctions 5 and 6. When the temperatures of junctions 1 and 4 are the same, we have an analogous situation to that when junctions 5 and 6 are at the same temperature.

If junctions 1 and 4 were at different temperatures, an extraneous emf would develop which would add to or substract from the emf due to a difference in temperature between junctions 5 and 6. This subtractive or additive difference would depend upon which leg of the thermocouple C–A was positive. The magnitude of the error would depend on how much of an emf develops in this thermocouple system per degree of temperature difference.

Qualitatively, the effects noted above may be defined mathematically as follows. Let us assume that the driving force at a thermocouple junction is proportional to the electronic work function difference $W_{xy}^{T_a}$ at a temperature T_a between the metals comprising the junction. Starting at junction 1 and continuing around the loop until we return to junction 1, the total emf for the system shown in Fig. 7 may be written

$$\text{emf}_t \propto W_{AC}^{T_3} + W_{CD}^{T_3} - W_{DC}^{T_3} - W_{CA}^{T_3} + W_{AB}^{T_2} - W_{BA}^{T_1} \propto W_{AB}^{T_2} - W_{BA}^{T_1}. \quad (22)$$

Note in Eq. (22) that in moving around the loop, the wire A makes contact with the wire C and that at a later point wire C makes contact with wire A, etc., for the other metals. The net result is to have emf's of equal magnitude but opposite sign generated if the junctions are at the same temperature. When these junctions are at the same temperature, it is equivalent to balancing the emf delivered by one battery with an emf delivered by a similar battery with reversed polarity.

Suppose now that the two junctions between the metals A and C, for example, are at different temperatures. The junctions 1 and 4 will generate a net emf, and depending upon whether A or B is the positive thermocouple leg and upon which of the junctions is at a higher temperature, the signal due to junctions 5 and 6 will be added to or substracted from the net emf generated by junctions 1 and 4.

In all normal thermocouple operation it is the case that we are measuring the differential output between two junctions, one of which (the reference junction) is at a constant temperature. In other words the temperature of one junction varies while that of the other (the reference junction) remains constant. Since the magnitude of this differential is generally in the millivolt range there is no need to amplify the differential signals. In general, one chooses as a thermocouple pair, metals whose output/° C differential is the greatest. This choice is, of course, also influenced somewhat by the temperature range of measurement, the requirements for inertness, etc. At moderate temperatures (up to several hundred degrees centigrade), copper–constantan thermocouples are frequently employed. For experiments up to 1200 or 1300°C, particularly in inert atmospheres, chromel–alumel thermocouples are used. Where oxidizing environments are involved and temperatures up to 1600–1700°C are to be measured or where precision grade thermocouples with notable long term stability are required, platinum–platinum 10 or 13% rhodium are preferred. These exhibit considerably smaller differential emf's/°C than the others and require more precise measuring equipment. For temperature measurements up to perhaps 2200°C, indium–indium rhodium couples are employed. The calibration curves for the these couples are not as precisely defined as those for the lower temperature thermocouples, since the calibrations are based solely on optical pyrometry together with a few poorly defined melting point standards.

In general, the differential signals for the three most frequently employed thermocouples are

Copper–Constantan ~ 0.04 mV/°C,
Chromel–Alumel ~ 0.04 mV/°C,
Platinum–Platinum 10% Rhodium ~ 0.005 mV/°C.

The temperature versus emf curves of thermocouples are not linear, so the above differentials, referenced to 0°C, vary somewhat depending on the temperature of the measuring junction.

If the reference junctions are maintained at 0°C, the sign of the emf will reverse when the measuring junction drops below 0°C in temperature. Tables of thermocouples are generally available below 0°C only for copper–constantan.

2. *Thermal Analysis*

If a binary mixture in a simple eutectic interaction is made molten and then cooled at a known, or better still, a uniform rate in an apparatus such as that depicted schematically in Fig. 8, a plot of the temperature of the melt versus time will have the appearance shown in Fig. 9.

Before discussing Fig. 9, it is of value to consider some of the details of Fig. 8. The furnace employed should have a large length-to-diameter ratio so that a flat temperature zone of sufficient length to encompass the sample is available. A ratio of 10/1 or more is generally needed. The furnace may advantageously have three separate windings (two outer and one inner) that may be separately controlled so as to permit tailoring of the gradient within the furnace. Each of these windings may be raised or lowered in temperature by a cam-controlled variable temperature controller or by a motor drive. The first method enables

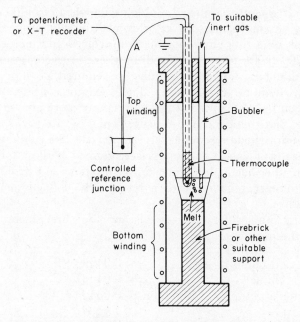

Fig. 8. A conventional thermal analysis experimental arrangement.

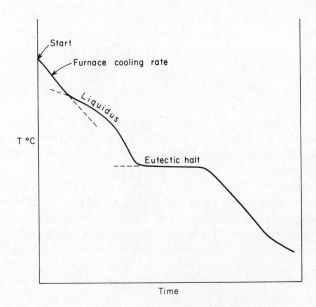

Fig. 9. A cooling curve in a binary eutectic system.

one to achieve linear temperature variation of the furnace temperature with time, the latter allows one to obtain a reproducible, but not necessarily uniform, temperature variation with time. It is useful, for certain purposes, to have a double wound furnace, i.e., an inner wound core and an outer wound core. Thus, if higher temperature experiments are to be performed, the inner core winding need not be overburdened. At other times, it is advantageous to have a furnace having a low heat capacity, i.e., one lightly lagged by the surrounding insulation, in order that rapid variations in temperature be possible. At still other times, it is useful to have heavily lagged furnaces. The number of furnace designs is large and it is best to have a few different types available.

The sample container is an important aspect of the experiment and must be chosen such that it will not be attacked by the molten sample. It may be of closed design or an open crucible depending again on the substance being studied. Quartz, platinum, platinum–rhodium, gold–palladium, gold, and silver receptacles have found extensive use.

In Fig. 8 it is seen that the measuring thermocouple is covered by a protective capsule affixed to a ground connection. This protective capsule prevents contact of the junction with the melt. It must be unreactive and have a high heat conductivity. The thermocouple support rod may be of ceramic or quartz, and special rods are available from a number of sources. These rods have one, two, or more wire channels in them, the thermocouple wires being available in diameters from about 1 mil up.

Time–temperature recorders (giving readings of emf) are readily available for recording data, although the more laborious manual recording of data may be employed with the use of a clock and a potentiometer. While the tendency to use rapid cooling or heating rates (approximately 5–10°C/min or more) is prevalent, the limited ability to establish quasi-equilibrium at these rates plus the attending temperature lag between sample and furnace walls will often yield inaccurate results. A good policy is to establish the maximum usable temperature-variation rate in preliminary experiments by decreasing the temperature variation rate until the recorded temperatures of given anomalies become constant.

Figure 8 shows a bubbler tube in the melt which provides agitation of the melt via gas stirring. The precautions to be observed in the choice of materials for this bubbler are obvious. It should be noted that sometimes an oxidizing or other environment must be maintained over the sample to prevent decomposition. The bubbling gas may serve a dual function in such instances.

With the preceding terse description of a few of the more important experimental aspects of a thermal analysis we can now focus our attention on Fig. 9, which shows schematically a time–temperature cooling curve in a binary eutectic system. At the start of the experiment, the sample is made completely molten. The furnace is then cooled at a known rate. So long as the sample remains completely molten, the temperature of the melt will follow the temperature of the furnace once it comes into steady-state balance with the latter. When the liquidus is reached, material starts to crystallize and latent heat of crystallization is evolved. At this point, some of the heat loss of the sample to the cooler furnace walls is counterbalanced by this evolved heat and the rate of temperature fall of the sample becomes less than the previous rate. During this liquidus crystallization the system is univariant, only sufficient crystallization occurring at each temperature to satisfy equilibrium requirements. Consequently, the temperature of the melt drops during liquidus crystallization. Finally, when the eutectic temperature is reached, both solid phases crystallize simultaneously and all of the liquid is consumed. This crystallization is isothermal in nature and the temperature of the system cannot drop until the liquid phase disappears. In practice, however, there may be three ways in which quasi-equilibrium may fail to be approximated during the cooling cycle. (1) The liquidus crystallization may not occur until the system has supercooled to a small or large extent, (2) the eutectic crystallization may supercool, and (3) the eutectic halt may not be isothermal if the heat loss to the walls greatly exceeds the kinetics of the eutectic heat of crystallization. A schematic representation of what might occur is shown in Fig. 10. The eutectic temperature may be evaluated in Fig. 10 by extrapolating the isothermal or near isothermal halt back to the base line curve, i.e., the dotted line. The start of the liquidus crystallization may also be approximated by extrapolating the steady state portion of the liquidus halt. A technique for avoiding undercooling is to premelt a sample of similar composition, allow it to crystallize, and then pulverize it, using the crushed material as seeds. These seeds are then dropped into the melt during the cooling cycle at frequent intervals.

After a cooling cycle is run, the same sample may then be examined in a heating cycle. The eutectic halt may be evaluated in the heating cycle, but is extremely difficult, if not impossible, to obtain information about the liquidus during heating since one would have to establish the point at which the last crystallite melts.

The determination of the binary eutectic phase diagram by the technique of thermal analysis involves obtaining traces of the types shown in Figs. 9 and 10 at a sufficient number of compositions so that the liquidi are unambiguously delineated. Figure 11 shows schematically

C. MEASUREMENT OF THERMAL ANOMALIES

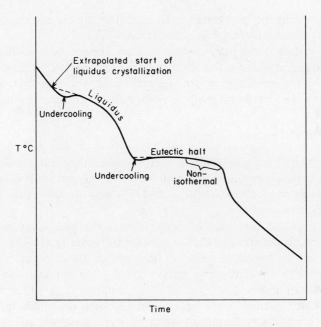

Fig. 10. Undercooling in the regions of crystallization.

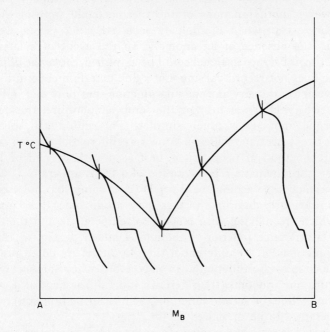

Fig. 11. The use of thermal analysis for defining a simple binary eutectic diagram.

a composite drawing of several idealized cooling curves and the construction of the phase diagram using the derived data.

It is seen that as the pure components are approached, cooling curves exhibit larger liquidus anomalies and smaller eutectic anomalies. The eutectic halt maximizes at the eutectic and the liquidus anomaly, in the form of an invariant melting point halt, maximizes at pure A or B.

3. *Differential Thermal Analysis*

Rather than continue a discussion of phase diagram resolution via thermal analysis approaches, it will better serve our purposes at this point if we introduce the technique of DTA and use this as the basis for further discussion.

It was pointed out that in conventional thermal analysis, one of the junctions is constant (stationary) while the other is varying (moving) with respect to temperature as a function of time. Since a slope is associated with the emf–time curve, even when no anomaly occurs, the onset of an anomaly is detected by a change in slope of the base-line curve. When the total latent heat associated with the anomaly is small, because either a small amount of crystallization is occurring, or the latent heats are small, the onset of an anomaly may be difficult to detect.

If, however, both junctions of the thermocouple were at the same temperature except when an anomaly occurred, a zero signal would be present in the absence of an anomaly. In the event of a latent heat anomaly, the unbalance in signal could be amplified. Since the differential technique does not record varying emf's due to a difference in temperature between a stationary and moving junction, but only emf differences between two junctions at near or the same temperature, the detection of a thermal anomaly is much simpler. This differential method of detecting latent heat anomalies is known as differential thermal analysis (DTA). Its method of operation is best seen with the aid of Fig. 12. The latter is a schematic representation of a DTA apparatus.

The thermocouple arrangement for DTA is somewhat more complex than was previously described in that there are in effect three junctions, two of which are moving with respect to temperature and the third of which is a reference junction. Junctions 4 and 5 of Fig. 12 represent the two junctions of the differential thermocouple. As noted previously all thermocouples are differential in nature but when measuring actual temperature one junction (the reference junction) is kept at a constant temperature while the second junction (the measuring junction) monitors the temperature to be measured.

When used as a "differential thermocouple," however, both the

C. MEASUREMENT OF THERMAL ANOMALIES

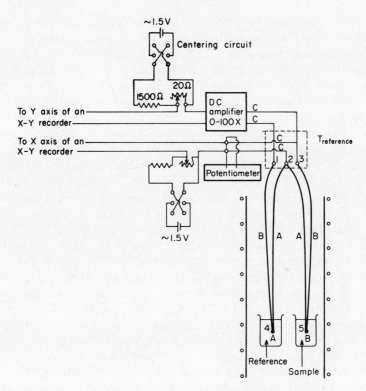

FIG. 12. A schematic representation of a DTA arrangement.

reference and measuring junctions monitor essentially the same temperature. So long as each portion of the system being monitored is at the same temperature, both junctions see this same temperature and the differential signal between them is zero. In Fig. 12, junctions 4 and 5 represent the differential junctions, the wire made of the metal B being interrupted by the voltage measuring circuit. The connections to the voltage measuring circuit are made by the arbitrary metal C (a copper wire, for example). So long as points 1 and 3 are at the same temperature, no extraneous emf is introduced into the signal developed by junctions 4 and 5 of the thermocouple A–B. For practical purposes, the differential signal between junctions 4 and 5 has a counter voltage circuit (the centering circuit) interrupting one of the thermocouple legs. The purpose of this circuit is to enable the adding or substracting of a fixed voltage to the differential voltage to enable centering of the signal at any arbitrary constant voltage. Thus, when the differential signal is introduced into one of the axes of an x–y recorder, the zero differential signal (base line) may be centered anywhere within the voltage range of the instrument. As shown, the centering voltage is capable of adding or subtracting approximately 20mV to the differential signal. Prior to introduction into the centering circuit, the differential signal is sent through a dc amplifier having the multiplicative range shown. This enables the use of small samples and/or samples exhibiting very small latent heat anomalies.

While the differential thermocouple enables one to detect the occurrence of an anomaly, it does not enable us to estimate the temperature at which the anomaly is occurring. This is where the added complexity arises in Fig. 12. The points 2 and 3 of Fig. 12, separated from one another by the third metal C represent one junction of a measuring thermocouple whose second junction is the point 5. Relative to the points 2–3, point 5 is the measuring junction of a conventional thermocouple while the points 2–3 together represent the reference junction of this conventional measuring thermocouple. The signal from this measuring thermocouple is sent to the second axis of an x–y recorder, its A leg being interrupted by a zero suppression circuit capable of adding or subtracting a fixed voltage during the course of an experiment. This circuit is identical to the centering circuit previously discussed except that its function is somewhat different. If our recorder has a useful voltage measuring range of 10 mV and we are measuring an emf of 13 mV, the signal is greater than can be accommodated. The zero suppression circuit enables us to introduce an opposing signal of 3 mV or more thereby reducing the signal into the recorder to 10 mV. The actual signal from the measuring thermocouple may be evaluated after the zero suppression signal is used to place the input signal within the recorder's range. The difference between the value of the suppressed signal on the recorder and the measured value at the potentiometer is the zero suppression voltage. This latter must be added to the value noted on the recording potentiometer to determine the precise emf at any time. Several means other than the simple, nonetheless effective, method for zero suppression shown in Fig. 12 are available. These are electronic in nature and offer the advantage of not relying on a battery to supply constant centering or suppression signals.

It is seen in Fig. 12 that two containers are present in the furnace, one of these containing a reference material, the other containing the sample. Upon heating or cooling the furnace, no differential signal is developed between junctions 4 and 5 so long as the material in the appropriate container is not undergoing an anomaly. If the reference material is inert, by which we mean that it does not melt or undergo a phase transformation, then it will always follow the furnace walls in temperature during either a heating or cooling cycle. If the sample undergoes some anomaly during heating or cooling, its temperature drop or rise will be perturbed when melting or undergoing some other types of phase transformations. At this point, its temperature will be lower or higher than that of the reference junction depending upon whether it is in a heating or cooling cycle, respectively, and a differential signal will develop between the junctions 4 and 5. Figure 13 shows a

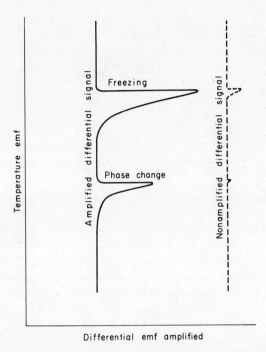

FIG. 13. Freezing and phase change in a pure substance.

schematic representation of such a differential trace for a cooling curve in which the freezing of a substance followed by a phase transformation occurs. Depending upon how one connects the differential thermocouple to the recorder, the exothermic phenomena depicted, may occur to the right or the left. As depicted, a heating cycle would show endotherms occurring to the left of the base line. For comparison, the same trace, as it might appear if the differential signal were not amplified, is shown by the dotted line in Fig. 13.

Figure 14 shows schematically a DTA cooling trace of a system exhibiting a liquidus crystallization, an incongruent transformation, a eutectic halt, and a phase transformation. In the phase diagram,

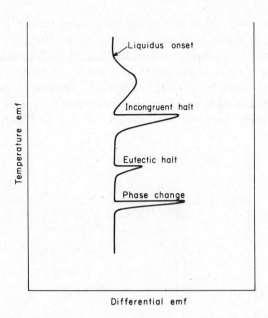

Fig. 14. Cooling cycle of a sample exhibiting a variety of latent heat anomalies.

the location of the incongruently melting compound would be determined by noting where the incongruent halt maximizes in size and where the eutectic halt tends toward zero. Since maintenance of equilibrium conditions in such a system is sometimes difficult, because a reaction must occur at the incongruent temperature, cooling curves alone may not suffice. If possible, samples should be solid-state reacted or liquid-phase quenched, followed by long term annealing, and heating curves should be run. These will be effective in determining the first composition at which a eutectic halt is observed near the incongruently melting compound location. These same samples prior to being examined by heating curves may be used for X-ray examination to determine at what composition a second solid phase appears to either side of the incongruently melting composition. Similarly, cooling and heating curves will also reveal the absence of a eutectic to the left or the right of a particular composition, the latter representing the stoichiometry of the incongruently melting compound. Figure 15 depicts a system exhibiting one intermediate compound which melts incongruently. Differential thermal analysis traces at different compositions are superimposed on the phase diagram. An indication of behavior of the type shown in Fig. 15 is that the liquidus halt is large near pure A (Trace I), decreases in size as the incongruent point i is reached (Trace II),

C. MEASUREMENT OF THERMAL ANOMALIES

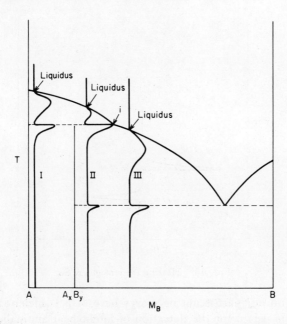

FIG. 15. DTA traces at several compositions in an incongruently melting system.

and becomes large again when the liquidus for A_xB_y is encountered (Trace III).

Since all equilibrium heating anomalies are accompanied by an absorption of heat while cooling anomalies are associated with exothermic phenomena, contrary behavior is indicative of metastable-to-stable processes taking place. For example, a substance such as selenium tends to crystallize as a glassy phase that is amorphous to X rays. Since the sensitivity of X rays is limited, the possibility exists that the glassy phase comprises crystallites of periodic microscopic nature with a particle size too small for X-ray detection of crystallinity. A heating trace of such a glassy phase has the appearance shown in Fig. 16. The DTA circuitry used for obtaining Fig. 16 is such that endotherms occur to the left. It is seen that at 77°C, an exotherm occurs and that at 214°C an endotherm occurs. This latter represents the melting point of Se. The 77°C effect represents the metastable phase transformation of amorphous to crystalline Se. Above 77°C but below 214°C, X-ray examination shows a crystalline pattern. Below 77°C, X-ray examination shows an amorphous pattern. The X-ray data coupled with the DTA data suggest strongly that the glassy phase is indeed amorphous.

While the number of examples of the use of DTA, coupled with other techniques for phase diagram resolution is numerous, further

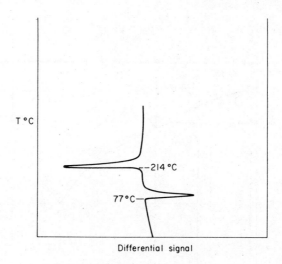

FIG. 16. Heating trace of glassy Se.

discussion on our part is not necessary here. It is sufficient to point out that DTA is useful for the detection of latent heat anomalies of almost any kind. Since the area under a DTA blip is proportional to the quantity of heat involved or absorbed, the method may also be employed as a calorimeter, and such usage has not been uncommon.

D. Other Useful Tools

While the brunt of the investigation of the condensed-phase behavior of a pure substance or mixture may be carried by DTA, this alone, except in simple instances, is not sufficient to delineate unambiguously the regions of a heterogeneous system. In conjunction with DTA, the use of solid state reacted or melt reacted samples for X-ray measurements are of great value. From the latter type of measurement one can define phase boundaries, define the number of phases, and detect nonequilibrium phases and the course of a reaction. Powder and single-crystal density measurements are useful for pinpointing compound location and solid-solution compositions, respectively. Resolution of solid solution diagrams whether of the ascending type or in systems exhibiting limited solid solubility is difficult because of the inability of maintaining equilibrium during a crystallization process. Samples may be made molten, then quenched by cooling rapidly. This will tend to keep crystallite size small. These quenched samples may then be pulverized and heat treated at temperatures below the solidus. Prelimi-

D. OTHER USEFUL TOOLS

nary experimentation is necessary to fix approximate solidus temperatures so that the annealing treatment may be conducted at as high a temperature as possible. X-ray examination of annealed samples after arbitrary time intervals is useful for determining whether equilibrium has been achieved in the annealing cycle. DTA heating curves of these annealed samples coupled with cooling curve studies and room or higher temperature X-ray analysis are sufficient in most instances for examining solid-solution systems.

A useful adjunct to differential thermal analysis, particularly in hydrate systems and the like, is the technique of thermogravimetric analysis, TGA. In its most efficient form TGA and DTA are run simultaneously on the same sample, with the TGA being used to monitor weight changes as a function of time. Using controlled environments where, for example, the aqueous tension is controlled, the variation in vapor pressure of a hydrate system with temperature may be determined, since dissociation of the hydrate system will not occur until its vapor pressure exceeds that of the environment.

Many phase transformations in pure materials tend to be sluggish. The fact that such transformations do not occur spontaneously makes the establishment of the temperature of transformation difficult. It is also a problem, in the cases where phase transformations are sluggish, to ascertain whether the transformation is reversible (an equilibrium phenomenon) or irreversible (thermodynamically spontaneous). The ultimate test is whether or not the transformation may be effected in two directions, i.e., starting with either of the phases, and producing the other. The technique most useful for identifying temperatures of transformation where latent heat anomalies cannot be detected because of the slowness of transformation is to heat treat samples at successively higher temperatures for varying lengths of time, followed by quenching them. Once the approximate temperature interval of transformation has been established, the increments in temperature of anneal may be reduced to define more precisely the temperature of transformation. Starting next with the higher temperature phase, annealings are conducted at temperatures below the previously established temperature of transformation to determine whether the process is reversible. In some instances annealing times of days, weeks, or months are necessary to acquire relevant data.

In cases where it is experimentally inconvenient to conduct elaborate DTA and related studies to establish the diagram of state or where only minute quantities of samples are available, so-called strip furnace techniques may be employed. Here a strip of inert, resistive material is used as a microfurnace, and a few milligrams of samples are placed on it. The temperature of the strip is raised while being monitored by an optical pyrometer and a microscope. At the points where meltings are observed to commence and terminate, the temperatures are noted. The system must be calibrated by the use of known melting point standards, preferably melting in the same temperature range as the unknown samples. Accuracies of the order of perhaps 10–20°C are obtainable by this method.

Appendix: References and Additional Remarks for Chapters 1–36, General Bibliography

In the following are provided additional comments which it was felt should be isolated from the body of the text because they would tend to interfere with continuity or because a separate mention would provide some emphasis. Not all chapters are commented upon, but this does not imply that comments are unneeded. Where comments are not given, it is anticipated that the reader will make use of the General Bibliography.

REFERENCES AND ADDITIONAL REMARKS

Chapter 3

For excellent introductory texts on probability and statistics the reader is referred to [1, 2]. Probability and distributions as the cornerstones of statistical mechanics are treated in context in [3, 4]. Errors in physical measurements are treated in [5, Chap. 2].

REFERENCES

1. H. L. Adler and E. B. Roessler, "Introduction to Probability and Statistics." Freeman, San Francisco, 1964.

2. F. Mosteller, R. E. K. Rourke, and G. B. Thomas, Jr., "Probability and Statistics." Addison-Wesley, Reading, Massachusetts, 1961.
3. J. E. Mayer and M. G. Mayer, "Statistical Mechanics." Wiley, New York, 1940.
4. G. S. Rushbrooke, "Introduction to Statistical Mechanics." Oxford Univ. Press, London, 1949.
5. N. H. Cook and E. Rabinowicz. "Physical Measurement and Analysis." Addison-Wesley, Reading, Massachusetts, 1963.

Chapter 4

For alternative presentations of the derivation of the phase rule, the reader is referred to the list of General References at the end of this Appendix, most of which contain such a treatment. Particularly, the reader is referred to Gibbs "Collected Works," Vol. I, cited among these General References and specifically to pp. 96–100 and 354–360.

The classic work on relating vapor pressures to evaporation rates and condensation coefficients is due to Langmuir and was based on studies of evaporation rates of wires in vacuum [1–3]. Wilkens modified Langmuir's approach to account for uncertainties about the area from which vaporization was occurring [4]. Questions concerning the evaporation phenomenon and deviations from the rate expected by applying Knudsen's equation [5] are discussed lucidly in [6]. An equally interesting paper on condensation coefficients is given in [7]. An extensive review entitled "Evaporation Mechanism of Solids" is cited in [8]. The reader is referred for additional pertinent information to the discussion and references relating to Chapter 36.

REFERENCES

1. I. Langmuir, *J. Am. Chem. Soc.* **35**, 931 (1913).
2. I. Langmuir and G. M. Mackay, *Phys. Rev.* **2**, 329 (1913); **4**, 377 (1914).
3. I. Langmuir, *J. Am. Chem. Soc.* **38**, 2221 (1916).
4. F. J. Wilkins, *Nature* **125**, 236 (1930).
5. M. Knudsen, *Ann. Physik* **28**, 75 (1909).
6. R. Littlewood and E. Rideal, *Trans. Faraday Soc.* **52**, 1598 (1956).
7. J. H. Stern and N. W. Gregory, *J. Phys. Chem.* **61**, 1226 (1957).
8. G. A. Somorjai, Evaporation Mechanism of Solids, *Progr. Solid State Chem.* **4**, 1 (1967).

Chapter 5

In conjunction with Section C, an excellent thoroughgoing treatment of free energy functions is given in [1]. The use of critical point phenomena, alluded to in Section D, particularly with water, for the growth of high melting materials at relatively low temperatures, was first described for the synthesis of silicates in 1913 [2] (see also [3]). A significant advance in the use of the technique came about when internally heated systems were designed [4]. From that time on the applications to a wide range of synthesis and single crystal growth, notably quartz, have been broad. An excellent review was published in [5]. The use in quartz crystal growth is described in detail in [6], and the application of the technique to more esoteric compounds, i.e., II–VI compounds, is presented in [7]. A recent brief introduction to the use of hydrothermal techniques for a wide variety of materials preparations is provided in [8].

An introduction to the general topic of phase transformations (Section H) is given in [9] (particularly Chapters 1, 5, 9, 11, and 14) where both first- and second-order transitions are considered. The mathematical treatment of order–disorder phenomena and phase changes based on the Ising problem are presented in [10]. The classic paper discussing second-order transitions, a topic which we have not considered, was written by Ehrenfest [11] to explain the liquid He I–He II transition. Hurst [12] subsequently claimed the Ehrenfest treatment to be a limiting case of phase transformations in general, rather than to represent a separate case. The topic of phase changes, primarily liquid–solid, is treated comprehensively in [13]. The topic of metastability and the nucleation process is treated elegantly by LaMer [14], and the melting phenomenon itself is addressed in a series of articles by Ubbelohde and co-workers [15–21]. Solid–solid and solid–liquid transitions as they relate to crystal structure are discussed in an excellent introductory article by Ubbelohde [22]. The solid–solid transformation was reviewed in a paper by Ubbelohde [23] in which second-order transformations are dealt with also. An excellent recent review article appears in [24, Chap. 4].

REFERENCES

1. J. L. Margrave, *J. Chem. Education* **32**, 520 (1955).
2. G. W. Morey and P. Niggli, *J. Am. Chem. Soc.* **35**, 1086 (1913).
3. G. W. Morey, *J. Am. Ceram. Soc.* **36**, 279 (1953).
4. F. H. Smyth and L. H. Adams, *J. Am. Chem. Soc.* **45**, 1172 (1923).
5. R. Roy and O. F. Tuttle, Investigations Under Hydrothermal Conditions. *In* "Physics and Chemistry of the Earth," Vol. I, Chap. 6, p. 138. Pergamon, New York, 1956.
6. A. C. Walker, *J. Am. Ceram. Soc.* **36**, 250 (1953).
7. A. Kremheller, A. K. Levine, and G. Gashurov, *J. Electrochem. Soc.* **107**, 12 (1960).
8. R. S. Bradley and D. C. Munro, "High Pressure Chemistry," p. 126. Pergamon, New York, 1965.
9. R. Smoluchowski, J. E. Mayer, and W. A. Weyl, "Phase Transformations in Solids." Wiley, New York, 1957.
10. H. S. Green and C. A. Hurst, "Order. Disorder Phenomena." Wiley (Interscience), New York, 1964.
11. P. Ehrenfest, *Leiden Comm.* Suppl. No. 75b, 1933.

12. C. A. Hurst, *Proc. Phys. Soc. (London)* **68**, 521 (1955).
13. D. Turnbull, *Solid State Phys.* **3**, 225 (1956).
14. V. K. LaMer, *Ind. Eng. Chem.* **44**, 1270 (1952).
15. J. N. Andrews and A. R. Ubbelohde, *Proc. Roy. Soc.* **228**, 435 (1955).
16. G. J. Landon and A. R. Ubbelohde, *Trans. Faraday Soc.* **52**, 647 (1956).
17. D. W. Plester, S. E. Rogers, and A. R. Ubbelohde, *Proc. Roy. Soc.* **235**, 469 (1956).
18. A. R. Ubbelohde, *Sci. J. Roy. Coll. Sci.* **26**, 1 (1956).
19. E. McLaughlin and A. R. Ubbelohde, *Trans. Faraday Soc.* **53**, 628 (1957).
20. G. J. Landon and A. R. Ubbelohde, *Proc. Roy. Soc.* **240**, 160 (1957).
21. E. McLaughlin and A. R. Ubbelohde, *Trans. Faraday Soc.* **54**, 1804 (1958).
22. A. R. Ubbelohde, *Brit. J. Appl. Phys.* **7**, 313 (1956).
23. A. R. Ubbelohde, *Quart. Rev.* **11**, 246 (1957).
24. C. N. R. Rao and K. J. Rao, Phase Transformations in Solids, *Progr. Solid State Chem.* **4**, 131 (1967).

Chapter 6

The topic of metastability and metastable transformations is the subject of a vast literature. To point up the problems encountered in ascertaining the number of phases exhibited by a compound and differentiating between stable and metastable phases for convenience we focus upon Nb_2O_5. A review is provided in [1] (See also [2-4]).

REFERENCES

1. A. Reisman and F. Holtzberg, The Structure and Physical Properties of Nb_2O_2 and Ta_2O_2. *In* "High Temperature Oxides" (A. M. Alper, ed.), Vol. 2, Chap. 7. Academic Press, New York, 1970.
2. F. Holtzberg, A. Reisman, M. Berry, and M. Berkenblit, *J. Am. Chem. Soc.* **79**, 2039 (1957).
3. A. Reisman and F. Holtzberg, *J. Am. Chem. Soc.* **81**, 3182 (1959).
4. H. Schäfer, R. Gruehn, and F. Schulte, *Angew. Chem.* **5**, 40 (1966).

Chapter 7

P. W. Bridgman is the acknowledged father of high pressure research, and the techniques he pioneered are described in great depth in his book [1]. Other excellent comprehensive treatments of the subject are offered in [2-4]. A review of the recent literature dealing with the effects of structure on single element and compound solids is given in [5].

REFERENCES

1. P. W. Bridgman, "The Physics of High Pressures." Macmillan, New York, 1952.
2. W. Paul and D. M. Warschauer, "Solids Under Pressure." McGraw-Hill, New York, 1963.
3. R. S. Bradley (ed.), "High Pressure Physics and Chemistry," Vol. 2. Academic Press, New York, 1963.

4. R. S. Bradley and D. C. Munro, "High Pressure Chemistry." Pergamon, New York, 1965.
5. W. Klement, Jr. and A. Jayaraman, Phase Relations and Structure of Solids at High Pressures. *Prog. Solid State Chem.* **3**, 289 (1967).

Chapter 15

The reader is referred to Mathot for an exposition of how the shapes of freezing point curves were used to infer information about the entropies of fusion of solvents.

REFERENCE

V. Mathot, *Bull. Soc. Chim. Belg.* **59**, 137 (1950).

Chapter 22

Prediction of compound formation, fields of stability, and the relationship to metastability is at a primitive stage. Interesting qualitative and semiquantitative approaches based on studies of dissociation phenomena, cationic field strengths, free energies of formation, positive and negative deviations from ideal solution behavior, and ionicity variations in homologous series have been used as the basis for understanding compound formation behavior better. The reader is referred to the following references for the several approaches mentioned.

REFERENCES

1. J. Kendall, A. W. Davidson, and H. Adler, *J. Am. Chem. Soc.* **43**, 1481 (1921).
2. J. Kendall and J. E. Booge, *J. Chem. Soc.* **127**, 1768 (1925).
3. J. Kendall and C. V. King, *J. Chem. Soc.* **127**, 1778 (1925).
4. J. D. M. Ross and I. C. Somerville, *J. Chem. Soc.* **128**, 2770 (1926).
5. V. M. Goldschmidt, *Skrifter Norske Videnskaps-Akad. Oslo I. Mat. Naturv. Kl.* **8**, 7 (1926).
6. A. Dietzel, *Z. Elektrochem.* **48**, 9 (1942).
7. C. Wagner, *Acta Met.* **6**, 309 (1958).
8. T. Penkala, *Bull. Acad. Polon. Sci.* **3**, 277 (1955); **4**, 615 (1956); **4**, 619 (1956).
9. E. Högfeldt, *Rec. Trav. Chim. Pays Bas* **75**, 790 (1956).
10. A. Reisman and F. Holtzberg, *J. Am. Chem. Soc.* **77**, 2115 (1955).
11. F. Holtzberg, A. Reisman, M. Berry, and M. Berkenblit, *J. Am. Chem. Soc.* **78**, 1536 (1956).
12. A. Reisman, F. Holtzberg, M. Berkenblit, and M. Berry, *J. Am. Chem. Soc.* **78**, 4514 (1956).
13. A. Reisman, F. Holtzberg, and E. Banks, *J. Am. Chem. Soc.* **80**, 37 (1958).
14. A. Reisman and F. Holtzberg, *J. Am. Chem. Soc.* **80**, 6503 (1958).
15. A. Reisman and F. Holtzberg, *J. Phys. Chem.* **64**, 748 (1960).
16. A. Reisman and J. Mineo, *J. Phys. Chem.* **65**, 996 (1961).
17. A. Reisman, *J. Phys. Chem.* **66**, 15 (1962).
18. A. Reisman and J. Mineo, *J. Phys. Chem.* **66**, 1181 (1962).

Chapter 26

Liquid miscibility gaps were first examined mathematically by van Laar [1]. An interesting commentary on miscibility gap phenomena as determined by values of the free energy as a function of composition is given in [2]. Additional insight into the occurrence of critical phenomena is to be gained from Prigogine and D. Defay cited among the General References, specifically Chapter 15, Section 5, and Marsh, also cited among the General References, pp. 54–66. The study of the system Cd–Se referred to is described in [3].

REFERENCES

1. J. J. van Laar, Z. Phys. Chem. **63**, 216 (1908); **64**, 257 (1908).
2. J. L. Meijering, N. V. Philips Report No. 2790, 1958.
3. A. Reisman, M. Berkenblit, and M. Witzen, J. Phys. Chem. **66**, 2210 (1962).

Chapter 27

Based on a graphical approach, Seltz [1] derived solidus and liquidus equations for ascending systems which are identical to those derived herein. Wagner [2] derived formulas for determining excess free energies of the coexisting solid–liquid phases and for evaluating the maximum width of the solidus–liquidus gap.

An interesting aspect of solid solution, particularly the variation of lattice constant with composition in cubic systems is an empirical rule attributed to Vegard and Dale [3]. This rule asserts that the lattice constant should vary linearly with composition in an ideal system and has been used to test many systems. Unfortunately, as pointed out by Zen [4], even in ideal systems the lattice constant should vary with the cube root of the composition. Alternatively, we might expect the unit cell volume or density to vary linearly with composition. This fact which might be anticipated intuitively, and which was demonstrated mathematically by Zen can be the basis for erroneous conclusions drawn from experimental results.

Interesting aspects of this question are seen with reference to the systems $KTaO_3$–$KNbO_3$ and $NaNbO_3$–$KNbO_3$ [5]. Neither of these systems is cubic, but one might expect continuous density variations with composition since both systems show complete miscibility in the solid state. Both show similar phase transformations of their end members. In the first system the phase transformation temperatures exhibit a linear variation with composition and a linear variation of density with composition. The second system shows nonlinear variation of transformation temperature with composition and a minimum in the density–composition curve.

The purification of materials (Section D) has, in recent years, achieved great attention because of the need for materials of ultra-high purity for electronic applications. Associated with this need is the growth of bulk single crystals of material for which techniques similar to those used for materials purification are frequently employed. Purification and growth techniques by vapor-phase methods are referenced in Chapter 35. Below we address ourselves to a few areas relating to purification and growth of bulk materials. A recent review of this area, particularly in terms of the kinetics of growth processes is given in [6]. Detailed descriptions of the many available techniques are to be found in [7–10]. The very widely used technique of zone refining (zone melting) was stimulated by the work of Pfann [11–12] and Reiss [13]. Application of the method to materials containing volatile solutes or which vaporized incongruently was pursued by Boomgaard and co-workers [14–16]. The thermodynamics, materials preparation, and crystal growth of II–VI compounds is reviewed in detail by Lorenz [17] and a few examples of p–T–X studies, upon which the technique for zone melting of decomposable materials rests, are to be found in [18–21]. The use of zone melting in a continuous process is described in [22]. The basic approach to the growth of ingots from the melt is that of Bridgman [23], and is referred to as the Bridgman technique. However, the method that is actually employed is that of Teal and Little [24]. The method is attributed to Czochralski [25] and frequently the term "Czochralski-grown crystals" is employed. Two other approaches of particular value where containers cannot be employed due to either the high melting point, or reactivity of the material, or both, are referred to as the Verneuil (flame fusion) [26] and the floating zone methods [27].

REFERENCES

1. H. Seltz, *J. Am. Chem. Soc.* **56**, 307 (1934).
2. C. Wagner, *Acta Met.* **2**, 242 (1954).
3. L. Vegard and H. Dale, *Z. Krist.* **67**, 148 (1928).
4. E-An Zen, *Am. Mineral.* **41**, 523 (1956).
5. A. Reisman and E. Banks, *J. Am. Chem. Soc.* **80**, 1877 (1958).
6. K. A. Jackson, Current Concepts in Crystal Growth from the Melt. *Progr. Solid State Chem.* **4**, 53 (1967).
7. H. S. Peiser, "Crystal Growth." Pergamon, New York, 1967.
8. J. C. Brice, "The Growth of Crystals from the Melt." North Holland Publ., Amsterdam, 1965.
9. J. J. Gilman, "The Art and Science of Growing Crystals." Wiley, New York, 1963.
10. W. D. Lawson and S. Nielsen, "Preparation of Single Crystals." Academic Press, New York, 1958.

11. W. G. Pfann, *J. Metals* **4**, 1, 747, 861 (1952); *Phys. Rev.* **89**, 322 (1953); *AIME Trans.* **203**, 961 (1955).
12. W. G. Pfann, "Zone Melting." Wiley, New York, 1958.
13. H. Reiss, *J. Metals* **6**, 1053 (1954).
14. J. van den Boomgaard, *Philips Res. Rept.* **10**, 319 (1955); **11**, 91 (1956).
15. J. van den Boomgaard, F. A. Kröger, and H. J. Vink, *J. Electron.* **1**, 212 (1955).
16. J. van den Boomgaard, *Philips Res. Rep.* **11**, 27 (1956).
17. M. R. Lorenz, IBM Res. Rept. RC-1390, Yorktown Heights, N.Y., 1965.
18. J. Bloem and F. A. Kröger, Philips Reprint No. 2379, Eindhoven, 1956.
19. J. van den Boomgaard and K. Schol, *Philips Res. Rep.* **12**, 127 (1957).
20. D. de Nobel, Thesis. Univ. of Leiden, May 1958 (see also *Philips Res. Rep.* **14**, 361 (1959)).
21. A. Muan, *Am. J. Sci.* **256**, 171 (1958).
22. K. J. Kennedy, *Rev. Sci. Instr.* **35**, 25 (1964).
23. P. W. Bridgman, *Proc. Am. Acad. Arts Sci.* **60**, 305 (1925).
24. G. Teal and R. Little, *Phys. Rev.* **77**, 647 (1950).
25. J. Czochralski, *Z. Physik. Chem.* **92**, 219 (1918).
26. A. Verneuil, *Compt. Rend.* **135**, 791 (1902).
27. P. H. Keck and M. E. Golay, *Phys. Rev.* **89**, 1297 (1953).

Chapter 30

The experimental data for the phase diagram Na_2CO_3–K_2CO_3 are presented in the following reference.

REFERENCE

A. Reisman, *J. Am. Chem. Soc.* **81**, 807 (1959).

Chapter 32

The phenomenon discussed in this chapter, particularly the discussion focused upon in Sections D–F leads to the formation of the so-called nonstoichiometric or defect compound, namely, one exhibiting a region of variable composition. The latter two descriptions are perhaps preferable since the concept of a nonstoichiometric compound might tend to imply that the combining weights of the components cannot be described in terms of small integral numbers. At any rate, where only one of the components in an incongruently vaporizing compound exhibits appreciable volatility, i.e. As in GaAs, O_2 in Nb_2O_5, H_2O in $CuSO_4 \cdot 5H_2O$, the variation of vapor pressure in the region of variable composition is most easily described by employing mass-action equations which include lattice vacancy concentrations and the dependence of the partial pressure of the volatile end member on the vacancy concentration. Consider, for example, the case of the incongruently vaporizing compound AB where B is the volatile end member and exhibits a valency of -2. When a B atom volatilizes from the lattice, it leaves behind a vacancy containing two electrons. The concentration of this unionized vacancy in the lattice may be denoted by [B^{II}]. Unionized vacancies may lose one or both of the electrons leading to concentrations of singly ionized vacancies [B^{I}], or doubly ionized vacancies [B^{0}]. For simplicity let us assume that one of two boundary states prevails, namely, that all vacancies are present in either the singly or doubly ionized state.

Single Vacancy Ionization State. The equilibria leading to such a situation are given by

$$B^= \rightleftarrows B^{II} + \tfrac{1}{2}B_2 \text{ (vapor)}, \tag{1}$$

$$B^{II} \rightleftarrows B^{I} + e^-. \tag{2}$$

Conservation of the total vacancy concentration [B^{II}] leads to

$$[B^{II}] = [B^{I}]. \tag{3}$$

From Eqs. (2) and (3) it is seen also that

$$[e^-] = [B^{I}]. \tag{4}$$

Furthermore, as [B$^=$], the total concentration of B atom lattice sites, is much greater than [B^{II}], the former is essentially constant. Utilizing this latter conclusion as well as Eqs. (3) and (4) enables us to write for the equilibrium constants for Eqs. (1) and (2), the relationships

$$K_1 = (p_{B_2})^{1/2}[B^{I}], \tag{5}$$

$$K_2 = \frac{[B^{I}]^2}{[B^{I}]}. \tag{6}$$

Solving Eqs. (5) and (6) for p_{B_2} in terms of [B^{I}] and combining constants gives

$$p_{B_2} = \frac{C}{[B^{I}]^4}. \tag{7}$$

Double Vacancy Ionization State. The pertinent equilibria for the case where all vacancies are present in the doubly ionized state are given by

$$B^= \rightleftarrows B^{II} + \tfrac{1}{2}B_2 \text{ (vapor)}, \tag{8}$$

$$B^{II} \rightleftarrows B^{0} + 2e^-. \tag{9}$$

The total vacancy conservation equation then takes the form

$$[B^{II}] = [B^{0}]. \tag{10}$$

From Eqs. (9) and (10) it follows then that

$$2[B^0] = [e^-]. \tag{11}$$

Again, utilizing the approximation that $[B^=]$ is constant and substituting for $[e^-]$ and $[B^{II}]$ in Eqs. (8) and (9), the equilibrium constants for Eqs. (8) and (9) may be written as

$$K_1 = (p_{B_2})^{1/2}[B^0], \tag{12}$$

$$K_2 = 4[B^0][B^0]^2/[B^0]. \tag{13}$$

Solving Eqs. (12) and (13) for p_{B_2} in terms of $[B^0]$ and combining constants yields

$$p_{B_2} = \frac{C}{[B^0]^6}. \tag{14}$$

Equations (7) and (14) indicate that the partial pressure of B is inversely proportional to the 4th or 6th power of the vacancy concentration, respectively. For the cases treated then it is seen that a small change in vacancy concentration leads to a proportionately enormous change in equilibrium vapor pressure. The inverse nature of this relationship should have been anticipated intuitively since, when the vacancy concentration is largest, the system concentration is closest to the three-phase line in which the volatile component concentration is smallest. Consequently, the volatile component partial pressure should be small. When the vacancy concentration is smallest, the concentration of the volatile component in the region of variable composition is largest and its vapor pressure should be larger.

Probably, the electron ionization state will not be at a boundary level and one might expect to see a vacancy concentration dependency varying up to the 6th power for bivalent volatile components. A case which does appear to lie at one of the boundaries, however, is that of nonstoichiometric Nb_2O_5. Thus, Brauer [1] found that the oxygen content of Nb_2O_5 could be varied between Nb_2O_5 and $Nb_2O_{4.8}$. Greener et al. [2] and others [3–5] demonstrated subsequently that the partial pressure of oxygen in the region of variable composition varies inversely with either the 4th or 6th power of the oxygen lattice vacancy concentration, depending upon the type of measurement used.

Additional insight into the topic of defect structures may be found in [6–11] with comprehensive discussions being given in [7, 8, and 12].

Comprehensive reviews on nonstoichiometry in chemical compounds are to be found in [13–14]. A review of the experimental techniques useful for defining p–T–X relationships for III–V compounds is given in [15]. The effect of complex gaseous molecules on vapor pressure measurements in high temperature systems is given in [16], and the mechanism of oxide vaporizations is described in [17]. Studies of the system $CuSO_4-H_2O$ are presented in [18–20].

REFERENCES

1. G. Brauer, *Naturwiss.* **28**, 30 (1940).
2. E. H. Greener, D. H. Whitmore, and M. E. Fine, *J. Chem. Phys.* **34**, 1017 (1961).
3. P. Kofstad and P. B. Anderson, *J. Phys. Chem. Solids* **21**, 280 (1961).
4. R. N. Blumenthal, J. B. Moser, and D. H. Whitmore, *J. Am. Ceram. Soc.* **48**, 617 (1965).
5. W. K. Chen and R. A. Swalin, *J. Phys. Chem. Solids* **27**, 57 (1966).
6. R. F. Brebrick, *J. Phys. Chem. Solids* **14**, 190 (1958); **18**, 116 (1961).
7. T. J. Gray, "The Defect Solid State," p. 1. Wiley (Interscience), New York, 1957.
8. F. A. Kröger and H. J. Vink, *Solid State Phys.* **3**, 310 (1956).
9. G. Mandel, *Phys. Rev.* **134**, A1073 (1964); **136**, A826 (1964).
10. G. Mandel, IBM Res. Rept. RC-633, Yorktown Heights, N.Y., 1962.
11. A. Reisman, M. Berkenblit, and M. Witzen, *J. Phys. Chem.* **66**, 2210 (1962).
12. R. A. Swalin, "Thermodynamics of Solids," p. 272. Wiley, New York, 1962.
13. G. G. Libowitz, Non-Stochiometry in Chemical Compounds, *Progr. Solid State Chem.* **2**, 216 (1966).
14. R. F. Brebrick, Non-Stochiometry in Binary Semiconductor Compounds. *Progr. Solid State Chem.* **3**, 213 (1967).
15. K. Weiser, p–T–X Diagrams of III–V Compounds. IBM Res. Rept. RC-529, Yorktown Heights, N.Y., 1961.
16. L. Brewer and J. S. Kane, *J. Phys. Chem.* **59**, 105 (1955).
17. L. Brewer, *Acta Chem. Scand.* **6**, 1 (1952).
18. H. J. Borchardt and F. Daniels, *J. Phys. Chem.* **61**, 917 (1957).
19. A. Reisman and J. Karlak, *J. Am. Chem. Soc.* **80**, 6500 (1958).
20. A. Reisman, *Anal. Chem.* **32**, 1566 (1960).

Chapter 35

Chemical vapor transport reactions are employed for the preparation of many materials and rely on one of the following three phenomena: (1) disproportionation of vapor phase species, (2) pyrolytic reduction of vapor-phase species, and (3) pyrolysis of vapor-phase species. The major use, particularly of the second method, is for the growth of single-crystal layers upon single-crystal substrates for electronics applications. This procedure is termed epitaxial growth, a subject that, in less than a decade, has received attention in hundreds if not thousands of literature articles.

A general introduction to the phenomenon of vapor-transport reactions is given in [1]. A thermodynamic approach to understanding multispecies vapor-transport phenomena is given in [2], and the work of Mandel [3, 5, 7] and Lever [4, 6] is to be recommended for the treatment of mass transport phenomena. Kinetics of the Ge–I_2 reaction and the general kinetics of open tube systems is described in [8], and the behavior of Si tetrahalide reduction systems is described in [9]. The use of transport reactions for the crystal growth of transition metal niobates is given in [10], and a thermodynamic treatment of the vapor phase Si–H–Cl system is presented in [11] with an analogous study for GaAs with H_2O and HCl being given in [12]. Other articles relating to the topic (in which extensive other references are contained) are given in [13–25].

REFERENCES

1. H. Schafer, "Chemical Transport Reactions." Academic Press, New York, 1964.
2. A. Reisman and S. A. Alyanakyan, *J. Electrochem. Soc.* 111, 1154 (1964).
3. G. Mandel, *J. Phys. Chem. Solids* 23, 587 (1962).
4. R. F. Lever and G. Mandel, *J. Phys. Chem. Solids* 23, 599 (1962).
5. G. Mandel, *J. Chem. Phys.* 37, 1177 (1962).
6. R. F. Lever, *J. Chem. Phys.* 37, 1078, 1174 (1962).
7. G. Mandel, *J. Chem. Phys.* 40, 683 (1964).
8. A. Reisman and M. Berkenblit, *J. Electrochem. Soc.* 113, 146 (1966).
9. T. O. Sedgwick, *J. Electrochem. Soc.* 111, 1381 (1964).
10. F. Emmenegger and A. Petermann, *J. Crystal Growth* 2, 33 (1968).
11. R. F. Lever, *IBM J. Res. Develop.* 8, 460 (1964).
12. M. Michelitsch, W. Kappallo, and G. Hellbardt, *J. Electrochem. Soc.* 111, 48 (1964).
13. J. C. Marinace, *IBM J. Res. Develop.* 4, 248 (1960).
14. A. Reisman, M. Berkenblit, and S. A. Alyanakyan, *J. Electrochem. Soc.* 112, 241 (1965).
15. A. Reisman and M. Berkenblit, *J. Electrochem. Soc.* 112, 315, 812 (1965).
16. S. A. Papazian and A. Reisman, *J. Electrochem. Soc.* 115, 961 (1968).
17. M. Berkenblit, A. Reisman, and T. B. Light, *J. Electrochem. Soc.* 115, 966 (1968).
18. T. B. Light, M. Berkenblit, and A. Reisman, *J. Electrochem, Soc.* 115, 969 (1968).
19. M. Farber and A. J. Darnell, *J. Phys. Chem.* 59, 156 (1955).
20. B. A. Joyce and R. R. Bradley, *J. Electrochem. Soc.* 110, 1236 (1963).
21. R. R. Monchamp, W. J. McAleer, and P. I. Pollak, *J. Electrochem. Soc.* 111, 879 (1964).
22. I. L. Kalnin and J. Rosenstock, *J. Electrochem. Soc.* 112, 329 (1965).
23. K. E. Haq, *J. Electrochem. Soc.* 113, 817 (1966).
24. V. J. Silvestri, *J. Electrochem. Soc.* 116, 81 (1969).
25. T. J. LaChapelle, A. Miller, and F. L. Morritz, Silicon Heteroepitaxy on Oxides by Chemical Vapor Deposition, *Progr. Solid State Chem.* 3, 1 (1967).

Chapter 36

The classic paper on the use of effusion techniques for the measurement of vapor pressures is that of Knudsen [1]. Equation 1, Chapter 36, represents the ideal situation, but in practice a correction term due to Clausing [2] is introduced. This factor is the probability that a molecule, having entered one end of a hole of finite length, will escape from the opposite end. This leads to the modified Knudsen equation

$$p_A = [l/(a_0 t)][2\pi(RT/\text{Mol. Wt.}_A)]^{1/2}. \tag{1}$$

One further significant modification is concerned with the Knudsen assumption that the loss of vapor through the orifice does not affect the equilibrium vapor pressure within the cell itself. This has been taken into account by several authors [3–6], and leads to a further modification,

$$p_A = p_{\text{equi}}/[1 + (a/A\alpha_K)], \tag{2}$$

where p_A is the vapor pressure measured by the Knudsen technique, p_{equi} is the true vapor pressure, a is cross-sectional area of orifice, $A =$ cross-sectional area of effusion cell, and α_K is the Knudsen coefficient.

In addition to the above, many further subtleties are introduced which we will not pursue, leaving to the interested reader the task of further exploration.

An excellent theoretical paper dealing with Knudsen flow is given in [7], and other aspects relating to orifice size, evaporation and condensation coefficients, and effusion cell design are dealt with in [8–10]. Molecular beam phenomena are dealt with in [11], and the mass spectrometer used with the effusion cell for identification of species is described in the extensive work of Goldfinger et al. [12] and elsewhere [13].

The torsion balance modification of the Knudsen technique was first described by Volmer [14] and then employed by others [15–18]. Its main use has been in the determination of molecular weights in conjunction with the more classical Knudsen approach. The method is not as widely used as might have been anticipated, perhaps because of the additional difficulty involved, and also because it is not readily used with mass spectrometric techniques. Searcy [19] has, however, pioneered its use in the United States, and modifications of the approach utilized by him have found considerable application among his students. More recent discussions of torsion techniques as applied to effusion and Langmuir-type approaches (see below) are given in [20–22].

In our discussion of vapor-pressure measurement we have, of necessity, restricted the number of techniques covered. We would, however, be remiss if we did not at this time mention the Langmuir free-evaporation technique, primarily because the concept of the accommodation coefficient or coefficient of evaporation α, or condensation coefficient, which is the subject of many debates on vapor pressure measurements first, arises in the context of his work. Langmuir's approach [23, 24] is based on measurements of weight loss from a heated filament of known surface area in a vacuum. It is assumed that the area of the filament is so small that the probability of an emitted molecule striking the filament is insignificant. The weight loss is related to the equilibrium vapor pressure by

$$p_L = \alpha_L p_{equi} = l/at(2\pi RT/\text{Mol. Wt.})^{1/2}, \tag{3}$$

where p_L is the calculated vapor pressure based on the measurement, l/at is the weight loss per unit area per unit time, where l is the total measured weight loss, a is the filament area, and t is the time. The method has applicability to measurements of very low vapor pressures and has been used extensively, particularly in conjunction with Knudsen experiments. The quantity α, which is called many things, is, effectively, the ratio of the number of molecules condensing on the surface to the number striking this area per unit time. It is, therefore, related to the probability factor we have called k_c in Chapter 4, Section D. It will be noted in that chapter that we showed that when k_c is invariant for materials, the equilibrium vapor pressure is proportional to the evaporation rate coefficient and therefore to the evaporation rate. This is, of course, the same argument employed by Langmuir, Eq. (3). Unfortunately, not only must α be the same for most materials, but also it must equal unity for Eq. (3) to provide actual equilibrium vapor pressures. For interesting discussions on α and cases where it is unity see [23, 25–28], and for cases where it is not unity see [29]. The quantity α must be determined experimentally [28, 30], although partially successful theoretical approaches have been proposed [31, 32].

One interesting technique for evaluating α is that due to Bogdandy et al. [33] where α is evaluated by determining the time for the equilibrium vapor pressure to be achieved, i.e.,

$$\ln(p_{sat} - p) = (-\alpha/V) S(kT/2\pi m)^{1/2} t + \text{const}, \tag{4}$$

where p_{sat} = equilibrium vapor pressure, p is the pressure at time t, S is the area of evaporating surface, V is the volume of the glass bulb, m is the mass of a molecule, and k is the Boltzman constant.

The transpiration method for evaluating vapor pressures is extensively used and dates back to 1913 or earlier [34]. For applications, the reader is referred to [35–38].

Distendable diaphragm gauges also have a long history [39–41], with numerous modifications having been reported. Elegant applications to the study of dissociative equilibria and reactions may be found in [42–48]. An automated version of this apparatus was described by Hochberg [49].

For a more comprehensive discussion of vapor-pressure measurement techniques, the reader is referred to [9].

For further details on temperature measurement, the reader is referred to the excellent treatments in [50–52]. For general techniques for measuring material properties at very high temperature, the work of Kingery [53] is recommended. For comprehensive discussions on temperature scales, the reader is referred to [54, 55].

An excellent introduction to differential thermal analysis is provided in [56], and interesting papers on applications and equipment design for the differential thermal analysis method are to be found in [57–73]; and, on a companion approach, thermogravimetric analysis in [74].

The use of strip furnace techniques is that due to Morey [75], and its application to phase-equilibria studies is described in [76].

The use of electrical conductivity changes associated with solid–solid and solid–liquid phase changes is found in [77, 78].

Additional insight into approaches other than those we have considered are to be obtained from the references already cited as well as from a number of the General References on phase diagrams and phase equilibria.

REFERENCES

1. M. Knudsen, *Ann. Physik* **28**, 1002 (1909).
2. P. Clausing, *Ann. Physik* **12**, 961 (1932).
3. R. Speiser, H. L. Johnston, and P. E. Blackburn, *J. Am. Chem. Soc.* **72**, 4142 (1950).
4. C. I. Whitman, *J. Chem. Phys.* **19**, 744 (1951); **20**, 161 (1952).
5. M. G. Rossman and J. Yarwood, *J. Chem. Phys.* **21**, 1406 (1953).
6. J. D. McKinley, Jr., and J. E. Vance, *J. Chem. Phys.* **22**, 1120 (1954).
7. W. C. Marcus, Union Carbide Nuclear Co., Rept. No. K1302, Parts 1 and 2, Oak Ridge, Tennessee, 1956.
8. E. Rutner, P. Goldfinger, and H. Rickert, "Condensation and Evaporation of Solids." Gordon and Breach, New York, 1964.
9. A. N. Nesmeianov, "Vapor Pressure of Chemical Elements," Chap. 1. Elsevier, Amsterdam, 1963.
10. E. W. Clower, Observations on Molecular Effusion. Thesis, Univ. of Illinois, Dept. of Chemistry, 1954.
11. R. C. Miller and P. Kusch, *J. Chem. Phys.* **25**, 860 (1956).
12. P. Goldfinger, M. Ackerman, and M. Jeunehomme, "Vaporization of Compounds and Alloys at High Temperature." Univ. Libre de Bruxelles, Lab. de Chim. Phys. Moleculaire, 1959.
13. P. Goldfinger, Mass Spectrometric Investigations of High Temperature Equilibria. *In* "Mass Spectrometry" (R. I. Reed, ed.). Academic Press, New York, 1965.
14. M. Volmer, *Z. Physik Chem.* (Bodenstein-Festband), 863 (1931).

15. K. Neumann and E. Volker, *Z. Physik. Chem.* **A161**, 33 (1932).
16. K. Niwa and M. Yosiyama, *J. Chem. Soc. Japan* **61**, 1055 (1940).
17. E. W. Balson, *Trans. Faraday Soc.* **43**, 54 (1947).
18. G. Wessel, *Z. Physik.* **12**, 961 (1951).
19. A. W. Searcy and R. D. Freeman, *J. Am. Chem. Soc.* **76**, 5229 (1954).
20. R. D. Freeman, "The Characterization of High Temperature Vapors," Chap. 7. Wiley, New York, 1967.
21. W. T. Lee and Z. A. Munir, *J. Electrochem. Soc.* **114**, 1236 (1967).
22. R. C. Blair and Z. A. Munir, *J. Phys. Chem.* **72**, 2434 (1968).
23. H. A. Jones, I. Langmuir, and G. M. Mackay, *Phys. Rev.* **30**, 201 (1927).
24. R. G. Johnson, D. E. Hudson, W. G. Caldwell, F. H. Spedding, and W. R. Savage, *J. Chem. Phys.* **24**, 917 (1956).
25. I. Langmuir, *Phys. Rev.* **2**, 329 (1913).
26. I. Langmuir and G. M. Mackay, *Phys. Rev.* **4**, 377 (1914).
27. A. L. Marshall, R. W. Dornte, and F. J. Norton, *J. Am. Chem. Soc.* **59**, 1161 (1937).
28. R. B. Holden, R. Speiser, and H. L. Johnston, *J. Am. Chem. Soc.* **70**, 3897 (1948).
29. S. Wexler, *Rev. Mod. Phys.* **30**, 402 (1958).
30. H. L. Johnston and A. L. Marshall, *J. Am. Chem. Soc.* **62**, 1382 (1940).
31. J. P. Hirth and G. M. Pound, *J. Chem. Phys.* **26**, 1216 (1957).
32. J. P. Hirth and G. M. Pound, *J. Phys. Chem.* **64**, 619 (1960).
33. L. V. Bogdandy, H. G. Klust, and O. Knacke, *Z. Elektrochem.* **58**, 460 (1955).
34. H. von Wartenberg, *Z. Elektrochem.* **19**, 482 (1913).
35. K. A. Sense, M. J. Snyder, and J. W. Clegg, *J. Phys. Chem.* **58**, 223 (1954).
36. A. Reisman, M. Berkenblit, and S. A. Alyanakyan, *J. Electrochem. Soc.* **112**, 241 (1965).
37. V. J. Silvestri, *J. Electrochem. Soc.* **112**, 748 (1965).
38. M. Berkenblit and A. Reisman, *J. Electrochem. Soc.* **113**, 93 (1966).
39. A. Ladenbing and E. Lehman, *Verhandl. Deut. Phys. Ges.* **8**, 20 (1906).
40. F. M. G. Johnson and D. McIntosh, *J. Am. Chem. Soc.* **31**, 1138 (1909).
41. G. Preuner and K. S. Schupp, *Z. Physik Chem.* **68**, 129 (1910).
42. V. J. Lyons and V. J. Silvestri, *J. Phys. Chem.* **64**, 266 (1960).
43. V. J. Silvestri, *J. Phys. Chem.* **64**, 826 (1960).
44. G. A. Somorjai, *J. Phys. Chem.* **65**, 1059 (1961).
45. V. J. Silvestri and V. J. Lyons, *J. Electrochem. Soc.* **109**, 963 (1962).
46. R. F. Lever, *J. Electrochem. Soc.* **110**, 775 (1963).
47. T. O. Sedgwick, *J. Electrochem. Soc.* **112**, 496 (1965).
48. T. O. Sedgwick and B. J. Agule, *J. Electrochem. Soc.* **113**, 54 (1966).
49. F. Hochberg, Electrochem. Soc. Mtg., N.Y., September 29–October 3, 1963, Abstract No. 139.
50. W. F. Roeser and S. T. Lonberger, Methods of Testing Thermocouples and Thermocouple Materials. Nat Bur. Stand. Circular 590, February 1958.
51. D. I. Finch, General Principles of Thermoelectric Thermometry. Tech. Publ. D1-100, Leeds and Northrup Co., 1962.
52. J. A. Hall, "The Measurement of Temperature." Chapman & Hall, London, 1966.
53. W. D. Kingery, "Property Measurements at High Temperatures." Wiley, New York, 1959.
54. H. F. Stimson, *J. Res. Nat. Bur. Stand.* **42**, 209 (1949).
55. R. B. Sosman, *Am. J. Sci.* Bowen Vol., Part II, 517 (1952).
56. W. J. Smothers and Y. Chiang, "Differential Thermal Analysis." Chem. Publ. Co., New York, 1958.
57. L. G. Berg, *Compt. Rend.* **49**, 648 (1945).

58. D. McConnell and J. W. Earley, *J. Am. Ceram. Soc.* **34**, 183 (1951).
59. J. A. Pask and M. F. Warner, *Bull. Am. Ceram. Soc.* **33**, 168 (1954).
60. F. W. Wilburn, *Trans. Soc. Glass Tech.* **38**, 371 (1954).
61. A. Reisman and F. Holtzberg, *J. Am. Chem. Soc.* **77**, 2115 (1955).
62. H. E. Kissinger, *J. Nat. Bur. Stand.* **57**, 217 (1956).
63. H. J. Borchardt and F. Daniels, *J. Am. Chem. Soc.* **79**, 41 (1957).
64. F. Holtzberg, A. Reisman, M. Berry, and M. Berkenblit, *J. Am. Chem. Soc.* **79**, 2039 (1957).
65. P. D. Garn and S. S. Flaschen, *Anal. Chem.* **29**, 271 (1957).
66. C. Groot and V. H. Troutner, *Anal. Chem.* **29**, 835 (1957).
67. A. Reisman, *J. Am. Chem. Soc.* **80**, 3558 (1958).
68. A. Reisman and J. Karlak, *J. Am. Chem. Soc.* **80**, 6500 (1958).
69. C. Campbell, S. Gordon, and C. L. Smith, *Anal. Chem.* **31**, 1188 (1959).
70. A. Reisman, F. Holtzberg, and M. Berkenblit, *J. Am. Chem. Soc.* **81**, 1292 (1959).
71. A. Reisman and F. Holtzberg, *J. Am. Chem. Soc.* **81**, 3182 (1959).
72. M. Markowitz and D. A. Boryta, *J. Phys. Chem.* **66**, 1477 (1962).
73. A. Reisman and M. Berkenblit, *J. Phys. Chem.* **67**, 22 (1963).
74. A. Reisman, *Anal. Chem.* **32**, 1566 (1960).
75. H. S. Roberts and G. W. Morey, *Rev. Sci. Instr.* **1**, 576 (1930).
76. A. Reisman, F. Holtzberg, M. Berkenblit, and M. Berry, *J. Am. Chem. Soc.* **78**, 4514 (1956).
77. A. Reisman, S. Triebwasser, and F. Holtzberg, *J. Am. Chem. Soc.* **77**, 4228 (1955).
78. P. Garn and S. S. Flaschen, *Anal. Chem.* **29**, 268 (1957).

GENERAL BIBLIOGRAPHY

General References

When a broad area has been treated by several authors one finds, in addition to differences in emphasis and content, unique flavors of presentation that the reader finds more or less rewarding, depending upon his own background and needs. "Phase equilibria" has been the subject of a number of specialized books each having much to offer in its own way. Since a different way of turning a phrase frequently enables one to grasp what he could not from some other source, the reader is urged to consult several of the books listed below for special areas of interest to him.

The following tabulation of works published in English, while not overly comprehensive, was chosen because it provides a wide gradation in technique ranging from purely descriptive overviews to those that are somewhat rigorous. The variation in topics addressed is large and the pedagogical methods vary enormously.

Included also, for the same reasons as above, are selected treatments on thermodynamics that the author has found useful. Some of these have major inputs on phase equilibria, based on the general approach involving "activity" and the special case of "regular" solutions, which the present author has chosen not to employ. For this reason the reader, in order to complete his formalism, is urged to consult these manuscripts, particularly those by Prigogine and Defay, and Guggenheim and Denbigh.

REFERENCES

S. T. Bowden, "The Phase Rule and its Applications." Macmillan, London, 1950.
E. F. Caldin, "An Introduction to Chemical Thermodynamics." Oxford Univ. Press, London and New York, 1958.
A. H. Cottrell, "Theoretical Structural Metallurgy." Arnold, London, 1951.
L. S. Darken and R. W. Gurry, "Physical Chemistry of Metals." McGraw-Hill, New York, 1953.
K. Denbigh, "The Principles of Chemical Equilibrium." Cambridge Univ. Press, London and New York, 1955.
P. S. Epstein, "Textbook of Thermodynamics." Wiley, New York, 1937.
A. Findlay, A. N. Campbell, and N. O. Smith, "The Phase Rule." Dover, New York, 1951.
J. W. Gibbs, "The Collected Works," Thermodynamics. Vol. I, Yale Univ. Press, New Haven, Connecticut, 1957.
S. Glasstone, "Thermodynamics for Chemists." Van Nostrand, Princeton, New Jersey, 1949.
P. Gordon, "Principles of Phase Diagrams in Materials Systems." McGraw-Hill, New York, 1968.
E. A. Guggenheim, "Thermodynamics." North Holland Publ., Amsterdam, 1949.
W. Hume-Rothery, J. W. Christian, and W. B. Pearson, "Metallurgical Equilibrium Diagrams." London Institute of Physics, London, 1952.
L. Kaufman and H. Bernstein, "Computer Calculations of Phase Diagrams." Academic Press, New York, 1970.
I. M. Klotz, "Chemical Thermodynamics." Prentice-Hall, Englewood Cliffs, New Jersey, 1953.
O. Kubaschewski and E. LL. Evans, "Metallurgical Thermochemistry." Pergamon Press, New York, 1958.
W. N. Lacey and B. H. Sage, "Thermodynamics of One-Component Systems." Academic Press, New York, 1957.
G. N. Lewis and M. Randall, "Thermodynamics." McGraw-Hill, New York, 1923 (see also the revised edition: K. S. Pitzer and L. Brewer, "Thermodynamics," 2nd ed. McGraw-Hill, New York, 1961).
J. S. Marsh, "Principles of Phase Diagrams." McGraw-Hill, New York, 1935.
G. Masing, "Ternary Systems." Dover, New York, 1944.
L. K. Nash, "Elements of Chemical Thermodynamics. "Addison-Wesley, Reading, Massachusetts, 1962.
L. S. Palatnik and I. Landau, "Phase Equilibria in Multicomponent Systems." Holt, New York, 1964.
F. M. Perel'mann, "Phase Diagrams of Multicomponent Systems." Consultants Bureau, New York, 1966.
A. Prince, "Alloy Phase Equilibria." Elsevier, Amsterdam, 1966.
F. N. Rhines, "Phase Diagrams in Metallurgy." McGraw-Hill, New York, 1951.
J. E. Ricci, "The Phase Rule and Heterogeneous Equilibria." Van Nostrand, Princeton, New Jersey, 1951.

D. R. Stull and G. C. Sinke, "Thermodynamic Properties of Elements," Vol. 18. Am. Chem. Soc., Washington, D. C., 1956.
R. A. Swalin, "Thermodynamics of Solids." Wiley, New York, 1962.
A. R. Ubbelohde, "An Introduction to Modern Thermodynamical Principles." Oxford Univ. Press, London and New York, 1952.

Compilations of Phase Diagrams

REFERENCES

R. P. Elliott, "Constitution of Binary Alloys," Suppl. 1. McGraw-Hill, New York, 1965.
M. Hansen, "Constitution of Binary Alloys." McGraw-Hill, New York, 1958.
E. M. Levin, H. F. McMurdie, and F. P. Hall, "Phase Diagrams for Ceramists." Am. Ceram. Soc., Columbus, Ohio, 1956.
E. M. Levin and H. F. McMurdie, "Phase Diagrams For Ceramists," Part II. Am. Ceram. Soc., Columbus, Ohio, 1959.
E. M. Levin, C. R. Robbins, and H. F. McMurdie, "Phase Diagrams for Ceramists" (M. K. Reser, ed.). Am. Ceram. Soc., Columbus, Ohio, 1964.

Subject Index

A

Accommodation coefficient, 529
Activity
 in equilibrium constants, 124
 relation to fugacity, 123
Allotrope, 52
Alloy systems, 396, 405
Ascending solid solutions, 284
 tie lines in, 290
Association
 degree of, 185
 in eutectic systems, 185, 188, 191
Avogadro constant, 19
Azeotrope, 362

B

Binary systems, see Eutectic systems, Solid solution systems
Boiling point
 diagrams, 341
 equations, 343
 in eutectic systems, 241
 normal, 44
 plates, 365

Bourdon gauge vapor pressure measurements, 488, 496
 gas phase reaction behavior deduced from, 497
Bridgman method, 523

C

Chemical, definition, 25
Chemical potential, 30
 of components and species, 109
 equivalency in coexisting phases, 110
Chemical vapor deposition, 477
Chemical vapor transport reactions, 477, 527
Chemical variable, 20
Clapeyron equation, 6, 40
Clausius–Clapeyron equation, 41
Condensed-vapor phase diagrams, 341
Condensation, 34
 sticking coefficients and, 36
Congruent melting, 167
 pressure effects, 264
 relation to unary solid-liquid equilibria, 267

SUBJECT INDEX 537

Congruently vaporizing systems, tie lines in, 371
Conjugate solutions, 273
Constituent, 26
Consulate point, 273
Common species effects, 177, 181, 184
Component, 20
    conservation of, 24
    of mole terms, 127
    minimum number of, 24
Compound
    definition of, 26
    flattening of melting peak, 211, 216
    formation in alloy systems, 396
        in ternary systems, 432, 434
    incongruently melting, 257, 264
Compressibility relation, 43
Computer analysis
    of ascending solid solutions, 301 ff.
    of condensed-vapor phase binary systems 348 ff.
    of condensed-vapor phase multicomponent systems, 477 ff.
    of eutectic systems, 204 ff.
    of minimum type solid solutions, 330 ff.
Coordinate transformations, 151, 222, 226, 231
Critical point, 44
Critical solution temperature, 273
Critical temperature, 45

D

Defect compounds, 524
Degrees of freedom, 7, *see also* Variance
Deliquescence, 384
Density
    of coexisting phases in unary systems, 53
    measurements in deducing phase diagrams, 514
Deviations
    dependence on number of particles, 18
    probability and, 15
Differential thermal analysis, 499, 508
Dimer-monomer behavior in solid solutions, 317
Disproportionation reactions, 478
    $Ge-I_2-H_2$ system, 484
Dissociation
    degree of, 172
    effects on binary solubility curves, 168, 173, 176, 197, 202, 203, 315, 333
Distributions, 12 ff.
    probability and, 15

E

Effluorescence, 385
Effusion method, 489, 528
Enantiomorphs, 52, 71
Enthalpy
    changes in phase transformations, 56
    diagrams, 86
    of fusion, constancy of, 128
    pressure effects, 87
    temperature effects, 88
Entropy
    diagrams, 89
    of fusion
        effects on solubility curves, 157
        table of compounds, 162, 163
        of elements, 159
Epitaxial growth, 478
Equidistribution, 16
Equilibrium, criteria for, 30
Equilibrium constant
    based on activities, 124
    on fugacities, 119
    concentration dependence, 103
    heterogeneous, 108, 116
    homogeneous, 95
    pressure dependence, 99
    temperature dependence, 105
Eutectic systems, 128
    binary
        association effects, 185, 188, 191
        boiling points in, 241
        common species effects, 177, 181, 184
        compound formation in, 217, 229
        computer analyses of, 204 ff.
        condensed-vapor phase diagrams for, 341
        cooling paths in, 141, 219, 259, 277
        coordinate transformations in, 151, 222 ff., 226, 231
        dissociation effects, 168, 176, 197, 202, 203
        equivalent species and mole terms, 125

eutectic point in, 144
flattening of compound maxima, 211, 216
graphical description, 140
incongruent vaporization in, 367
incongruently melting compounds in, 257
limited solid solubility in, 389
melting points of compounds in, 230
metastability in, 133
miscibility gaps in, 271, 273, 282
phase changes in, 255
solid-gas compositions, 244, 249
space models, 234, 264
species interactions in, 192, 196
stepwise variation of vapor pressure in incongruently vaporizing, 376
tie lines in, 141
univariant melting point equation, 129 ff.
vapor pressures in, 132
ternary, 415
cooling paths in, 419
compound formation in, 432, 434
incongruent phenomena in, 448
isothermal sections, 423, 451
Eutectoid systems, 399
Evaporation rate, 33, 518
Experimental noise, 11
deviations and, 12
Extensive properties, 9, 29

F

Fractional crystallization, 291
Fractional distillation, 363 ff.
Free energy, 29
changes, 98, 100
functions, 42
standard states, 98, 101
Fugacity, 118
Fusion Processes, 130

G

Gas saturation method, 492
Gaussian error function, 18
Gibbs free energy, 29

H

Heat of fusion
constancy of, 128
effect on ascending solid solutions, 303
on minimum solid solutions, 332
sign of, 130
Heating curve of Se, 513
Heterogeneous equilibria, 2
Hydrate systems, 377, 379
stability of, 386
Hydrogen reduction reactions, 378
Hydrothermal synthesis, 45

I

Ideal behavior, 93
Ideal gas equation, 94
Impurities in pure phases, 291
Incongruent melting, 167, 257
effects of pressure on, 264
relation to unary solid-liquid curves, 267
Incongruent vaporization, 367, 524
Incongruently vaporizing systems
hydrates, 377
tie lines in, 370
Independent chemical variable, 20, *see* Component
Inert gas, effects in unary systems, 39, 79, 81
Intensive properties, 2
Isothermal distillation, 359
Isotherms
in condensed-vapor phase diagrams, 344, 351
in eutectic space model, 252
in solid solution systems, 299
in ternary eutectic systems, 423

K

Kinetics, 4
Knudsen method, *see* Effusion method

L

Langmuir evaporation, 518, 529
Lattice defect equilibria, 525
Lever arm principle
applications, 149

coordinate transformations, 151
  derivation, 146
  tie line construction, 142
Lewis and Randall treatment of fugacities, 120
Liquidus, 131
Liquidus curves, see Solubility curves
Liquidus equations
  for simple eutectic systems, 129
  for vapor phase composition
    in simple eutectic systems, 138
    in solid solutions, 289
  from vapor pressure data, 132

M

Macroscopic properties, 12
Mass
  constancy of in systems, 10
  fraction, 22
Melting point, maximum in binary eutectic systems, 130
Melting point curve
  effect of inert gas pressure on, 82
  slope of in unary systems, 45, 47
Metastability
  in binary systems, 133, 144
  in Nb_2O_5, 72
  in unary systems, 48
  vapor pressure in unary systems and, 49
Metastable phase changes, 54, 72
  enthalpic behavior in, 56
Microscopic properties, 12
Minimum type solid solutions, 313
  analytical relationships for, 320, 321, 323, 326
  dissociation effects in, 333
  graphical description, 315
  hypothetical examples in condensed-vapor phase systems, 352 ff.
  isobaric univariance in, 318
  K_2CO_3–Na_2CO_3 system, 335
  space models, 327
Miscibility gaps, 271
  cooling behavior in, 277
  effect of pressure on, 283
  in solid regions, 282
  tie lines in, 273
Mole fraction, 22, 126
  in thermodynamics, 23

Molecularity, 21
Monomer-dimer behavior in solid solutions, 317
Monotropy, see Phase changes
Multicomponent systems, phase diagrams, 483

N

Noise, experimental, 11
Nonstoichiometric compounds, 524, see Incongruently vaporizing systems, Variable composition

O

One component systems, see Unary systems
Open tube vapor transport systems, 479

P

Parameters
  extensive, 9, 29
  intensive, 2
Partial molar free energy, see Chemical potential
Partial molar volume, 112
Peritectic systems
  binary, 255, 257, 405
  ternary, 436 ff., 447
Phase(s)
  definition, 10
  equipotential, 47
  maximum number of, 24, 39
Phase changes
  in binary eutectic systems, 255
  enantiomorphic, 52, 71
  1st and 2nd order, 52
  metastable, 54
  monotropic, 63, 69, 72, 74
  pressure effects on, 61, 75
  study of, 515
  theory of, 519
Phase diagrams, experimental determination of, 512
Phase equilibria, 2
Phase rule, 5
  derivation of, 31 ff.
  reduced, 32
Polymorphs, 52

SUBJECT INDEX

Pressure
  effect on compound melting, 264, 267
    on miscibility gaps, 283
    on unary solid-liquid transformations, 44, 47
Probability, 15
  equidistribution, 16
Projections of equipotential lines in unary systems, 47
Properties
  extensive, 9
  intensive, 2
  macroscopic, 12
  microscopic, 12

Q

Quadruple point, 238

R

Raoult's Law, 93
  approximate nature of, 135
  use in determining solubility curves, 133

S

Segregation coefficients in solid solutions, 293
Single crystal growth, 478, 523
Singular point between congruency and incongruency, 266
Solid solubility
  limited, 389, 397
  in ternary systems, 459, 472
Solid solutions
  analytical relationships for ideal, 289
  for K_2CO_3–Na_2CO_3 system, 335
  for minimum type, 320, 321, 323, 326
  concentration triangles, 464
  dimer-monomer behavior, 317
  dissociation effects, 315
  graphical description, 289, 301
  hypothetical examples, 301 ff., 330 ff., 352 ff.
  isotherms in ascending, 299
  in condensed-vapor phase regions, 351
  liquidus and solidus contours (table), 311
  minimum type, 314, 330

normal sublimation temperature of, 296
peritectic series of, 395
quenched, 294
space model, 295
structure of, 284
subsolidus behavior, 397
substitutional, 285
ternary, 459 ff., 472 ff.
Solubility curves
  ascending solid solution melting temperature range, 290
  association effects, 185, 188, 191
  common species effects, 177, 181, 184
  contours of in ascending solid solutions, 311
  dissociation effects, 168, 173, 176, 197, 202, 203
  entropy of fusion effects, 157
  eutectic, 129 ff.
  flattening of at compound maxima, 211, 216
  inflected, 153, 156, 271
  interaction effects, 192, 196
  minima in, 313
  relation to unary solid-liquid curves, 267
  theoretical, 204 ff.
  vapor phase composition along, 138
  vapor pressure along, 131
Space models
  eutectic, 234, 236, 264, 273, 278
    isotherms in, 252
  solid solution, 295, 327
  systems with limited solid solubility, 401, 405
Species, 5, 20
  conservation of mole terms, 127
  effects on solubility curves, 167
  ideal gas equation and, 94
  mole of, 15
  solid phase, 316
Standard states, 98, 101
  unit conversions, 102
Sticking coefficient, 36
Strip furnace, 515
Stoichiometry, 19
Structure changes, see Phase changes
Surface area effects on vapor pressure, 82
Sublimation point diagrams, 341
Sublimation temperature, 44
Systems, 10

SUBJECT INDEX

T

Ternary systems
  concentration triangles, 464
  cooling paths, 419 ff.
  graphical representation of, 408, 411
  isothermal gradient lines in, 417
  isothermal sections, 451
  peritectic reactions in, 436 ff., 448
  projection diagrams, 410
  solid solubility in, 459
  solid-vapor equilibria in, 477
    Ge–I$_2$–H$_2$, 484
  subsystems in, 434
  variance in, 407
Thermal analysis, 499
Thermal anomalies, measurement of, 503
Thermodynamic concentrations and pressures, 93, 118, 124
Thermodynamics, 4
  concept of species, 5
  mole fraction in, 23
  thermogravimetry, 515
  thermometry, 499
Tie lines
  in eutectic systems, 141
  in incongruently vaporizing systems, 370, 371
  in miscibility gaps, 273
  solid-liquid ratios from, 142
  in solid solution systems, 290
Time constant, 36
Transformations
  coordinate, see Coordinate transformations
  structural, see Phase changes
Transpiration method, 488, 491

U

Unary systems
  enthalpy diagrams, 86
  equipotential projections, 47
  inert gas effects, 39, 79, 81
  maximum number of phases in, 39
  melting point curve slope, 45, 47
  metastability in, 48, 49, 51
  metastable phase changes in, 54
  monotropy in, 63, 69, 72, 74
  phase changes in, 52, 54, 61
  surface area effects, 82
  univariance in, 40
Univariance
  achievement of in multicomponent systems, 483 ff.
  in binary systems, 128, 286
  isobaric in minimum type solid solutions, 318
  isothermal, 43

V

van 't Hoff equations, 130
Vapor phase, composition of, 138, 244, 249
Vapor pressure, 33
  deduction of liquidus curves from, 132
  effects of inert gas pressure on, 79, 81
    of surface area on, 82
  equilibrium, 35
  measurement of, 488 ff.
  in region of variable composition, 375, 384
  stepwise variation of in incongruently vaporizing systems, 376
  triple point in binary dissociative systems, 368
  variation with temperature along solubility curve, 131
Vaporization, incongruent, 367
Variable composition
  regions of, 380
  vapor pressure in region of, 375, 384
Variables, maximum number of, 7
Variance
  of homogeneous systems, 95
  of systems, 7
  in two component systems, 128
Vegard's law, 285, 522

Z

Zone refining, 292, 523
  floating, 293
  segregation coefficient and, 293

Physical Chemistry

A Series of Monographs

Ernest M. Loebl, Editor

Department of Chemistry, Polytechnic Institute of Brooklyn, Brooklyn, New York

1. W. Jost: Diffusion in Solids, Liquids, Gases, 1952
2. S. Mizushima: Structure of Molecules and Internal Rotation, 1954
3. H. H. G. Jellinek: Degradation of Vinyl Polymers, 1955
4. M. E. L. McBain and E. Hutchinson: Solubilization and Related Phenomena, 1955
5. C. H. Bamford, A. Elliott, and W. E. Hanby: Synthetic Polypeptides, 1956
6. George J. Janz: Thermodynamic Properties of Organic Compounds — Estimation Methods, Principles and Practice, revised edition, 1967
7. G. K. T. Conn and D. G. Avery: Infrared Methods, 1960
8. C. B. Monk: Electrolytic Dissociation, 1961
9. P. Leighton: Photochemistry of Air Pollution, 1961
10. P. J. Holmes: Electrochemistry of Semiconductors, 1962
11. H. Fujita: The Mathematical Theory of Sedimentation Analysis, 1962
12. K. Shinoda, T. Nakagawa, B. Tamamushi, and T. Isemura: Colloidal Surfactants, 1963
13. J. E. Wollrab: Rotational Spectra and Molecular Structure, 1967
14. A. Nelson Wright and C. A. Winkler: Active Nitrogen, 1968
15. R. B. Anderson: Experimental Methods in Catalytic Research, 1968
16. Milton Kerker: The Scattering of Light and Other Electromagnetic Radiation, 1969
17. Oleg V. Krylov: Catalysis by Nonmetals — Rules for Catalyst Selection, 1970
18. Alfred Clark: The Theory of Adsorption and Catalysis, 1970

Physical Chemistry
A Series of Monographs

19 ARNOLD REISMAN: Phase Equilibria: Basic Principles, Applications, Experimental Techniques, 1970
20 J. J. BIKERMAN: Physical Surfaces, 1970

In Preparation
E. A. MOELWYN-HUGHES: Statistics and Kinetics of Solution
J. C. ANDERSON: Chemisorption and Reaction on Metallic Films
S. PETRUCCI: Ionic Interactions (In Two Volumes)